U0282806

CISCO

思科数据中心

产品配置

薛润忠 韩大海 张国清◎主编

网络专家**精心撰写**，

全面展示数据中心配置

实践性强，

网络工程师快速入门的必备手册

北京邮电大学出版社
www.buptpress.com

内 容 简 介

当今云计算风起云涌，数据中心技术不断推陈出新，作为数据中心网络产品的业界龙头，思科公司一直引领变革。但最新技术的原版书都是用英文撰写的，对英文不是很精通，又想快速掌握其产品的工程师来说无疑是个难题。本书就是一本可以给读者快速熟悉思科数据中心产品配置的工具书。

建立一个数据中心，按照从简单到复杂的步骤执行：先建设网络平台，再建设应用平台。建设网络平台先从配置网络交换机开始，然后是安装存储交换机，最后是配置路由器和防火墙。应用平台的建设是从搭建服务器开始的，之后是完成服务器虚拟化的配置。本书在写作上按照数据中心产品进行分类，每个章节后面给出该类产品的典型配置，供读者参考。

图书在版编目(CIP)数据

思科数据中心产品配置 / 薛润忠，韩大海，张国清主编. -- 北京：北京邮电大学出版社，2015.6

ISBN 978-7-5635-4308-3

Ⅰ.①思… Ⅱ.①薛…②韩…③张… Ⅲ.①数据库系统—研究 Ⅳ.①TP311.13

中国版本图书馆 CIP 数据核字(2015)第 054291 号

书　　名：思科数据中心产品配置

著作责任者：薛润忠　韩大海　张国清　主编

责 任 编 辑：徐振华　孙宏颖

出 版 发 行：北京邮电大学出版社

社　　址：北京市海淀区西土城路 10 号(邮编：100876)

发 行 部：电话：010-62282185　传真：010-62283578

E-mail：publish@bupt.edu.cn

经　　销：各地新华书店

印　　刷：北京源海印刷有限责任公司

开　　本：787 mm×1 092 mm　1/16

印　　张：16

字　　数：412 千字

版　　次：2015 年 6 月第 1 版　2015 年 6 月第 1 次印刷

ISBN 978-7-5635-4308-3　　　　定　价：35.00 元

· **如有印装质量问题，请与北京邮电大学出版社发行部联系** ·

序

数据中心一直是重要的企业资产，也是IT用以保护、优化和发展业务的战略性重点机构。由于盲目追求新技术和缺乏对整体发展的计划，使得目前许多企业出现了服务器、存储资源使用率低下，能源和人员成本占数据中心总运行成本过高的问题。如果一个企业把IT预算的70%都花费在维护方面，将会不可避免地给企业的竞争力造成影响。为了解决这一问题，许多公司都致力于改变数据中心设施的整合和虚拟化，优化现有技术，消除运营孤井，使IT效率、响应能力和永续性达到新的高度。

思科数据中心3.0改变了目前的数据中心域环境——服务器、存储和网络——作为独立孤井运行的情况，能够根据需要将位于任意物理地点的设备混合、匹配和配置到单一架构和一套共享网络服务中。由于网络能自动从虚拟化服务器、存储和网络服务池中部署基础设施，可以实现按需发现和配置资源，不仅缩短了响应时间，而且降低了错误率，使得企业可以将富有经验的IT人员重新分配到具有更高价值的工作岗位上，从根本上改变了IT运行其核心资源的方式，在效率和企业响应能力方面获得了突破性的优势。

本书作者长期从事运营网络设计和实施，研究的重点是核心网络和数据中心架构，有丰富的网络设计经验，出版了多本思科认证书籍和网络技术书籍，深受读者的喜爱。

本书通过对思科数据中心架构、数据中心交换机Nexus、思科服务器UCS统一计算系统、数据中心存储交换机等内容进行介绍，可使读者快速掌握思科数据中心系列产品的配置，帮助企业提高运营效率，消除运营孤井，顺利过渡到未来的虚拟数据中心。

牛少彰

前　　言

本书是一本介绍思科数据中心设备配置的书，配备了大量实际案例，通俗易懂，便于读者在实际工程中参考，是思科网络工程师和数据中心维护者难得的工具书。

全书分为8个章节，循序渐进地搭建了一个思科数据中心。第1章介绍了思科数据中心3.0架构，涉及了统一网络、统一计算和统一存储的基本概念和主要技术。第2章说明了数据中心Nexus交换机产品和经典的7-5-2架构，内容包含了数据中心交换机常用的技术和配置。第3章介绍的是思科MDC存储交换机，思科没有存储设备，但是存储交换机是思科统一阵列的下一个目标，本章给出了思科存储交换机的详细的配置步骤和案例。第4章阐述了思科数据中心边缘路由器ASR 9000的配置知识，由于ASR 9000设备主要在运营商数据中心使用，所以这方面的资料少之又少，作者在ASR 9000方面积累了大量的实际工程经验，相信在看过本书后对您的工程安装和实施有很大的帮助。第5章介绍的是思科防火墙设备，包括思科ASA防火墙的基本配置和介绍，章尾给出了数据中心出口的一个典型配置。第6章介绍的是思科UCS服务器的安装和配置，思科UCS产品是数据中心划时代的革新产品，如果跟思科的交换机产品配合使用更能发挥其强大的性能优势。第7章介绍了一个虚拟化安装实例，在电脑上就可以实现数据中心经典的Vmotion的功能。第8章描述的是思科将各种网络产品在虚拟世界的实现，包括虚拟交换机、虚拟防火墙、虚拟路由器和虚拟的负载均衡设备。

更多关于本书配置的理论部分，请参阅作者与思科工程师合作翻译的《Data Center Virtualization Fundamentals》，中文书名为《数据中心虚拟化技术权威指南》。

由于思科数据中心技术日新月异，加之作者水平有限，书中难免存在谬误，恳请读者指正。我们将密切跟踪思科数据中心新技术的发展，吸收您的意见，适时编撰书的升级版本。

目　　录

第1章 思科数据中心技术

1.1 思科数据中心 3.0 架构

典型的思科数据中心 3.0 架构如图 1-1 所示。

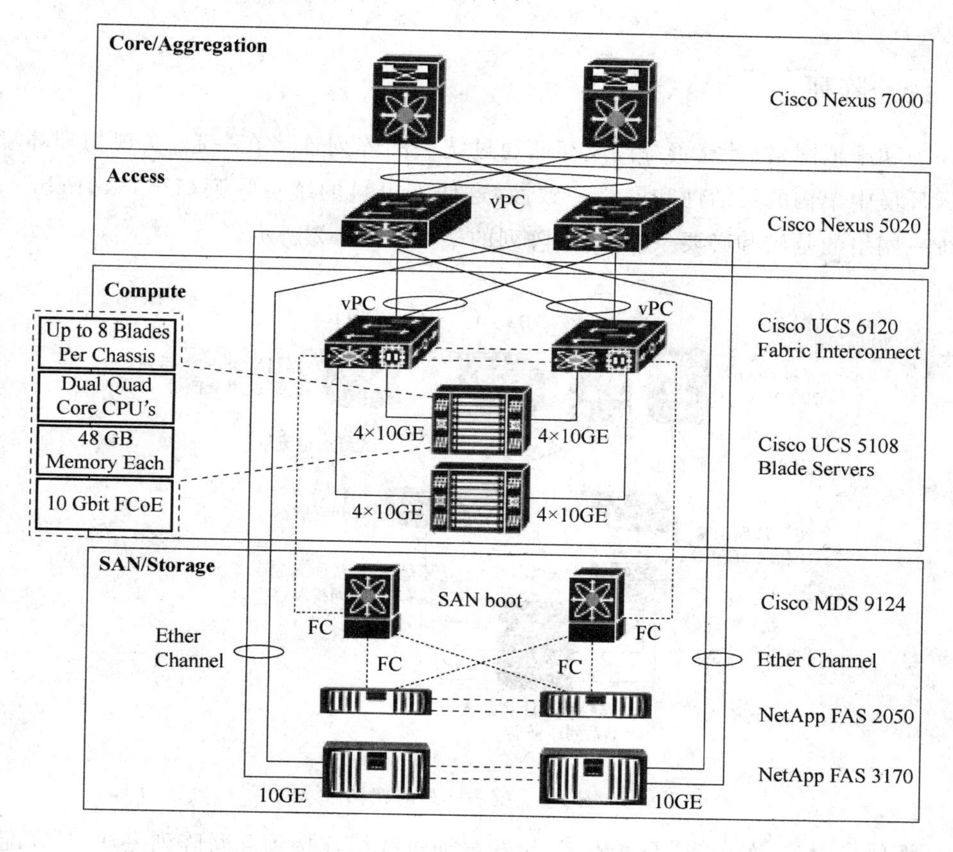

图 1-1　典型的数据中心架构

(1) 网络方面，两个 UCS 6120 阵列互联器可以将多达 40 个 UCS 5108 机箱上联到 Nexus 5000接入交换机，再通过核心/汇聚层(Core/Aggregation)的 Nexus 7000 交换机在数据中心之间或到数据中心之外交换流量，在 UCS 6120 和 Nexus 5000 之间，Nexus 5000 和 Nexus 7000 之间都通过虚拟端口通道(virtual Port Channel，vPC)绑定多条物理链路，从而形成一个高带宽和冗余的逻辑链路。

(2) 计算方面，每个 UCS 5108 机箱可插入 8 个半宽带片服务器，UCS 5108 系列通过 4×10GE万兆接口卡上联到 UCS 6120 阵列交换机，再通过统一的网络进行数据交换，这里的万兆接口卡其实是 UCS 2104 阵列互联设备。

这里的 UCS 6120 承担了架顶交换机的角色，数据中心典型的三层架构(接入层/汇聚层/核心层)也变成了二层架构(接入层/核心层)，大容量、可扩展和扁平化是对适应数据中心新的应用需求做出的改变。

存储设备 NetApp FAS2050 可以通过 FC 口接到思科存储交换机 MDS 9124，再通过 MDS 连到 UCS 6120 进入阵列交换(图 1-1 虚线部分)，也可以像 NetApp FAS3170 那样通过以太口直接上联到 Nexus 5020 交换机(图 1-1 实线部分)。

1.2 统一网络技术

1.2.1 统一阵列

将物理和虚拟网络、计算资源以及存储通过统一的阵列连接在一起，实现物理环境、虚拟环境和云环境中架构的灵活性和网络一致性，这便是思科的统一阵列(Unified Fabric)，统一阵列是统一网络的基础和必要条件，统一阵列的架构如图 1-2 所示。

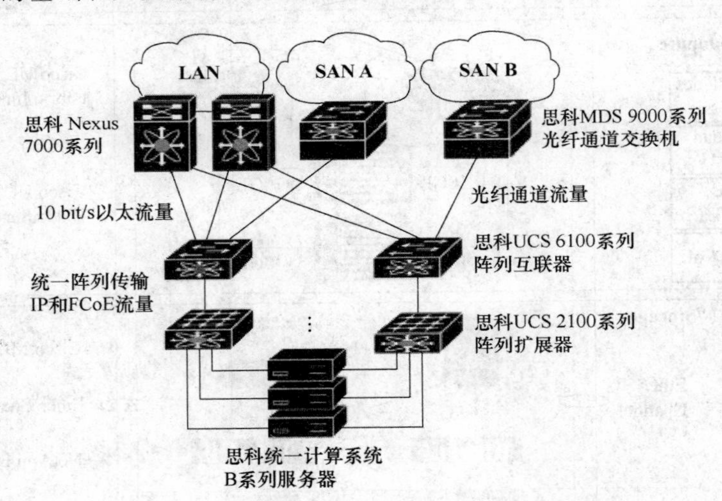

图 1-2 统一阵列的架构

统一阵列采用“一次布线”部署模式，机箱只通过线缆连接到互联阵列一次，I/O 配置的改变只需通过管理系统进行，而无须安装主机适配器以及对机架和交换机重布线。此统一阵列不再需要在每个服务器中部署冗余以太网和光纤通道适配器，也不必采用独立布线连接接入层交换机，并为每种网络媒体使用不同的交换机，因此大大简化了机架布线。所有流量都通过路由器互联到中央服务器，随后以太网和光纤通道流量可独立传输到本地非整合网络。

该统一阵列基于万兆以太网，采用标准扩展支持更多流量类型并优化管理。它支持以太网和以太网光纤通道(Fiber Channel over Ethernet，FCoE)，其管理特性使得以太网和 FCoE 等多种流量的管理能独立进行，支持带宽管理，且各流量级别间无干扰。

统一阵列为虚拟环境提供了创建基于统一 I/O 连接的大型服务器资源池的能力，通过编程，该资源池能够以与数据中心当前最佳实践相一致的方式运行。在虚拟化软件使用基于光纤通道的共享存储的环境里，就无须再部署冗余的 HBA 卡、收发器、电缆和上游交换机端口，这些成本相当于一个小型服务器。正如稍后所讨论的那样，Cisco VN-Link 技术支持每个虚拟机和互联阵列间的虚拟网络连接，简化了虚拟机及其网络的管理，包括轻松移动虚拟机，自动保持安全性等网络特性。

统一阵列在 UCS 这款产品内的实现：首先，每个刀片服务器都自带一个融合性网卡(Converged Network Adapter，CNA)，这块卡内置以太网的处理核心和以太网光纤通道的处理核心，能同时发送和接收数据流量及存储流量；然后，通过 Blade 机箱上的 Cisco UCS 2100 系列阵列扩展器(Fabric Extender)，将 CNA 直接和 Cisco UCS 6100 系列阵列互联(Fabric Interconnect)连接起来，这样就减少了原本在 Blade 机箱层所需要做的一次交换；最后阵列互联会根据流量转发给不同的上层交换机，例如，数据流量会转发给 Cisco Nexus 7000 系列网络交换机，存储流量会转发给 Cisco MDS 9000 系列光纤交换机。

1.2.2 虚拟链路

随着虚拟化技术在数据中心的大规模使用，虚拟机已经嵌入到主机当中，如何对数据流进行管理和流控以及安全问题就需要思考和解决的问题，思科的方案是把交换机 vDS(虚拟分布交换机)及 ASA 1000V 也嵌入到主机里，虚拟机通过 vDS，在数据从 VM 一出来时就打上标记，这样的数据流会被 vDS 以及可能的 ASA 1000V 控制，这样一个网络结构跟物理世界的网络完全一致，这个数据流的链路就叫虚拟链路(VN-LINK)。

在普通的刀片虚拟机环境下，有可能在网络 Access(接入)层出现 3 种交换机：其一是在 Access 层外置的交换机，如 Cisco Catalyst 6500 系列；其二是在 Blade 机箱层的刀片交换机，如 Cisco Nexus 4000 系列；其三是主机层的虚拟交换机(Virtual Switch)。但是在 UCS 上，Cisco 利用多种技术将这 3 种交换机简化至一种。

在 Access 层，UCS 提供 Cisco UCS 6100 系列阵列互联设备来支持 UCS 整个系统内所有虚拟机的流量，而且建立了一个单点控制和管理网络流量的接口，总体而言，可以认为它是一个增强版的 Access 层交换机。

在 Blade 机箱层，UCS 取消了刀片交换机，而是采用 Cisco UCS 2100 系列阵列扩展器。通过这个设备能够将 Blade 服务器所产生的所有流量直接传输给上层的阵列互联设备，这样能简化整个网络架构。

在主机层，放弃原先主机上的虚拟交换机，转而让虚拟机直接通过网络接口虚拟化技术来直接连接物理的网卡。

虚拟链路(VN-LINK)是一整套能在分布式虚拟技术环境下直接运行的网络方案，并且提供了与其他思科网络产品相似的让人很熟悉的功能集和运行模型。在设计方面，虚拟链路的核心思想就是让虚拟机在网络方面的使用和管理都尽可能地和现有的网络架构融合在一起。这样的设计理念不仅能很好地让虚拟环境更好地适应当今数据中心的网络环境，而且能维护思科在网络方面的地位。VN-LINK 包括两个子方案：其一是基于 Cisco Nexus 1000V 的解决方案，这个方案是以软件为主的实现；其二是基于网络虚拟化(Network Interface Virtualization，NIV)的解决方案，这个方案是硬件实现。

1. VN-LINK 软件方面——Cisco Nexus 1000V

Cisco Nexus 1000V 是基于 VMware vDS 的框架，所以其在总体的架构上也和 vDS 非常相近。它的做法也是分离了交换机的数据功能和控制功能，首先数据功能不是由主机上的虚拟交换机来处理，而是由安装在主机上的思科的虚拟太模块(Virtual Ethernet Module，VEM)来管理，而控制功能则被集中起来至主机之外，可以是安装在一个特制的虚拟机内(Cisco Nexus 1000V 使用这种做法)，也可以是在虚拟数据中心内或者是内嵌在一个特制的物理机(Cisco Nexus 1010)上。图 1-3 为 Cisco Nexus 1000V 的架构。

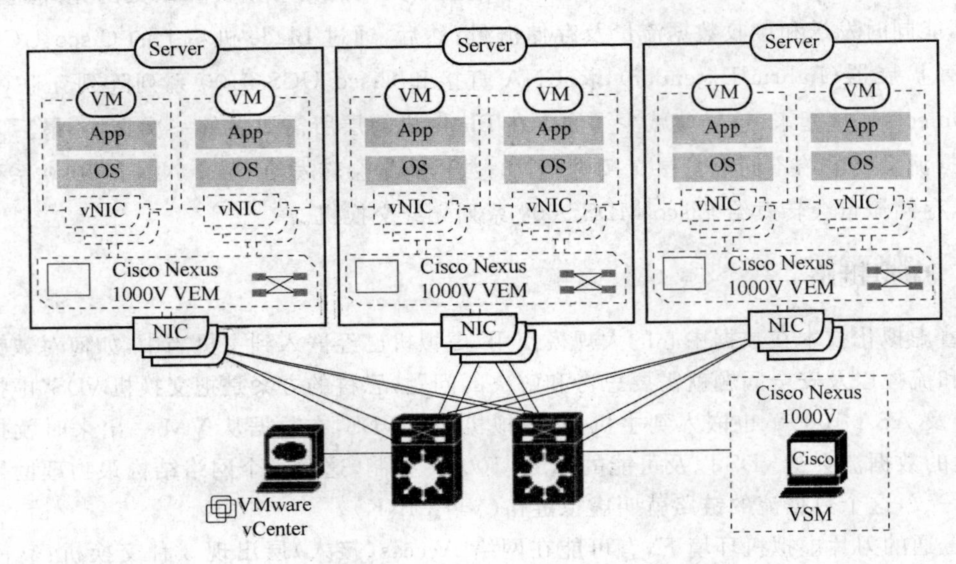

图 1-3 Cisco Nexus 1000V 架构

通过 VEM/VSM 的组合，除了非常支持原有 vDS 的功能集合，还能提供一些能更好地与现有企业网络架构融合的特性，例如，其自带的企业级 NX-OS 系统，堪称业界事实标准的 Cisco IOS CLI，Private VLAN，Encapsulated Remote SPAN 和 NetFlow v. 9 等。

2. VN-LINK 硬件方面——NIV＋Nexus7/5/2

NIV(Network Interface Virtualization)在技术层面上和 Cisco Nexus 1000V 相差很大，它主要通过引入支持 NIV 技术的新设备来避免使用如 Cisco Nexus 1000V 或 Virtual Switch 等软件交换技术，VN-LINK NIV 结构如图 1-4 所示。

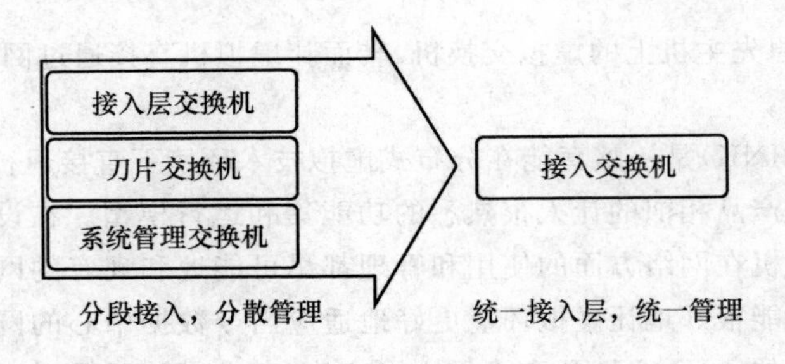

图 1-4 VN-LINK NIV 结构

通过整合上述 3 种交换机，能够让所有的流量都直接传送到阵列互联设备，这样不仅极大地简化了网络的架构，而且统一了整个网络的管理。

下面介绍网络接口虚拟化技术是如何让虚拟机直接连接物理网卡的。一台主机上运行的虚拟机的数量是远大于1的，所以虚拟机内置的虚拟网卡是远多于实际物理网卡的数量的。为了解决这个不对称的问题，思科在UCS中引入了支持网络接口虚拟化技术的网卡Cisco UCS M81KR，也称为"Palo"，因为网络接口虚拟化技术能够将一个物理的端口虚拟成多个虚拟的端口，所以Palo卡能虚拟多达256块虚拟网卡来对应虚拟机内置的网卡，而且加上Intel的Directed I/O技术，使得Palo卡在速度上也非常优秀。

从虚拟化的网络接口卡开始，数据包通过Nexus 2000/5000/7000经过一个完整的虚拟链路传输，大大提高了传输效率，这便是思科的VN-LINK。

1.2.3 虚拟端口通道

虚拟端口通道(virtual Port Channel，vPC)是Cisco NX-OS用于解决STP阻塞端口而使用的技术。通过将两台设备虚拟成一条链路，使得系统可以使用多个冗余链路转发数据，如图1-5所示。

传统端口信道通信的最大限制在于端口信道只能在两个设备之间运行。在大型网络中，设计中常常需要同时支持多个设备，来提供某些形式的硬件故障备用路径，这一备用路径的连接方式常常会导致环路，从而限制对单一路径实施端口信道技术的优势。为突破此限制，Cisco NX-OS软件平台提供了一种名为虚拟端口通道(即vPC)的技术。尽管对于与端口信道相连的设备来说，一对作为vPC对等终端的交换机就像是单一逻辑实体，但这两个作为逻辑端口信道终端的设备仍是两个或多个独立设备，端口信道绑定只能够在两个设备之间，而vPC的一端可以是多个物理设备，这便是最大的不同，如图1-5所示。vPC环境结合硬件冗余性和端口信道环路管理的优势，升级到一个完全基于端口信道的环路管理机制，所能获得的另一主要优势是，链路恢复速度大大加快。生成树协议从链路故障中恢复的时间大约为6 s，而完全基于虚拟端口捆绑组的解决方案则有可能在不到1 s的时间完成故障恢复。

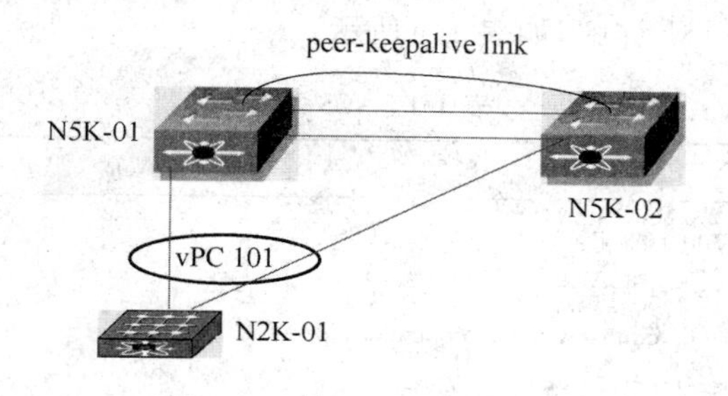

图1-5 vPC逻辑示意图

尽管vPC不是实施此解决方案的唯一技术，但其他解决方案都有很多缺陷，限制了它们的实际使用，特别是当部署在密集高速网络的核心或分布层时更是如此。所有的多机箱端口信道技术都仍需要在作为端口信道终端的两个设备间具有，该链路所占的带宽通常要远远小于与此终端相连的vPC的总带宽。思科对vPC技术进行专门设计，仅限于将此ISL用于交换机管理流量和偶尔来自于故障网络端口的流量。其他厂商在技术设计时并未以此为目标，所以实际上在扩展规模方面有很大限制，因为它们需要使用ISL处理控制流量以及对等设备间接近一半的数据吞吐率，对于小型环境来说，这种方法可能足够，但对于可能有数TB数据流量的环境来说，就无法满足需要。

vPC 为二层网络提供了重要优势，并借助第二层功能提供的优势，对第三层互联进行一系列改进。在第二层网络中，能够实现以下优势：

- 通过冗余系统提高系统可用性；
- 无须使用生成树协议，就能进行环路管理；
- 始终提供完全系统带宽可用性；
- 迅速恢复链路故障；
- 为任意支持 IEEE 802.3ad 的边缘设备提供端口信道连接。

此外，还支持第三层特性：

- 通过 HSRP 配置进行"主用-主用"第三层转发；
- 通过"主用-主用"HSRP 进行完全第三层带宽访问；
- 通过"主用-主用"PIM 指定路由器进行第三层迅速组播融合。

1.2.4 虚拟交换机系统

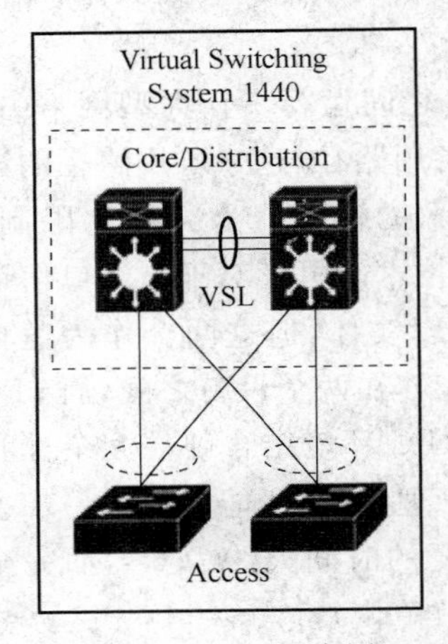

图 1-6 VSS 示意图

虚拟交换机系统（VSS）通过多机箱以太通道（MultiEther Channel）建立一个无环路的拓扑结构，VSS 系统对外体现为单一独立的设备，通过 STP 技术来防止内部环路的产生，图 1-6 是逻辑示意图，目前用在思科 4500,6500 和 6800 系列交换机中。

VSS 配置说明如下所示。

在两台交换机上启动 startup-config 配置文件的 VSS 配置必须一致，如果更新了交换机的优先级或者抢占了功能，只有在保存了配置文件和交换机重启后，配置才会生效。对于线路冗余，在虚拟交换链路（VSL）配置的时候为每个交换机配置至少两个端口，对于模块冗余，两个端口可以位于机箱的不同交换模块上。思科 Catalyst 65 系列交换机 VSS 虚拟交换系统配置详解，如表 1-1 所示。

表 1-1 VSS 配置

步 骤	6509 的配置
初始化 Catalyst 6509 交换机	6509-1 的配置： VSS-sw1 # conf t Enter configuration commands, one per line. End with CNTL/Z. 6509-2 的配置： VSS-sw2 # conf t Enter configuration commands, one per line. End with CNTL/Z.
配置 VSS Domain ID	6509-1 的配置： VSS-sw1(config) # switch virtual domain 100 Domain ID 100 config will take effect only after the exec command 'switch convert mode virtual' is issued

续表

步　骤	6509 的配置
	6509-2 的配置： VSS-sw2(config)＃switch virtual　domain 100 Domain ID 100 config will take effect only after the exec command ‘switch convert mode virtual’ is issued
设置交换机 Switch ID	6509-1 的配置： VSS-sw1(config-vs-domain)＃switch 1 VSS-sw1(config-vs-domain)＃ 6509-2 的配置： VSS-sw2(config-vs-domain)＃switch 2 VSS-sw2(config-vs-domain)＃
设置两台交换机的 priority	VSS-sw1＃(config)＃switch 1 priority110 VSS-sw2(config)＃switch 1 priority 100
配置 VSL	6509-1 的配置： VSS-sw1＃conf t VSS-sw1(config)＃interface portchannel 1 VSS-sw1(config-if)＃switch virtual link 1 VSS-sw1(config-if)＃no shut VSS-sw1(config-if)＃ 6509-2 的配置： VSS-sw2＃conf t VSS-sw2(config)＃interface portchannel 2 VSS-sw2(config-if)＃switch virtual link 2 VSS-sw2(config-if)＃no shut VSS-sw2(config-if)＃
物理接口划入端口组	6509-1 的配置： VSS-sw1(config)＃interface range tenGigabitEthernet 5/4-5 VSS-sw1(config-if-range)＃channel group 1 mode on VSS-sw1(config-if-range)＃no shut VSS-sw1(config-if-range)＃^Z VSS-sw1＃ Switch-1＃ platform hardware vsl pfc mode pfc3 6509-2 的配置： VSS-sw2(config)＃interface range tenGigabitEthernet 5/4-5 VSS-sw2(config-if-range)＃channelgroup 2 mode on VSS-sw2(config-if-range)＃no shut VSS-sw2(config-if-range)＃^Z VSS-sw2＃ Switch-2＃ platform hardware vsl pfc mode pfc3c

续 表

步　骤	6509 的配置
VSS 模式转换	6509-1 的配置： VSS-sw1＃switch convert mode　virtual This command will convert all interface names to naming convention“interface-type switch-number/slot/port”，save the running config to startupconfig and reload the switch. Do you want to proceed? [yes/no]：yes Converting interface names Building configuration… [OK] Saving converted configurations to bootflash … [OK] 6509-2 的配置： VSS-sw2＃switch convert mode　virtual This command will convert all interface names to naming convention“interface-type switch-number/slot/port”，save the running config to startupconfig and reload the switch. Do you want to proceed? [yes/no]：yes Converting interface names Building configuration… [OK] Saving converted configurations to bootflash… [OK]
重启交换机	6509-1 的配置： System detected Virtual Switch configuration… Interface TenGigabitEthernet1/5/4 is member of PortChannel 1 Interface TenGigabitEthernet1/5/5 is member of PortChannel 1 <snip> 00：00：26：%VSL_BRINGUP-6- MODULE_UP：VSL module in slot 5 switch 1 brought up Initializing as Virtual Switch Active 00：08：01：SW1_SP：Card inserted in Switch_number = 2 , physical slot 3， interfaces are now online VSS > VSS>en VSS＃

续 表

步 骤	6509 的配置
	6509-2 的配置： System detected Virtual Switch configuration… Interface TenGigabitEthernet2/5/4 is member of PortChannel 2 Interface TenGigabitEthernet2/5/5 is member of PortChannel 2 <snip> 00:00:26: %VSL_BRINGUP-6- MODULE_UP: VSL module in slot 5 switch 2 brought up Initializing as Virtual Switch standby 00:01:43: %CRYPTO-6-ISAKMP_ON_OFF: ISAKMP is OFF 00:01:43: %CRYPTO-6-ISAKMP_ON_OFF: ISAKMP is OFF VSS-sdby> Standby console disabled
检验 VSS 配置	VSS# show switch virtual Switch mode : Virtual Switch Virtual switch domain number : 100 Local switch number : 1 Local switch operational role: Virtual Switch Active Peer switch number : 2 Peer switch operational role : Virtual Switch Standby
保存配置	VSS# conf t Enter configuration commands, one per line. End with CNTL/Z. VSS(config)# write

1.2.5 虚拟设备环境

虚拟设备环境(Virtual Device Contexts，VDC)是 Cisco 基于操作系统级别的一虚多网络虚拟化技术，VDC 架构如图 1-7 所示，表 1-2 展示的是网络一虚多技术的区别。

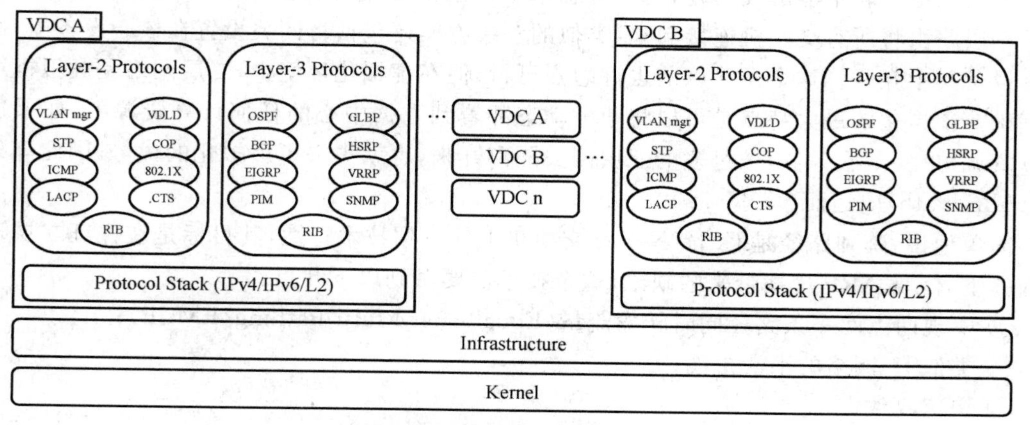

图 1-7 单个物理设备中的 VDC 架构

表 1-2　一虚多技术对比

技　术	逻辑虚拟化对象
VLAN	数据平面
VRF	数据平面＋少部分控制平面(路由协议)
虚拟防火墙	数据平面＋管理平面
VDC	数据平面＋控制平面＋管理平面＋系统资源

Cisco 目前公布的软件规格为每物理设备最多虚拟 4 个 VDC,并支持每 VDC 4k VLANs 和 200 VRFs 进行分层次虚拟化部署。按照云计算的需求来看,一般给每个用户分配的都是以带宽为度量的网络资源,不会以交换机为单位进行虚拟网络资源分配,所以 VDC 主要应用在企业级的环境。

1.2.6　阵列路径

当今的大部分数据中心网络都是遵循标准的层次化理念建设的,分为接入层和汇聚/核心层,接入层和汇聚层之间为二层链路,三层网关设在汇聚或核心,所有的二层链路上都运行生成树协议(STP),当任意两点间有一条以上路径可达时,STP 会阻塞多余的路径,以保证两点间只有一条路径可达,从而防止环路的产生。这种模式在过去很长一段时间被大规模采用,因为其部署起来非常简单,接入层设备不需要复杂的配置,大部分的网络策略只要在汇聚层集中部署就能分发到全网。但随着数据中心的规模不断扩张,这种模型逐渐显得力不从心。

首先,未来数据中心内部的横向流量将越来越大,新加入的设备同原有设备之间仍然要运行 STP,如果两台服务器之间只有一条链路可行,其余的万兆交换机端口全被阻塞,不但是投资的极大浪费,而且也无法支持业务的快速扩展;其次,当交叉链路数量增加时,二层网络的设计会变得非常复杂,哪条链路该保留哪条链路该阻断和三层网关设在何处等类似的问题会出现,这就失去二层网络配置简单的优势;最后,传统的二层 MAC 地址没有层次化的概念,同一个二层网络内的接入交换机上存储本网段所有设备的 MAC 地址,这很容易导致边缘设备的 MAC 地址空间耗尽,特别是在虚拟化的数据中心内,虚拟机的 MAC 地址数量可能 以千计。

如果二层互联不能解决问题,另一种思路是在汇聚交换机和接入交换机上设置 IP 网关,通过三层路由将所有交换机连接起来,类似的解决方案还包括将网关设置在核心设备上,通过核心设备集中互联。这个做法以前也许勉强可行,但在虚拟化环境中,二层网络是虚拟机迁移的基础。虚拟化的最大特点是可以将业务动态部署到数据中心的任何计算资源上,如果这些计算资源(也就是服务器)被过多的三层网关隔离开来,也就失去了虚拟化的优势,因此阵列路径(FabricPath)应运而生。

简单地说,阵列路径是 Cisco Nexus 交换机上的一项技术特性,其目标是在保证二层环境的前提下,修复 ARP 广播带来的缺陷,这个技术需要做到以下几点:

① 实现两点间多条路径同时转发流量(Equal Cost Multi Pathing,EMCP);

② 类似 IP 网络的平滑扩展;

③ 快速收敛;

④ 防止广播风暴;

⑤ 保持原有二层网络配置的简洁性。

更准确地说,阵列路径要摆脱传统二层"交换"的弊端,在二层环境中实现类似三层 IP 的"路由"行为。

既然二层网络的问题是控制平面的缺失,阵列路径的思路就清晰了,那就是重建一个控制平面。

为了能够高效地支持数据中心扩展,这个新的控制平面需要具备几个基本功能,包括如下几个方面:

① 主动建立邻居关系,并基于链路状态维护一个路由数据库;

② 支持等价路由;

③ 支持灵活的寻址方式;

④ 保留原有二层网络配置的简单风格。

为了构建这样一个控制平面,阵列路径主要做了以下两件事:

① 新增一个二层帧头;

② 增加一套简化的 IS-IS 路由协议。

图 1-8 显示阵列路径网络示意图。

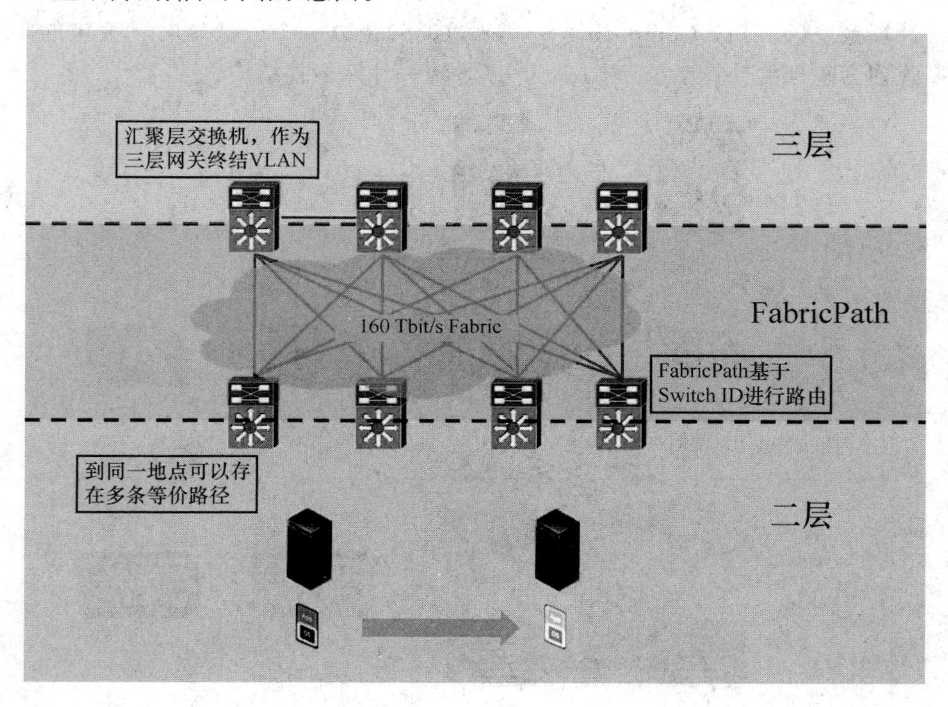

图 1-8 阵列路径网络示意图

图 1-8 是一个典型的阵列路径组网图,汇聚设备同接入设备之间为阵列路径网络,阵列路径网络内没有运行 STP,多条链路都能够转发数据,目前版本的阵列路径支持 16 条等价路由,也就是说在使用万兆链路的情况下,任意两点间的带宽可达到 160 Gbit/s(16 条等价链路结合,每条等价链路为 16 个万兆 port channel)。

接入的设备作为网关连接了传统以太网络同阵列路径网络,阵列路径网关上可以进行"基于会话的 MAC 地址学习",只有那些目的地址为本地设备的数据帧的源地址会被放入网关的 MAC 地址表,其他数据帧的源地址以及广播帧的源地址都不会被学习,这就保证了边缘网关设备的 MAC 地址表里只保存与本地有会话关系的 MAC 地址，这个举措能够大大缩小虚拟化数据中心内接入设备的 MAC 地址表体积。

基于 IS-IS 的特性，阵列路径网络设备的 Switch ID 可以动态修改，而不影响流量转发，当数据中心规模不断扩张时，可以利用阵列路径平滑地扩展。

阵列路径是 Cisco 近期在数据中心领域最重要的一个发布，同时也预示着基础网络向下一代模型转型的开始。数据中心内不断增长的横向流量推动了二层多路径技术的迅速发展，阵列路径是这股潮流的重要组成部分。

Cisco 已经发布了支持阵列路径的多个产品，并且承诺现有架构与 TRILL 标准兼容，当 TRILL 正式标准化之后，只需要升级现有设备的软件，就能够与标准的 TRILL 交换机互联互通。

1.3 统一计算系统

统一计算系统（Unified Computing System，UCS）是以一个以低延时无丢包的 10 Gbit/s 统一交换阵列为基础的计算系统，图 1-9 为 UCS 统一计算拓扑结构。

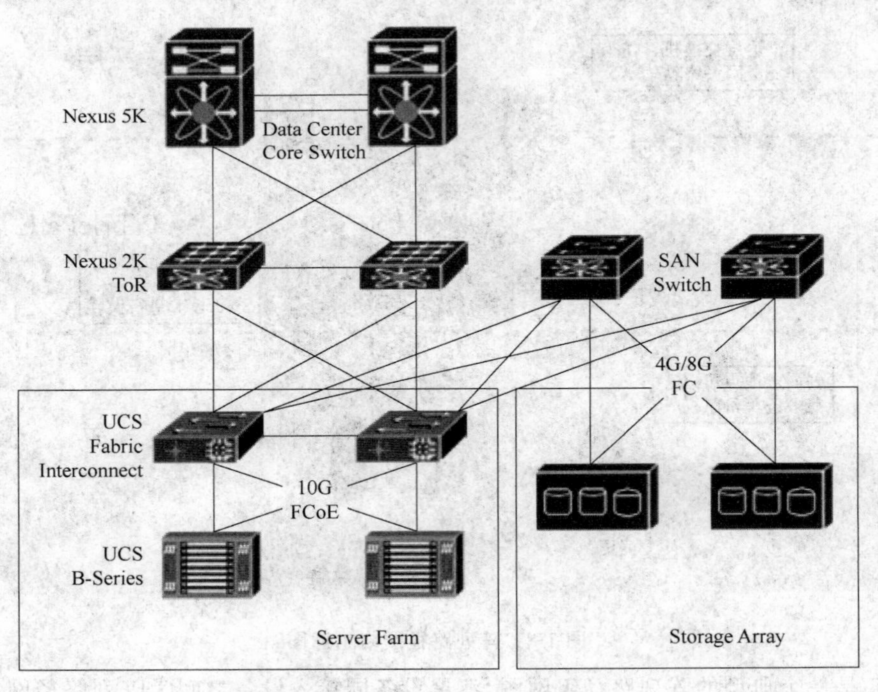

图 1-9 UCS 统一计算拓扑结构

思科统一计算系统的重要目标是实现传统数据中心最大程度的资源整合。一般在传统数据中心中存在两种网络：传统的数据局域网和使用光纤存储交换机的存储交换网络。统一交换平台将这两种网络实现在统一的传输平台上，即使用一种交换技术同时实现传统数据网络功能和远程存储。这样才能最大化地实现资源的整合，降低系统复杂程度，从而便于实现跨平台的资源调度和虚拟化服务，提高投资的有效性，同时还降低了管理成本。

刀片服务器群的 LAN 接入将直连数据中心有线网络的 ToR 接入设备，并在该统一计算系统内部实现自交换矩阵至虚拟服务器网卡的全链路层面的 FEX 贯通。最终在网络层面可以识别并可控到每一台虚拟机网卡的流量，并进而在其上实现可自定义的各类功能，这便是 UCS 的核心思想。

1.3.1 无状态计算

无状态计算（Stateless Computing）在维基百科上的定义是：一般指在计算主体上不存在任何状态信息或特定配置，各个主体都是无差别部署，这样的好处就是可快速定制和返还计算资源，通常可通过服务化等手段抽取状态。

既然单个计算主体上不再有任何状态信息，像 MAC，IP，WWNN 和 WWPN 等信息如何提供呢？那就是要介绍的另一个概念：服务配置文件。

思科统一计算系统是建立在“无状态”计算这一基础概念上的。基于这个设计理念，用户可以采用“一次布线”的方式，利用组件虚拟化的基础架构并灵活调用各种计算资源。这种架构使得用户可以轻松地从 Platform as a Service (PaaS) 和 Infrastructure as a Service (IaaS) 服务转入云计算模式。思科不但将服务器虚拟化，同时将输入输出的连接（如数据中心的交换和存储阵列本身）也虚拟化，将“无状态模型”提升到一个新的高度。

1.3.2 服务器配置文件

Cisco UCS Manager 使用服务配置文件（Service Profiles）来配置服务器及其 I/O 连接。服务配置文件由服务器、网络和存储管理员创建，并存储在 Cisco UCS 6100 系列互联阵列中。在当今的数据中心，服务器很难部署和改变使用目的，因为这通常要花费几天甚至几周的时间来实施。这一问题的出现是因为服务器、网络和存储团队需要仔细的人工协调，来确保其所有设备都能实现互操作。服务配置文件允许将思科统一计算系统中的服务器视作“裸计算能力”，在应用工作负载中进行分配和重新分配，从而能够更加动态、高效地发挥出当今数据中心内的服务器的处理能力。

服务配置文件由服务器及服务器所需的相关局域网和存储域网络连接组成。当为一台服务器部署了一个服务配置文件之后，Cisco UCS Manager 便可自动配置该服务器、适配器、扩展模块和互联阵列，以匹配该服务配置文件中所规定的配置。设备配置的自动化可减少配置服务器、网络接口卡（NIC）、主机总线适配器（HBAs）和局域网与 SAN 交换机等所需的人工步骤。人工步骤的减少可进一步降低人为错误的概率，改进一致性，缩短服务器部署时间。

服务配置文件可使虚拟化环境和非虚拟化环境同时受益。工作负载可能需要在服务器之间进行转移，以变更指定给一项工作负载的硬件资源或者使服务器离线进行维护或升级。服务配置文件可用于提高非虚拟化服务器的灵活性。它们还可与虚拟化集群一起使用，轻松增添新资源，来补充现有虚拟机的灵活性。同时，服务配置文件还可用于支持思科服务器的 VN-LINK 功能，以运行支持 VN-LINK 的虚拟机管理程序。

1.4 统一存储

要实现存储网络和传统数据网络的双网合一，思科统一交换平台实现二者的一体化交换。当前在以太网上融合传统局域网和存储网络的唯一成熟技术标准是以太网光纤通道，它已在标准上给出了如何把存储区域网络(Storage Area Network,SAN)的数据帧封装在以太网帧内进行转发的相关技术协议。

FCoE 技术标准可以将光纤通道映射到以太网,可以将光纤通道信息插入以太网信息包内,从而让服务器-SAN 存储设备的光纤通道请求和数据可以通过以太网连接来传输,而无须专门的光纤通道结构,进而可以在以太网上传输 SAN 数据。FCoE 允许在一根通信线缆上传输 LAN 和 FC SAN 通信,融合网络可以支持 LAN 和 SAN 数据类型,减少数据中心设备和线缆数量,同时降低供电和制冷负载,收敛成一个统一的网络后,需要支持的点也跟着减少了,这有助于降低管理负担。它能够保护客户在现有 FC-SAN 的投资基础上,提供一种以 FC 存储协议为核心的 I/O 整合方案,FCoE 协议结构如图 1-10 所示。

FCoE 是 2007 年国际信息技术标准委员会(INCITS)的 T11 委员会(和 FC 标准制定是同一组织)开始制定的标准,2009 年 6 月标准完成(FC-BB-5)。FCoE 基于 FC 模型而来,仍然使用 FSPF 和 WWN/FC ID 等 FC 的寻址与封装技术,只是在外层新增加了 FCoE 报头、Ethernet 报头封装和相应的寻址动作,可以理解为类似 IP 和 Ethernet 的关系。

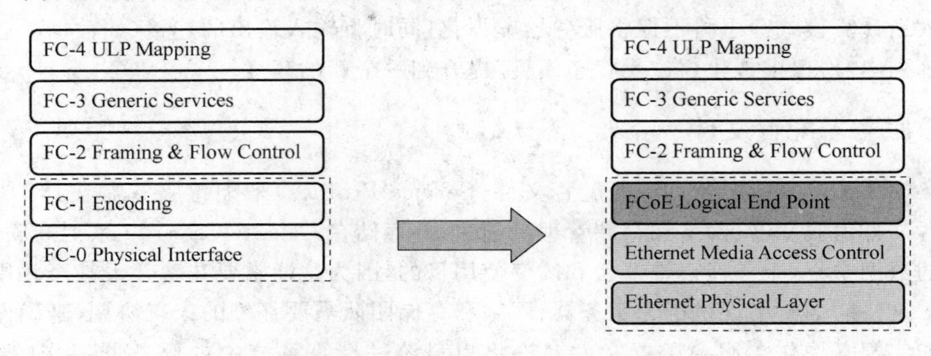

图 1-10　FCoE 协议结构

1.5 小　结

数据中心设备通过统一的矩阵互联,矩阵是物理层面的电路交换,其性能比链路交换和三层路由不只是快一个数量级。再通过 vPC,VDC,VSS,VN-tag 和 VN-LINK 等一系列技术解决链路层面的扩展问题,满足数据中心大二层的需求,这是思科统一的网络。

UCS 服务器通过扩展内存等技术提高了数据中心服务器的性能,利用阵列扩展(UCS 6000系列和 Nexus 2000 系列)将计算性能提升到最大,而无状态技术的配置文件在大量部署服务器时大大降低了复杂性,提高了可维护性,这是思科统一的计算,没有这些技术,UCS 就跟其他品牌数据中心服务器没有多大区别了。

物理层统一矩阵,同时提供 IP 和 FC 统一接口的交换机及 FCoE 等技术,使得存储 IP 化,这便是思科的统一存储。

第 2 章　数据中心交换机 Nexus

2.1　Nexus 系列产品介绍

Nexus 系列交换机是思科在数据中心解决方案中使用的主打产品，其 VN-tag、VN-LINK、阵列结构、FabricPath、FCoE 等一系列技术统一了网络、计算和存储资源，使得思科数据中心产品异常强大。Nexus 产品线涵盖了从 2000 到 9000 系列，本章主要介绍 Nexus 2000，5000 和 7000 产品及其详细的配置过程。

2.1.1　Nexus 2000

Nexus 2000 系列阵列扩展器是一种新型数据中心产品，可以满足高度可扩展的、灵活的服务器联网解决方案。它能够与 Cisco Nexus 5000 系列交换机协作，为服务器汇聚提供高密度、低成本的连接，图 2-1 列出了该系列产品。

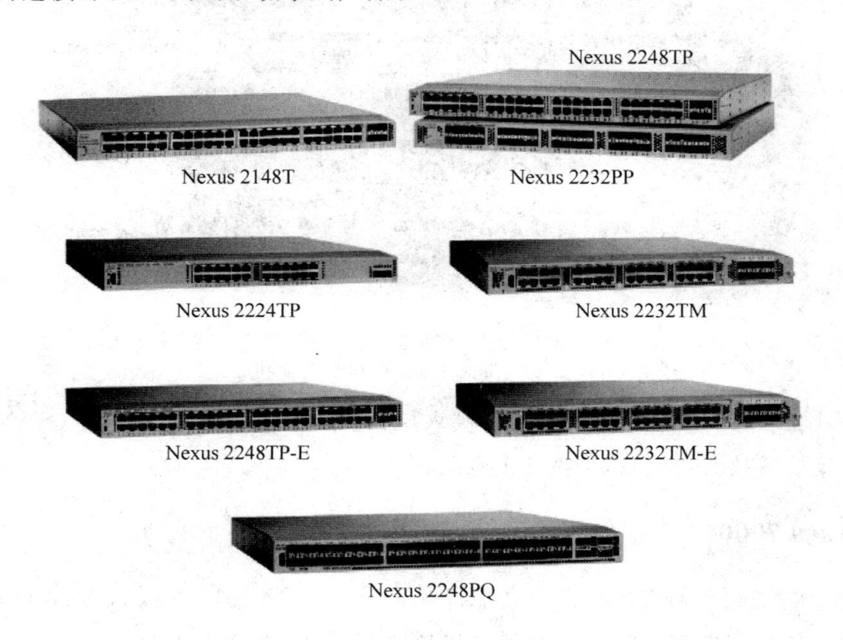

图 2-1　Cisco Nexus 2000 系列阵列扩展器

Cisco Nexus 2148T 阵列扩展器是该系列产品中的主力。它外形小巧，高为 1U 设备，可为服务器连接提供 48 个固定千兆以太网接口和最多 4 个万兆以太网上行链路接口。它的各种特性都为数据中心的网络设计，例如，前后贯通的冷却方式，位于后部靠近服务器端口的交换机端口，以及便于用户从前面板维修的组件。

Nexus 2000 缺省不带任何的 NX-OS 以及配置，每次启动的时候，都会与上层交换机(Nexus 5000 或 Nexus 7000)比对 NX-OS 版本和配置。如果版本和配置有变化，则强制与上级交换机同步。

Nexus 2000 相当于上联交换机的一块办卡，所以配置都在上联交换机上。连接 Nexus 2000 的交换机(N5K 或 N7K)需要使用 10GE 端口，并需要上联交换机进行一些配置，要说明的是在 Nexus 交换机里，10G 口都标以 Ethernet，而不再是 TenGigE。

2.1.2 Nexus 5000

在新一代的数据中心中，基础设施的范围将会大幅扩展，而用户对其负载能力也会提出越来越高的要求。作为 Cisco Nexus 系列数据中心级交换机的组成部分，Cisco NexusTM 5000 系列使用一种创新的架构，可以实现高性能、基于标准的以太网统一矩阵并简化数据中心的转型。该系列交换机可以将局域网、存储网和服务器集群整合到一个统一的矩阵之中。在众多业界领先的技术支持下，Cisco Nexus 5000 系列，如图 2-2 所示，能够克服下一代数据中心所带来的诸多挑战，包括高密度端口、多核心和针对虚拟机优化的服务等。

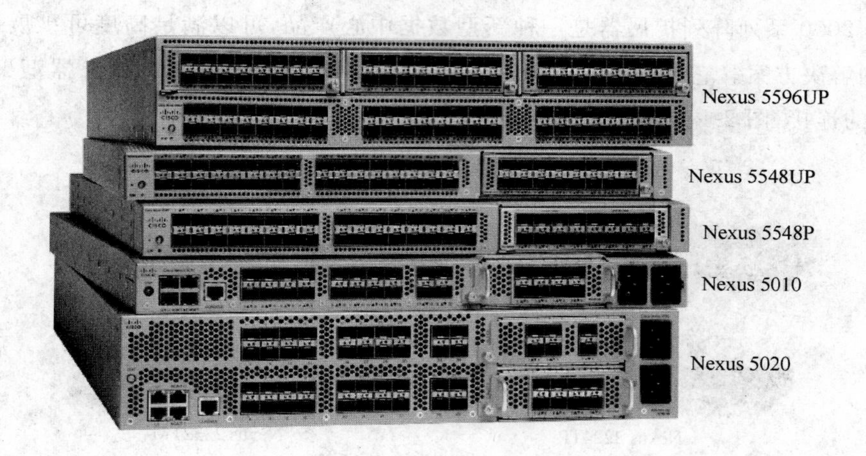

图 2-2 Cisco Nexus 5000 交换机

在典型网络结构中，Nexus 5000 位于接入层，连接服务器群、存储网络(SAN)和园区网(LAN)的汇聚层或核心层设备，如图 1-1 所示。

2.1.3 Nexus 7000

Nexus 7000 是此系列交换机中的高端产品(笔者编写此书时 Nexus 9000 已经发布)，它的每个插槽最多可支持 55 万兆位，而每个机箱则可支持 13.6 兆兆位。它具备高度的千兆、万兆、4 万兆和亿兆以太网扩展能力，并具备无中断的运行中软件升级 (In-Service Software Upgrade，ISSU) 功能，实现了高可用性。后面介绍的重叠传输虚拟化(Overlay Transport Virtualization，OTV)技术是在 Nexus 5000/7000 之间提供的数据中心互连 (Data Center Interconnected，DCI)解决方案技术，是叠加式数据中心和园区核心部署的理想之选。图 2-3 显示了 Nexus 7000 系列产品的外观。

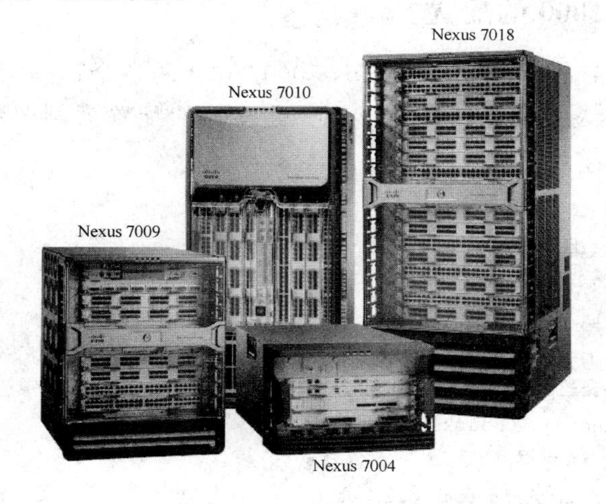

图 2-3 Nexus 7000 系列产品

2.2 Nexus 5000/7000 基础配置

介绍完思科 Nexus 系列的主要产品，从本节开始将介绍思科数据中心的产品：Nexus 交换机的配置。Nexus 所用的操作系统为 NX-OS，其配置格式与 IOS 基本一致。通常一台 Nexus 5000 或 7000 至少需要 VLAN 的配置，vPC、路由配置也是不可或缺的，其他特征可依据客户要求加以选择。

2.2.1 实验拓扑图

在正式配置之前，给出本书配置的总体拓扑图，如图 2-4 所示。

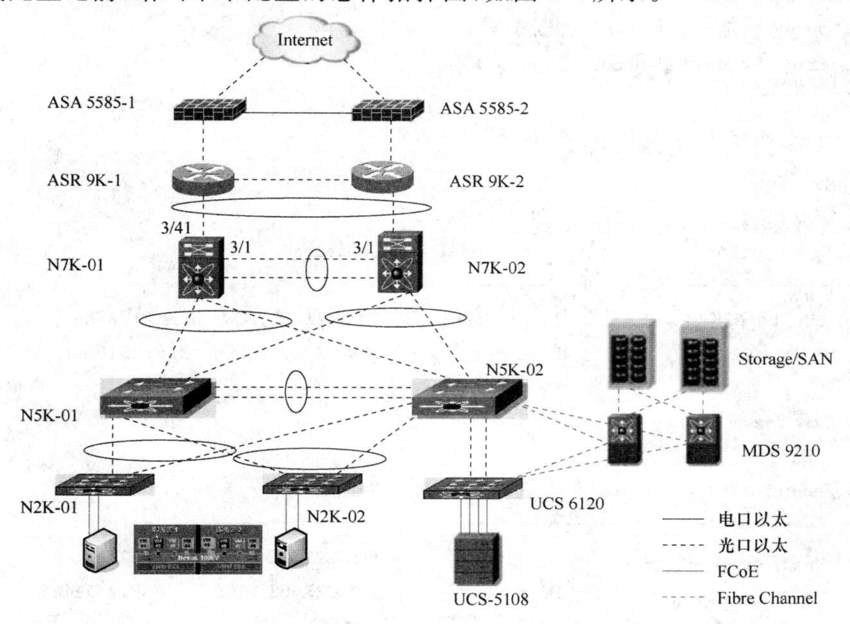

图 2-4 实验总体拓扑

2.2.2 连接到 Nexus 2000 的配置

Nexus 2000 无须做任配置，相当于 Nexus 5000/7000 的一个远程板卡，在 N5K 或 N7K 上做些配置后才能识别 Nexus 2000。下面以在 N5K 上的配置为例解释如何将 N2K 连接上来。

步骤一

```
N5K-01(config)#feature fex
! 开启 fex 特性
N5K-01(config)#fex 101
! 创建 fex
N5K-01(config-fex)#description cnnect_to_N2K
N5K-01(config-fex)#pinning max-links 1
!绑定成一条链路连到上联的交换机
Change in max-links will cause traffic disruption.
! 提示:更改 max-links 将引起数据中断

N5K-01(config-if-range)#interface port-channel 101
N5K-01(config-if)#switchport mode fex-fabric
! 端口模式改为 fex-fabric
N5K-01(config-if)#fex associate 101
! 将 port-channel 关联到相应的 FEX
N5K-01(config-if)#no shutdown
N5K-01(config-if)#end

interface Ethernet1/13
N5K-01(config-if) fex associate 101
! 指定关联的 Fabric Module 成为第 101 个关联的模块
N5K-01(config-if) switchport mode fex-fabric
! 指定接口的功能用于连接 Fabric Module
N5K-01(config-if) channel-group 101
```

步骤二

此时 N2K 需要下载镜像文件。

```
N5K-01#show fex
FEX      FEX                   FEX          FEX
Number   Description           State        Model      Serial
------------------------------------------------------------------------
100   connect_to_N2K-C2148          Image Download   N2K-C2148T-1GE   JAF1212AFXA
---      ---------                  Discovery        N2K-C2248TP-1GE   JAF1251BBKL

N5K-01#show fex
FEX      FEX                   FEX          FEX
Number   Description           State        Model      Serial
------------------------------------------------------------------------
100   connect_to_N2K-C2148          offline          N2K-C2148T-1GE   JAF1212AFX
---      ---------                  Discovery        N2K-C2248TP-1GE     JAF1251BBKL
```

步骤三

下载镜像文件后，设备重启，然后查看 N2K 状态，此时状态为 Online。通过一段时间的监测，上层交换机就可以发现并且配置 Nexus 2000。由于在上层交换机上看到的 Nexus 2000 上的端口都是本地端口，所以这个具有 fex-fabric 角色的端口算是一个功能很特殊的 Trunk。

同步完成之后，将可看到如下信息：

```
N5K-01 # show fex
FEX      FEX                  FEX        FEX
Number   Description          State      Model       Serial
------------------------------------------------------------------
100   connect_to_N2K-C2148        online    N2K-C2148T-1GE   JAF1212AFX
---      --------                 Discovery     N2K-C2248TP-1GE      JAF1251BBKL
```

一个阵列扩展器可以被处于 vPC 形态的多个上层交换机所识别，也可以被两侧交换机同时配置和管理。但是为了保证 Fabric Module 在系统切换时保持正确的形态，我们需要在两侧的上层交换机上同步配置。

2.2.3 Nexus 5000/7000 设备管理

(1) 配置主机名

```
vrf context management
ip route 0.0.0.0/0 192.168.10.254
! 网管的缺省网关
```

(2) 配置管理地址

```
interface mgmt0
ipaddress 192.168.10.172/24
```

(3) 查看相关特性

缺省不打开的特性都是关闭的，启用命令：feature fex。

```
show feature
! 查看打开的所有功能特性
```

2.2.4 VLAN 配置命令

(1) 创建一个 VLAN 并激活。

```
N7K# configure terminal
N7K(config)# vlan 2
N7K(config-vlan)# name test
N7K(config-vlan)# state active
N7K(config-vlan)# no shutdown
N7K(config-vlan)# exit
N7K(config)#
```

(2) 创建 PVLAN。

在 Private VLAN 的概念中，交换机端口有 3 种类型：Isolated port(隔离端口)、Community port(团体端口)和 Promiscuous port(混杂端口)。它们分别对应不同的 VLAN 类型：Isolated port 属于 Isolated PVLAN，Community port 属于 Community PVLAN，而代表一个 Private VLAN 整体的是 Primary VLAN。前面两类 VLAN 需要和它绑定在一起，同时它还包括 Promiscuous port。在 Isolated PVLAN 中，Isolated port 只能和 Promiscuous port 通信，彼此不能交换流量；在 Community PVLAN 中，Community port 不仅可以和 Promiscuous port 通

信，而且彼此也可以交换流量。Promiscuous port 与路由器或第三层交换机接口相连，它收到的流量可以发往 Isolated port 和 Community port。PVLAN 的应用对于保证接入网络的数据通信的安全性是非常有效的，用户只需与自己的默认网关连接，一个 PVLAN 不需要多个 VLAN 和 IP 子网就提供了具备第二层数据通信安全性的连接，所有的用户都接入 PVLAN，从而实现了所有用户与默认网关的连接，而与 PVLAN 内的其他用户没有任何访问。PVLAN 功能可以保证同一个 VLAN 中的各个端口相互之间不能通信，但可以穿过 Trunk 端口，这样即使同一 VLAN 的用户，相互之间也不会受到广播的影响。

PVLAN 当中使用的一些规则：

① 一个"Primary VLAN"当中至少有 1 个"Secondary VLAN"，没有上限；

② 一个"Primary VLAN"当中只能有 1 个"Isolated VLAN"，可以有多个"Community VLAN"；

③ 不同"Primary VLAN"之间的任何端口都不能互相通信（这里"互相通信"是指二层连通性）；

④ "Isolated port"只能与"promiscuous port"通信，除此之外不能与任何其他端口通信；

⑤ "Community port"可以和"promiscuous port"通信，也可以和同一"Community VLAN"当中的其他物理端口进行通信，除此之外不能和其他端口通信。

PVLAN 逻辑关系如图 2-5 所示。

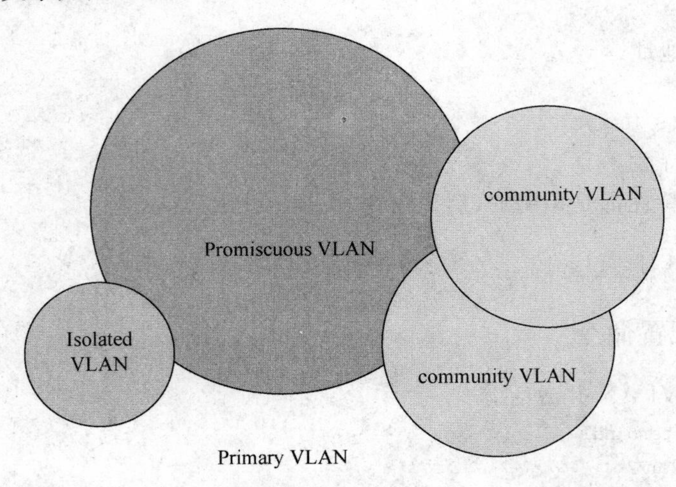

图 2-5　描述不同类型 PVLAN 之间的关系

```
N7K# configure terminal
N7K(config)# vlan 2
N7K(config-vlan)# private-vlan primary
N7K(config-vlan)# exit
N7K(config)# vlan 3
N7K(config-vlan)# private-vlan community
N7K(config-vlan)# exit
N7K(config)# vlan 4
N7K(config-vlan)# private-vlan isolated
N7K(config-vlan)# exit
```

(3) 关联 Secondary VLAN 到主 VLAN。

```
N7K(config)# vlan 2
N7K(config-vlan)# private-vlan association 3,4
N7K(config-vlan)# exit
```

(4) 创建一个 PVLAN 主机和混杂端口并把它们分配到正确的 VLAN。

```
N7K(config)# interface ethernet 1/11
N7K(config-if)# switchport
N7K(config-if)# switchport mode private-vlan host
N7K(config-if)# exit
N7K(config)# interface ethernet 1/12
N7K(config-if)# switchport
N7K(config-if)# switchport mode private-vlan promiscuous
N7K(config-if)# exit
```

(5) 配置接口可以访问的 VLAN。

```
N7K(config)# interface ethernet 1/11
N7K(config-if)# switchport private-vlan host-association 2 3
N7K(config-if)# exit
N7K(config)# interface ethernet 1/12
N7K(config-if)# switchport private-vlan mapping 2 3 4
N7K(config-if)# exit
```

(6) 创建一个 VLAN 接口和可以访问的 VLAN。

```
N7K(config)# interface vlan 2
N7K(config-vlan)# private-vlan mapping 3 4
N7K(config-vlan)# exit
N7K(config)#
```

2.2.5 vPC 配置

一个虚拟端口通道(vPC)允许带有 port channel 功能的单独的服务器或交换机连接到两台思科 Nexus 7000/5000 设备，就像连接到一台设备一样，增加了并行的带宽，同时获得了冗余的好处，从 NX-OS 5.1(4)开始，每台物理交换机设备支持至少 256 个 vPC。

1. vPC 配置

配置之前，先给出设备拓扑，如图 2-6 所示。

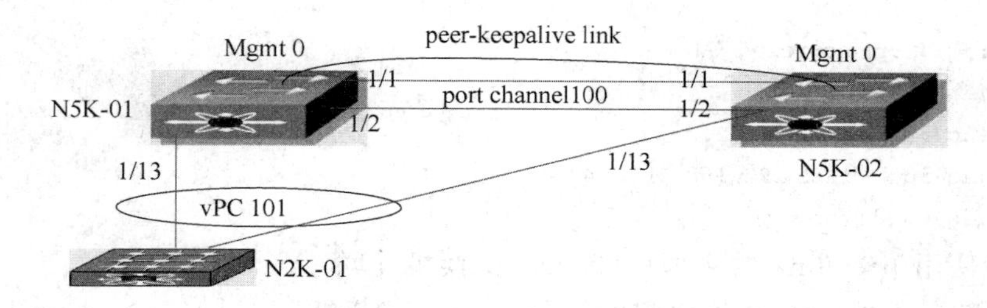

图 2-6 vPC 设备拓扑

(1) 开启 vPC 特性

```
N5K-1(config)# feature vpc
! 启用 vPC 功能
```

```
N5K-2(config)# feature vpc
N5K-1(config)# feature lacp
! 为链路汇聚控制协议
N5K-2(config)# feature lacp
```

(2) 建立 vPC domain 及增加 vPC peer-keepalive 链路

一台设备属于且只能属于一个 vPC Domain，一个 vPC Domain 有且只能拥有两个成员。Domain 在配置时，需要指定 vPC 对端设备地址用以传递两边的状态信息，且 vPC domain 号不能与 port channel number 一样。

```
N5K-1(config)# vpc domain 10
N5K-1(config-vpc-domain)# role priority 110
! 配置 N5K-1 为 vPC 主用，低值将被选举为 vPC 主交换机
N5K-1(config-vpc-domain)#  peer-keepalive destination 192.168.10.254 source 192.168.10.253
N5K-2(config)# vpc domain 10
N5K-2(config)#  role priority 120
N5K-2(config-vpc-domain)#  peer-keepalive destination 192.168.10.253 source 192.168.10.254
```

为 vPC peer-keepalive 链路创建独立的 VRF 接口来保证其通信，缺省将管理口设置成 VRF 接口，并且接入一个单独的交换机上，如果不使用缺省的管理口交换 peer-keepalive 信息，必须利用独立的物理口接口完成此功能，这样会占用一个物理端口，所以通常用管理口实现 peer-keepalive 功能，管理口做 peer-keepalive link 的拓扑，如图 2-6 所示。

用管理口做 peer-keepalive link 的配置：

```
N5K-1(config)# vrf context management
N5K-1(config)int mgmt0
N5K-1(config)ip address 192.168.10.253/24
N5K-2(config)# vrf context management
N5K-2(config)int mgmt0
N5K-2(config)ip address 192.168.10.254/24
```

如果不用管理口做 peer-keepalive 端口，命令格式参照如下：

```
N5K-1(config)# vrf context PKAL
N5K-1(config)int e1/1
N5K-1(config)vrf member PKAL
N5K-1(config)ip address 192.168.10.253/24
N5K-1(config)no shutdown

N5K-2(config)# vrf context PKAL
N5K-2(config)int e1/1
N5K-2(config)vrf member PKAL
N5K-2(config)ip address 192.168.10.254/24
N5K-2(config)no shutdown
```

(3) 配置使用 peer-link 的物理口(Nexus 交换机互联)

同一级 N5K 或同一级 N7K 之间的 port channel 链路称为 peer-link，peer-link 是 vPC 转发机箱间流量的链路，链路必须使用 10G 以太网，配置手册推荐使用至少 2 条 10G 以太网电缆进行捆绑。

注意：switchport mode fabricpath 隐含为 Trunk 模式。

配置 vPC peer-link：

```
N5K-1(config)# interface port-channel 100
N5K-1(config-if)# switchport mode trunk
N5K-1(config-if)# vpc peer-link
N5K-2(config)# interface port-channel 100
N5K-2(config-if)# switchport mode trunk
N5K-2(config-if)# vpc peer-link

N5K-1(config)# interface Ethernet1/1-2
N5K-1(config-if)# switchport mode trunk
N5K-1(config-if)# channel-group 100 mode on
N5K-2(config)#  interface Ethernet1/1-2
N5K-2(config-if)# switchport mode trunk
N5K-2(config-if)#  channel-group 100 mode on
```

(4) 配置 Nexus 2000 到 vPC 101

将一段设备连接到两侧设备链路推入各自的 EthernetChannel 组，并且将参加配置的 EthernetChannel 加入 vPC 组，保证对应的 EthernetChannel 在相同的转发 vPC 当中，便完成了该配置。

```
N5K-1(config)# interface port-channel 101
N5K-1(config-if)# switchport  mode fex-fabric
N5K-1(config-if)# fex associate 101
N5K-1(config-if)# vpc 101
! 这里将 EthernetChannel 加入 vPC 组
N5K-1(config)# interface Ethernet 1/13
N5K-1(config)# switchport mode fex-fabric
N5K-2(config)#  fex associate 101
N5K-1(config)# channel-group 101
! 这里将一个接口推入定义的一个组

N5K-2(config)# interface port-channel 101
N5K-2(config-if)# switchport  mode fex-fabric
N5K-2 (config-if)# fex associate 101
N5K-2(config-if)# vpc 101
N5K-2(config)# interface Ethernet 1/13
N5K-2(config)# switchport mode fex-fabric
N5K-2(config)# fex associate 101
N5K-2(config)# channel-group 101
```

vPC 配置的两端都必须是相同的 Trunk 配置，如 LACP(链路汇聚控制协议)或 no protocol(无协议)。

一致的 LACP System priority，有利于 vPC 状态下 LACP 的收敛，推荐的配置为 vPC 成员拥有相同的值。命令需要在全局和 vPC 配置模式下使用。

(5) 保存配置

```
N5K-1(config)#  copy running-config startup-config
N5K-2(config)#  copy running-config startup-config
```

2. 管理 vPC 常用命令

(1) 显示 vPC 简要信息

```
N5K-1(config)# show vpc brief
Legend:
                    (*) - local vpc is down, forwarding via vPC peer-link
vPC domain id                       : 10
Peer status                         : peer adjacency formed ok
vPC keep-alive status               : peer is alive
Configuration consistency status    : success
vPC role                            : primary
```

(2) 查看统计状态信息

```
N5K-1# show vpc statistics peer-link
 port-channel1 is up
   Hardware: Port-Channel, address: 000d.eca3.9da5 (bia 000d.eca3.9da5)
   MTU 1500 bytes, BW 160000000 Kbit, DLY 10 usec,
       reliability 255/255, txload 1/255, rxload 1/255
   Encapsulation ARPA
   Port mode is trunk
   full-duplex, 10 Gb/s
   Beacon is turned off
   Input flow-control is off, output flow-control is off
   Switchport monitor is off
   Members in this channel: Eth1/1, Eth1/2, Eth1/3, Eth1/4, Eth1/21, Eth1/22, Eth1/23,
Eth1/24, Eth1/25, Eth1/26, Eth1/27, Eth1/28, Eth1/29, Eth1/30, Eth1/31, Eth1/32
   Last clearing of "show interface" counters never
   30 seconds input rate 2645539776 bits/sec, 4842090 packets/sec
   30 seconds output rate 1184009112 bits/sec, 2177802 packets/sec
   Load-Interval #2: 5 minute (300 seconds)
     input rate 2.48 Gbps, 5.55 Mpps; output rate 1.22 Gbps, 2.24 Mpps
   RX
     356216924324 unicast packets  11366280385 multicast packets  0 broadcast packets
     367583204709 input packets  25317981509171 bytes
     0 jumbo packets  0 storm suppression packets
     0 runts  0 giants  0 CRC  0 no buffer
     0 input error  0 short frame  0 overrun  0 underrun  0 ignored
     0 watchdog  0 bad etype drop  0 bad proto drop  0 if down drop
     0 input with dribble  0 input discard
     81372137 Rx pause
   TX
     95309974177 unicast packets  2491785887 multicast packets  0 broadcast packets
     97801760066 output packets  6605708744882 bytes
     0 jumbo packets
     2 output errors  0 collision  0 deferred  0 late collision
     0 lost carrier  0 no carrier  0 babble
     377039096 Tx pause
   7 interface resets
```

注意:vPC 功能不需要额外的 License 许可。

其他常用命令如下所示。

show feature:显示 vPC 是否启用。

show port-channel capacity:显示配置多少 port-channel 及几条可用。

show running-config vpc:显示 vPC 运行的配置。

show spanning-tree summary:显示 spanning-tree 端口状态总结,包括 vPC peer 交换机状态。

show vpc brief:显示 vPC 简略信息。

show vpc consistency-parameters:显示穿过所有 vPC 接口相同的参数状态。

show vpc peer-keepalive:显示 peer-keepalive 信息。

show vpc role:显示 peer 状态、本地设备角色、vPC 系统 MAC 地址、系统优先级、MAC 地址及本地 vPC 设备的优先级。

show vpc statistics:显示 vPC 状态。

2.2.6 路由配置

1. 静态路由的配置

静态路由配置格式如下:

```
Nexus7000(config)#
Nexus7000(config)# vrf context production
Nexus7000(config-vrf)# ip route 0.0.0.0 0.0.0.0 192.168.10.2
```

检查静态路由情况命令:

```
Nexus7000# show ip route static
IP Route Table for vrf  "default"
'*' denotes best ucast next-hop          '**' denotes best mcast next-hop
'[x/y]' denotes [preference/metric]
192.168.2.1/32, 1 ucast next-hops, 0 mcast next-hops
    *via 192.168.10.2, Ethernet1/13, [1/0], 00:00:13, static
```

2. OSPF 路由的配置

OSPFv2 是一个在 RFC2328 中定义的标准路由协议。在启动 OSPF 进程或在端口下启动 OSPF 前,首先应该激活 OSPF 功能,OSPF 应该在接口下配置而不是在进程中配置。ip ospf passive-interface 命令应该应用在接口下,注意不同类型路由器或不同厂家对接的 hello 间隔和 dead 间隔可能不通,下面是配置的例子。

(1) 配置 OSPF 命令如下:

```
Nexus7000(config)# feature ospf
Nexus7000(config)# router ospf 10
Nexus7000(config-router)# router-id 192.168.1.1
Nexus7000(config)# interface ethernet 1/13
Nexus7000(config-if)# ip address 192.168.10.1 255.255.255.0
Nexus7000(config-if)# ip router ospf 10 area 0
```

(2) 验证 OSPFv2 路由表：

```
Nexus7000＃ show ipospf route
Nexus7000＃ show route
IP Route Table for VRF "default"
'*' denotes best ucast next-hop
'**' denotes best mcast next-hop
'[x/y]' denotes [preference/metric]

0.0.0.0/0, ubest/mbest: 1/0
 *via 192.168.10.254, Eth2/1, [1/0], 00:17:48, static
1.1.1.1/32, ubest/mbest: 1/0
 *via 192.168.10.93, Eth2/1, [110/20], 00:00:32, ospf-10, type-2
```

(3) 验证 OSPF 邻居表：

```
Nexus7000＃
Nexus7000＃ show ospf neighbors
OSPF Process ID 10 VRF default
Total number of neighbors: 1
Neighbor ID Pri State Up Time Address Interface
6.6.6.7 1 FULL/BDR 00:13:02 192.168.10.93 Eth2/1
Nexus7000＃ restartospf 10
! 重启进程
Nexus7000＃ clear ipospf 10 neighbor ?
! 清除邻居关系
   *                   Clear all neighbors
   A.B.C.D             Source IP address, or router ID of the neighbor
ethernet               Ethernet IEEE 802.3z
loopback               Loopback interface
port-channel           Port Channel interface
```

2.3 Nexus 高级配置选项

2.3.1 VDC 配置

VDC 是 Nexus 7000 系列的特色功能。通过将物理机箱划分为多个逻辑交换机，核心交换机区域将可以获得多台物理隔离的高性能交换机。VDC 具有完全隔离的路由表、VRF 和接口，因此可以获得真实交换机属性的配置。

VDC 的资源是占用全局机箱的，因此在必要的时候，需要通过调整 VDC 资源配置来进行 VDC 功能和性能的调整。所有进入 VDC 的接口和资源都不能被其他 VDC 或缺省 VDC 使用。

(1) 分配 VDC 资源。

```
N7K＃ vdc vdc2_1 id 2
  allocate interface Ethernet1/13-24
  allocate interface Ethernet2/1-3
```

```
boot-order 1
limit-resource vlan minimum 16 maximum 4094
limit-resource monitor-session minimum 0 maximum 2
limit-resource monitor-session-erspan-dst minimum 0 maximum 23
limit-resource vrf minimum 2 maximum 1000
limit-resource port-channel minimum 0 maximum 768
limit-resource u4route-mem minimum 8 maximum 8
limit-resource u6route-mem minimum 4 maximum 4
limit-resource m4route-mem minimum 8 maximum 8
limit-resource m6route-mem minimum 2 maximum 2
```

(2) 通过命令 switchto vdc,可以查看当前 VDC 的数量和状态。系统机箱本身默认为 VDC1,最多可以建立 3 个另外的 VDC。登录到系统默认的 switch 下,可以通过 switch to vdc 命令在不同的 VDC 之间跳转。

```
N7K# switch to vdc vdc2_1
Last login: Thu Nov 25 16:40:19 UTC 2010 on ttyS0
Last login: Thu Nov 25 17:06:47 on ttyS0
Cisco Nexus Operating System (NX-OS) Software
TAC support: http://www.cisco.com/tac
Copyright (c) 2002-2010, Cisco Systems, Inc. All rights reserved.
The copyrights to certain works contained in this software are
owned by other third parties and used and distributed under
license. Certain components of this software are licensed under
the GNU General Public License (GPL) version 2.0 or the GNU
Lesser General Public License (LGPL) Version 2.1. A copy of each
such license is available at
http://www.opensource.org/licenses/gpl-2.0.php and
http://www.opensource.org/licenses/lgpl-2.1.php
switch-vdc2_1#
```

当位于其他 VDC 当中时,无法通过 switch to vdc 的方式进行 VDC 的跳转。系统保存配置和 reload 都有针对单独 VDC 的配置。

不同 VDC 的名称,除了在 VDC 命令中直接指定,还可以进入到 VDC 配置界面后,直接用 hostname 命令进行更改。

(3) 新建一个 VDC 命令。

```
N7000(config)# vdc VDC02
N7000(config-vdc)# allocate interface Ethernet 1/1-8
! 给 VDC 分配物理接口
NEXUS7000#switchto vdc VDC02
! 登录到 VDC02 上
```

(4) 检查当前 VDC 状态。

```
switch(config)# show vdc

vdc_idvdc_name state mac type lc
------   --------   -----   ----------   ---------   ------
1 switch active 00:26:98:0d:01:41 AdminNone
```

```
2 vdc2 active 00:26:98:0d:01:42 Ethernet m1 f1 m1xl m2xl
3 vdc3 active 00:26:98:0d:01:43 Ethernet f2
4 new-vdc active 00:26:98:0d:01:44 Ethernet m1 f1 m1xl m2xl
switch(config)#
```

2.3.2 FabricPath 配置

在数据中心的每个要运行 FabricPath 的设备上输入下列命令，即可开启数据中心的 FabricPath。思科的 FabricPath 与 Trill 基本相同，并做了很多改进。图 2-7 为 Nexus 7000 和 Nexus 5000 4 台设备之间配置 FabricPath 的情形。

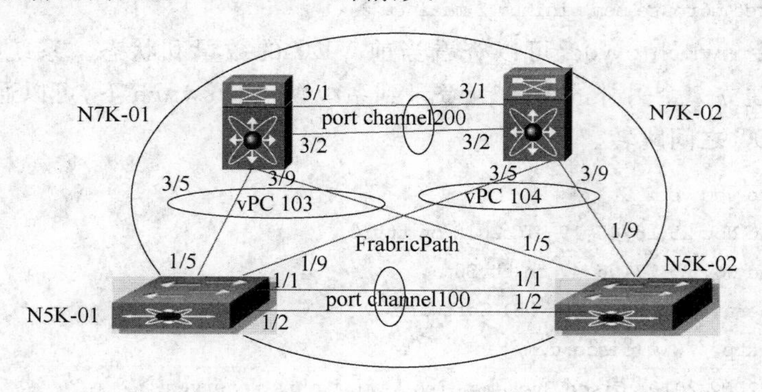

图 2-7　某数据中心 FabricPath 的实现

(1) 启用 FabricPath 命令。

```
N7K(config)# install feature-set fabricpath
N7K(config)# feature-set fabricpath

N7K(config)# interface port-channel201
N7K(config-if)# switchport mode fabricpath

N7K (config)# interface ethernet 3/5
N7K (config-if)# switchport mode fabricpath
N7K (config-if)# channel-group 201 mode active
N7K (config-if)#
```

以下为可选配置：

```
N7K (config)# fabricpath topology 1
! 指定 FabricPath 号，类似 OSPF 自治区域号
N7K (config-fabricpath)# member vlan 10

vlan 10
! 参与 FabricPath 的 VLAN
  mode fabricpath
  name Network-Deivce-Mgmt

interface port-channel100
! 参与 FabricPath 的 port-channel
  switchport
  switchport mode fabricpath
```

```
  medium p2p
  vpc peer-link
  fabricpath topology-member 1

fabricpath switch-id 201
fabricpath domain default
  maximum-paths 128
  root-priority 110
```

注意:FabricPath 需要相应的 License。

(2) 为传统(CE)VLAN 配置 MAC 学习模式(可选配置项)。

```
mac address-table learning-mode conversational vlan vlan-id
```

传统 VLAN 用传统的 MAC 地址学习方式,你可以修改成 conversational MAC 地址学习模式。

(3) 配置远端 MAC 学习模式(可选配置项)。

```
mac address-table fabricpath remote-learning
```

缺省 MAC 地址学习是 enable,可以在管理的板卡内 disable。

(4) 为 Core Port 配置 MAC 学习模式(可选配置项)。

```
hardware fabricpath mac-learning module module_number {port-group port_group}
```

缺省 MAC 地址学习是 enable,可以在管理的 VDC 内 disable。

(5) 配置 Switch ID(可选配置项)。

```
fabricpath switch-id value
```

缺省情况下由 FabricPath 给每个 FabricPath 设备分配 ID,修改交换机 ID 不会中止流量的损失。

(6) 配置 FabricPath 计时器(可选配置项)。

```
fabricpath timers {allocate-delay seconds | linkup-delay seconds seconds | linkup-dalay always |
transition-delay seconds}
```

这 4 个计时器如果修改的话,请注意保持全网的一致性。

(7) 关闭 FabricPath Link-Delay(可选配置项)。

```
fabricpath linkup-delay
```

此命令加速网络收敛,如果网络静态配置交换机 ID,在没有冲突的情况下且无须检测冲突时,可以在全网交换机上做此配置,正常情况下强烈建议不要关闭 linkup-delay。

(8) 关闭 FabricPath Graceful Merge(可选配置项)。

```
fabricpath graceful-merge disable
```

如果关闭此选项需要关闭 FabricPath 网络里所有的交换机,缺省是 enable, 且在 disable 时会引起丢包。

(9) 为单播和组播配置 TTL(可选配置项)。

```
fabricpath ttl unicast
fabricpath ttl multicast
```

缺省情况下,FabricPath 会分配一个 TTL 值,但是可以修改它,它在分组从边缘端口进入时进行标记,在穿过核心端口时减掉。

(10) 强制链路启用(可选配置项)。

```
fabricpath force link-bringup
```

此命令只在网络出现故障及 FabricPath 链路没有正常启动时使用，正常时请不要使用，该命令在 copy running-config startup-config 时不会保存。

(11) 检查命令。

检查开启的功能模块命令：

```
Nexus7000# show feature-set
Feature Set Name      ID      State
--------------------  ------  --------
fabricpath            2       enabled
virtualization        5       installed

Nexus7000# show fabricpath topology detail
topo-Description                       Topo-ID    Topo-State
-------------------------------------  ---------  ------------------
0                                      0          Up     Reason : --
1                                      1          Up     Reason : --
```

检查阵列路径交换 ID：

```
Nexus7000# show fabricpath switch-id table
Legend: '*'- this system
==========================================================================
SWITCH-ID     SYSTEM-ID         FLAGS         STATE       STATIC  EMULATED
-----------+----------------+-----------+------------+--------------------
101           002a.6af8.2841    Primary    Confirmed     Yes      No
*102          002a.6af8.2741    Primary    Confirmed     Yes      No
201           e4c7.2212.8cc1    Primary    Confirmed     Yes      No
202           5087.8940.5dc1    Primary    Confirmed     Yes      No
1000          5087.8940.5dc1    Primary    Confirmed     No       Yes
1000          e4c7.2212.8cc1    Primary    Confirmed     No       Yes
1100          002a.6af8.2741    Primary    Confirmed     No       Yes
1100          002a.6af8.2841    Primary    Confirmed     No       Yes
Total Switch-ids: 8
```

2.3.3 FCoE 的配置

FCoE 是在以太网上提供存储 FC 流量的技术，以太网需要提供一个全双工、无丢包的环境，无丢包的行为是在以太网中运行一个基于优先级的流量控制(Priority Flow Control, PFC)的机制，保证在数据包冲突时防止丢失数据。

FCoE 功能思科只在 10G 端口上支持，且需要 License 许可，没有 License 情况下仍然可以启用 FCoE，但是重启后所有 FCoE 命令自动消失。通常情况下，Nexus 交换机接口与融合网络适配器(Converged Network Adapter, CNA)网卡之间会自动协商运行参数，如果协商不成则手动配置。

配置步骤如下所示。

(1) 启用 FCoE，命令如下：

```
N7K# configure terminal
N7K(config)# feature fcoe
2008 Nov 11 20:43:38 switch %$ VDC-1 %$ %PFMA-2-FC_LICENSE_DESIRED:
FCoE/FC feature will be enabled after the configuration is saved followed by a reboot
```

注意:FCoE 启用后,需要重启 Nexus 5000/7000 交换机才能生效。

(2) 在接口启用 FCoE 功能,命令如下:

```
N7K# configure terminal
N7K(config)# interface ethernet 1/4
N7K(config-if)# fcoe mode on
```

(3) PFC 功能在自能协商时会启用,也可以强制启用或关闭,命令如下:

```
N7K# configure terminal
N7K(config)# interface ethernet 1/2
N7K(config-if)# priority-flow-control mode on
N7K(config-if)#no  priority-flow-control mode on
```

(4) 检查 FCoE 是否开启,命令如下:

```
N7K# show fcoe
FCoE/FC feature is desired.
```

2.3.4 OTV 配置

重叠传输虚拟化(Overlay Transport Virtualization,OTV)是思科为了解决在三层网络中提供二层互通的问题提出的解决方案,这个技术主要应用在解决多个云计算数据中心跨站点的问题。

1. OTV 基本配置

OTV 术语

- Edge Device:负责执行 OTV 功能的设备,并且将站点连接到 WAN/MAN 互联网。
- Internal Interface:ED 设备中面向站点内部的二层接口,会接收 STP BPDU 信息。
- Join Interface:ED 设备中面向核心网络的三层可路由的接口,实际上用它转发 OTV 数据包给远程站点的 ED 设备。
- Overlay Interface:ED 设备上的虚拟接口,封装二层帧到 IP 单播或组播包中,构成 OTV 数据包。
- Multicast Control-Group:所有的 ED 设备使用一个相同的组播地址发送以及交换 OTV 控制协议的更新。
- Multicast Data-Group:处理组播数据流量地址,配合子网掩码,是一组专用 IP 范围。
- Authoritative Edge Device:同一站点如果有多个 ED 设备,只有一个授权 ED 处理 VLAN 内 MAC 地址转发,每个 VLAN 可以指定不同的 AED。

下面以图 2-8 为例,介绍 OTV 的基本配置。

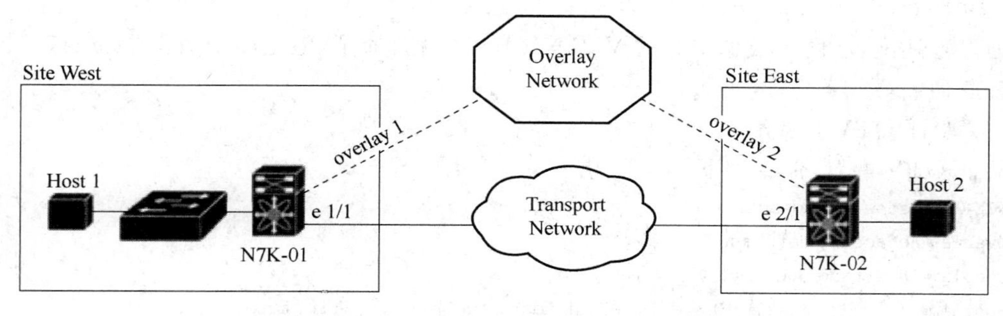

图 2-8 OTV 组网模型

(1) 在两边数据中心都有的配置

开启 OTV 功能,命令如下:

```
featureotv
```

两边需要打通的 VLAN,命令如下:

```
vlan 101-104
```

两边 OTV 协议通信的 VLAN,命令如下:

```
otv site-vlan100
```

(2) 数据中心 Site West 配置

配置 OTV 标识,命令如下:

```
N7K-01(config)# otv site-dentifier WestSize
N7K-01(config)# interface Ethernet1/1
! 两端需要 OTV 连接的端口
N7K-01(config-if)# ip address 10.1.1.1/24
N7K-01(config-if)# ip igmp version 3
N7K-01(config-if)# no shutdown
N7K-01(config)# interface  Overlay 1
N7K-01(config-if)description site sitewest
N7K-01(config-if)otv join-interface Ethernet1/1
N7K-01(config-if)otv control-group 239.1.1.1
N7K-01(config-if)otv data-group 232.1.1.0/28
N7K-01(config-if)otv extend-vlan 101-104
N7K-01(config-if)no shutdown
```

(3) 数据中心 Site East 配置

配置 OVT 标识,命令如下:

```
N7K-02(config)# otv site-dentifier EastSize
N7K-02(config)interface Ethernet2/1
N7K-02(config-if)ip address 10.1.1.2/24
N7K-02(config-if)ip igmp version 3
N7K-02(config-if)no shutdown
N7K-02(config)interface Overlay2
N7K-02(config-if)description site siteeast
N7K-02(config-if)otv join-interface Ethernet2/1
N7K-02(config-if)otv control-group 239.1.1.1
N7K-02(config-if)otv data-group 232.1.1.0/28
N7K-02(config-if)otv extend-vlan 101-104
N7K-02(config-if)no shutdown
```

通过上述配置,两个数据中心的 VLAN101-104 就打通了,但是两个数据中心的三层首先要保证联通。

(4) 验证 OTV 配置

验证 OTV 配置,命令如下:

```
N7K-02# show otv adjacency
Overlay Adjacency database
Overlay-Interface Overlay1 :
Hostname         System-ID         Dest Addr      Up time    Adj-state
DC-B-N7010-VDC   0026.51c3.af44    10.1.1.2       04:20:22   UP
```

注意：OTV 功能有 License 要求，配置前确认设备有 OTV 功能许可。

2. OTV 高级选项配置

(1) OTV 底层依靠 IS-IS 建立连接，进入 OTV IS-IS VPN 配置模式。

```
N7K-02# configure terminal
N7K-02(config)#otv-isis default
N7K-02(config-router)#vpn Test1
N7K-02(config-router-vrf)#
```

(2) 配置 Edge Devices 认证，认证可以是明文也可以是 MD5 方式。

```
N7K-02# configure terminal
N7K-02(config)# interface overlay 1
N7K-02(config-if-overlay)# otv isis
N7K-02(config-if-overlay)# otv isis authentication-type md5
N7K-02(config-if-overlay)# otv isis authentication keychain OTVKeys
```

(3) 配置 OTV 控制平面协议数据单元(PDU)认证。

```
N7K-02# configure terminal
N7K-02(config)# otv-isis default
N7K-02(config-router)# vpn Marketing
N7K-02(config-router-vrf)# authentication-check
N7K-02(config-router-vrf)# authentication-type md5
N7K-02(config-router-vrf)# authentication keychain OTVKeys
```

(4) 配置 OTV 相邻服务器功能选项。

```
N7K-02# configure terminal
N7K-02(config)# interface overlay 1
N7K-02(config-if-overlay)# otv adjacency-server unicast-only
! 只允许 OTV 相邻服务器单播功能
N7K-02(config-if-overlay)# otv use-adjacency-server 192.0.2.1 unicast-only
```

(5) 配置 OTV 邻居 ARP 发现超时值。

```
N7K-02# configure terminal
N7K-02(config)# interface overlay 1
N7K-02(config-if-overlay)# otv arp-nd timeout 70
! 单位秒，缺省值 480 s
```

(6) 关闭 ARP 邻居发现缓存。

```
N7K-02# configure terminal
N7K-02(config)# interface overlay 1
N7K-02(config-if-overlay)# no otv suppress-arp-nd
```

(7) 配置指定设备单播泛洪。

```
N7K-02# configure terminal
N7K-02(config)# otv flood mac 0000.ffaa.0000 vlan 328
```

(8) 配置 OTV VLAN 映射。

当两个数据中心 VLAN 不一致时，做 VLAN 映射。

```
N7K-02# configure terminal
N7K-02(config)# interface overlay 1
N7K-02(config-if-overlay)# otv vlan mapping 1-5 to 7-11
```

(9) 配置 OTV 网络专门的组播。

```
N7K-02# configure terminal
```

```
N7K-02(config)# interface overlay 1
N7K-02(config-if-overlay)# otv broadcast-group 225.1.1.10
```

(10) 配置 OTV 快速收敛。

```
Feature OTV
Feature BFD
N7K-02# configure terminal
N7K-02(config)# feature interface-vlan
N7K-02(config)# interface vlan 2500
N7K-02(config-if)# no ip redirects
N7K-02(config-if)# ip address 172.1.2.1 255.255.255.0
N7K-02(config-if)# no shutdown
```

(11) 配置快速故障检测。

```
Feature OTV
Feature BFD
N7K-02# configure terminal
N7K-02(config)# otv-isis default
N7K-02(config-router)# track-adjacency-nexthop
N7K-02(config-router)# exit
N7K-02(config)# otv site-vlan 10
N7K-02(config-site-vlan)# otv isis bfd
```

(12) 配置 MAC 地址重分布到不同的数据中心。

```
N7K-02# configure terminal
N7K-02(config)# otv-isis default
N7K-02(config-router)# vpn Marketing
N7K-02(config-router-vrf)# redistribute filter route-map otvFilter
```

(13) OTV 负载均衡。

一个中心在有多个 AED 且具有相同 VLAN 的情况下,配置成负载均衡模式。

```
N7K-02# configure terminal
N7K-02(config)# otv site-vlan 10
```

(14) OTV 参数调整。

注意:下列参数为缺省值,一般不作修改。

```
N7K-02# configure terminal
N7K-02(config)# interface overlay 1
N7K-02(config-if-overlay)# otv isis csnp-interval 100
N7K-02(config-if-overlay)# otv isis hello-interval 30
N7K-02(config-if-overlay)# otv isis hello-multiplier 30
N7K-02(config-if-overlay)# otv isis hello-padding
N7K-02(config-if-overlay)# otv isis lsp-interval 30
N7K-02(config-if-overlay)# otv isis metric 30
N7K-02(config-if-overlay)# otv isis lsp-interval 30
```

(15) 关闭带 IP 地址池的 Tunnel 极化。

```
N7K-02# configure terminal
N7K-02(config)# feature otv
N7K-02(config)# otv depolarization disable
N7K-02(config)# exit
```

3. OTV 检查命令

表 2-1 列出了常用检查 OTV 运行状态的命令。

表 2-1　OTV 常用命令

命　令	含　义
show running-configuration otv [all]	显示当前运行的 OTV 配置
show otv overlay [*interface*]	显示 overlay 接口信息
show otv adjacency [detail]	显示 overlay 网络相邻信息
show otv [overlay *interface*] [vlan [*vlan-range*] [authoritative | detail]]	显示与 overlay 接口关联的 VLAN 信息
show otv isis site [database | statistics]	显示本地和邻接 edge device 的 BFD 配置
show otv site [all]	显示本地 OTV 信息
show otv [route [*interface* [neighbor-address *ip-address*]] [vlan *vlan-range*] [*mac-address*]]	显示 OTV 路由信息
show otv mroute vlan *vlan-id* startup	显示来自 ORIB(OTV Routing Information Base) VLAN 的组播信息
show forwarding distribution otv multicast route vlan *vlan-id*	显示指定 VLAN FIB(Forwarding Information Base) OTV 组播信息
show otv vlan-mapping [overlay *interface-number*]	显示本地和远端 VLAN 的转换映射
show mac address-table	显示 MAC 地址信息
show otv internal adjacency	显示 overylay 网络附加信息

2.4　Nexus 交换机日常维护命令

(1) 查看版本信息。

```
N7K-02# show version
! 使用 show version 命令以获取软件和硬件的信息
Cisco Nexus Operating System (NX-OS) Software
TAC support: http://www.cisco.com/tac
Copyright (c) 2002-2008, Cisco Systems, Inc. All rights reserved.
The copyrights to certain works contained in this software are
owned by other third parties and used and distributed under
license. Certain components of this software are licensed under
the GNU General Public License (GPL) version 2.0 or the GNU
Lesser General Public License (LGPL) Version 2.1. A copy of each
such license is available at
http://www.opensource.org/licenses/gpl-2.0.php and
http://www.opensource.org/licenses/lgpl-2.1.php
Software
  BIOS:      version 3.15.0
loader:      version N/A
```

```
kickstart:  version 4.0(1)
system:     version 4.0(1)
BIOS compile time:    03/04/08
kickstart image file is: bootflash:/n7000-s1-kickstart.4.0.1.bin
kickstart compile time: 3/6/2008 2:00:00 [04/02/2008 08:12:57]
system image file is:  bootflash:/n7000-s1-dk9.4.0.1.bin
! 当前使用的 OS 版本
system compile time:  3/6/2008 2:00:00 [04/02/2008 08:58:14]
Hardware
cisco N7K-02 C7010 (10 Slot) Chassis ("Supervisor module-1X")
Intel(R) Xeon(R) CPU  with 4129620 kB of memory.
Processor Board ID JAB114000CC
Device name: n7000
bootflash: 2030616 kB
  slot0:              0 kB (expansion flash)
Kernel uptime is 0 day(s), 15 hour(s), 9 minute(s), 39 second(s)
```

(2) 验证阵列模块。

```
N7K-02# show module fabric
Xbar  Ports   Module-Type                                     Model              Status
---   -----   ----------------------------------------------- ------------------ ----------
1     0       Xbar                                            N7K-C7010-FAB-1    ok
! 阵列模块之一
2     0       Xbar                                            N7K-C7010-FAB-1    ok
3     0       Xbar                                            N7K-C7010-FAB-1    ok
Xbar  Sw              Hw      World-Wide-Name(s) (WWN)
---   --------------  ------  --------------------------------------------------
1     NA              0.404   --
2     NA              0.404   --
3     NA              0.404   --
Xbar MAC-Address(es)                                          Serial-Num
---  ----------------------------------------                 ----------
1     NA                                                      JAB114700X6
2     NA                                                      JAB114700WQ
3     NA                                                      JAB114700WZ
```

(3) 查看模块状态。

```
N7K-02# show module
! 使用 show module 的命令,以确定每个模块状态
Mod  Ports   Module-Type                          Model              Status
---  -----   -----------------------------------  ------------------ ----------
1    48      10/100/1000 Mbps Ethernet Module     N7K-M148GT-11      ok
2    48      10/100/1000 Mbps Ethernet Module     N7K-M148GT-11      ok
3    32      10 Gbps Ethernet Module              N7K-M132XP-12      ok
4    32      10 Gbps Ethernet Module              N7K-M132XP-12      ok
5    0        Supervisor module-1X                N7K-SUP1           active *
6    0        Supervisor module-1X                N7K-SUP1           ha-standby
Mod  Sw              Hw      World-Wide-Name(s) (WWN)
---  --------------  ------  --------------------------------------------------
```

```
1      4.0(1)        0.902      --
2      4.0(1)        0.902      --
3      4.0(1)        0.504      --
4      4.0(1)        0.504      --
5      4.0(1)        0.801      --
6      4.0(1)        0.801      --
Mod    MAC-Address(es)                           Serial-Num
---    --------------------------------------    ----------
1      00-1b-54-c1-00-38 to 00-1b-54-c1-00-6c    JAB114100WE
2      00-1b-54-c0-fe-cc to 00-1b-54-c0-ff-00    JAB114100WK
3      00-1b-54-c1-0b-cc to 00-1b-54-c1-0b-f0    JAB114602F7
4      00-1b-54-c1-0a-64 to 00-1b-54-c1-0a-88    JAB114602FD
5      00-1b-54-c0-fe-b8 to 00-1b-54-c0-fe-c0    JAB114000CC
6      00-1b-54-c0-ff-18 to 00-1b-54-c0-ff-20    JAB114000R1
```

（4）查看风扇运行状态。

```
N7K-02#  show environment fan
！使用 show environment fan 命令来验证机箱风扇和电源风扇的状态
Fan:
------------------------------------------------------------
Fan              Model              Hw        Status
------------------------------------------------------------
Fan1(sys_fan1)   N7K-C7010-FAN-S    0.409     Ok
Fan2(sys_fan2)   N7K-C7010-FAN-S    0.409     Ok
Fan3(fab_fan1)   N7K-C7010-FAN-F    0.209     Ok
Fan4(fab_fan2)   N7K-C7010-FAN-F    0.209     Ok
Fan_in_PS1       --                 --        Ok
Fan_in_PS2       --                 --        Ok
Fan_in_PS3       --                 --        Absent
Fan Air Filter : Absent
```

（5）查看功率。

```
N7K-02# show environment power
！检查电源功率的峰值及实际功耗
Power Supply:
Voltage: 50 Volts
Power                                 Actual      Total
Supply    Model                       Output      Capacity    Status
                                      (Watts)     (Watts)
1         N7K-AC-6.0KW                   0 W         0 W      Shutdown
2         N7K-AC-6.0KW                 712 W      3000 W      Ok
3         N7K-AC-6.0KW                1539 W      6000 W      Ok
                                      Actual      Power
Module    Model                       Draw        Allocated   Status
                                      (Watts)     (Watts)
-------   -------------------------   ---------   ---------   ------------
1         N7K-M132XP-12               N/A          750 W      Powered-Up
2         N7K-M132XP-12               N/A          750 W      Powered-Up
3         N7K-M148GT-11               N/A          400 W      Powered-Up
```

```
4      N7K-M148GT-11        N/A      400 W    Powered-Up
5      N7K-SUP1             N/A      210 W    Powered-Up
6      N7K-SUP1             N/A      210 W    Powered-Up
Xb1    N7K-C7010-FAB-1      N/A       60 W    Powered-Up
Xb2    N7K-C7010-FAB-1      N/A       60 W    Powered-Up
Xb3    N7K-C7010-FAB-1      N/A       60 W    Powered-Up
Xb4    xbar                 N/A       60 W    Absent
Xb5    xbar                 N/A       60 W    Absent
fan1   N7K-C7010-FAN-S      N/A      720 W    Powered-Up
fan2   N7K-C7010-FAN-S      N/A      720 W    Powered-Up
fan3   N7K-C7010-FAN-F      N/A      120 W    Powered-Up
fan4   N7K-C7010-FAN-F      N/A      120 W    Powered-Up
N/A - Per module power not available
Power Usage Summary:
--------------------
Power Supply redundancy mode (configured)            PS-Redundant
Power Supply redundancy mode (operational)           Non-Redundant
Total Power Capacity (based on configured mode)          9000 W
Total Power of all Inputs (cumulative)                   9000 W
Total Power Output (actual draw)                         2251 W
Total Power Allocated (budget)                           4700 W
Total Power Available for additional modules             4300 W
```

（6）查看环境温度。

```
N7K-02# show environmental temperature
```

！使用 show env temp 命令来检查每个模块温度值，如果 minor 或 major 的数值相同，那么 SNMP 将会发送报警消息

```
Temperature:
--------------------------------------------------------------------
Module   Sensor          MajorThresh   MinorThres   CurTemp     Status
                         (Celsius)     (Celsius)    (Celsius)
--------------------------------------------------------------------
1        Crossbar(s5)    105           95           35          Ok
1        CTSdev1 (s6)    115           105          67          Ok
1        CTSdev2 (s7)    115           105          59          Ok
1        CTSdev3 (s8)    115           105          54          Ok
1        CTSdev4 (s9)    115           105          50          Ok
1        CTSdev5 (s10)   115           105          47          Ok
1        CTSdev6 (s11)   115           105          51          Ok
1        CTSdev7 (s12)   115           105          46          Ok
1        CTSdev8 (s13)   115           105          48          Ok
1        CTSdev9 (s14)   115           105          43          Ok
1        CTSdev10(s15)   115           105          42          Ok
1        CTSdev11(s16)   115           105          39          Ok
1        CTSdev12(s17)   115           105          40          Ok
1        QEng1Sn1(s18)   115           110          44          Ok
1        QEng1Sn2(s19)   115           110          42          Ok
1        QEng1Sn3(s20)   115           110          40          Ok
```

```
1          QEng1Sn4(s21)   115             110         42         Ok
```

(7) 查看系统冗余状态。

```
N7K-02# show system redundancy status
! 用 show system redundancy status 命令来查看"Active/Standby"和高可用性(HA) 管理状态
Redundancy mode
---------------
      administrative:   HA
         operational:   HA
This supervisor (sup-1)
-----------------------
    Redundancy state:   Active
    Supervisor state:   Active
      Internal state:   Active with HA standby
Other supervisor (sup-2)
------------------------
    Redundancy state:   Standby
    Supervisor state:   HA standby
    Internal state:     HA standby
```

(8) 复制文件。

从根目录复制 samplefile 到目录 mystorage 下,命令如下:

```
N7K-02# copy slot0:samplefile slot0:mystorage/samplefile
```

从当前目录复制,命令如下:

```
N7K-02# copy samplefile mystorage/samplefile
```

复制激活引擎下的文件到备用引擎的 bootflash 里,命令如下:

```
N7K-02# copy bootflash:system_image bootflash:! sup-2/system_image
```

(9) 覆盖当前 NVRAM 中的配置。

```
N7K-02# copy nvram:snapshot-config nvram:startup-config

Warning: this command is going to overwrite your current startup-config:
Do you wish to continue? {y/n} [y] y
```

(10) 备份当前运行的配置文件到 bootflash。

```
N7K-02# copy system:running-config bootflash:my-config
```

(11) 备份启动文件到 TFTP 服务器。

```
N7K-02# copy startup-config tftp://192.168.10.100/my-config
```

(12) 将 License 文件上传到文件夹 bootflash,然后执行安装。

```
N7000#install license bootflash:license_file.lic
```

(13) 验证安装的许可证。

```
N7000#show license usage
```

(14) 当大量的接口需要分配相同的参数时,port-profile 是非常有用的。

```
N7000(config)# port-profile type Ethernet

N7000(config)# port-profile type ethernet Email-Template
! port-profile 的类型为"ethernet"
N7000(config-ppm)# switchport
```

```
N7000(config-ppm)# switchport access vlan 10
N7000(config-ppm)# spanning-tree port type edge
N7000(config-ppm)# no shutdown
N7000(config-ppm)# description Email Server Port
N7000(config-ppm)# state enabled
! 激活 port-profile

N7000(config)# interface ethernet 2/1-2
! 在端口上应用 port-profile
N7000(config-if-range)# inherit port-profile Email-Template
```

2.5 Nexus 交换机配置案例

如图 2-10(2.3.2 节)所示的案例的核心是两台 N7K-02 交换机，每台 N7K-02 交换机连接两台汇聚的 Nexus 5000 交换机，Nexus 之间互联都为万兆接口，N7K-02 之间多个接口绑定在 port channel200，Nexus 5000 之间多个万兆接口绑定在 port channel100，N5K-4 到 Nexus 7000 配置成 vPC104。

下面按照测试拓扑介绍相关配置，多余的配置为了不占篇幅已经去掉。

2.5.1 Nexus 5000-1 的配置

以下是 Nexus 5000-1 的设备配置：

```
N5K-01# sh run
! Command: show running-config
! Time: Sun Mar 29 06:01:45 2009

version 5.2(1)N1(4)
install feature-set fabricpath
feature-set fabricpath
hostname N5K-01

feature privilege
feature telnet
cfs eth distribute
feature interface-vlan
feature lacp
feature vpc
feature lldp
feature fex

username admin password 5 $1$Ool/4wdY$cWu8eARlRiHB99Rmv.hyH/
role network-admin
no password strength-check

banner motd #Nexus 5000 Switch
```

```
ip domain-lookup
class-map type qos class-fcoe
class-map type queuing class-fcoe
  match qos-group 1
class-map type queuing class-all-flood
  match qos-group 2
class-map type queuing class-ip-multicast
  match qos-group 2
class-map type network-qos class-fcoe
  match qos-group 1
class-map type network-qos class-all-flood
  match qos-group 2
class-map type network-qos class-ip-multicast
  match qos-group 2
fex 101
  pinning max-links 1
  description "FEX0101"
fex 103
  pinning max-links 1
  description "FEX0103"
fex 105
  pinning max-links 1

  description "FEX0105"
fex 107
  pinning max-links 1
  description "FEX0107"
fex 109
  pinning max-links 1
  description "FEX0109"
fex 111
  pinning max-links 1
  description "FEX0111"
fex 113
  pinning max-links 1
  description "FEX0113"
snmp-server user admin network-admin auth md5 0xd03c0e3b749e4e706da7c8093bdb4e67
priv 0xd03c0e3b749e4e706da7c8093bdb4e67 localizedkey

vrf context management
vlan 1
  mode fabricpath
vlan 10
  name Network-Device-Mgmt
  mode fabricpath
vlan 100
```

```
  name server1
vlan 112
  name iLO
  mode fabricpath
vpc domain 10
  role priority 110
  peer-keepalive destination 192.168.10.254 source 192.168.10.253
  peer-gateway
  auto-recovery
  fabricpath switch-id 1100
port-profile default max-ports 512

interface Vlan1

interface port-channel100
  description *****VPC+Peer-Link-To-N5K_02*****
  switchport mode fabricpath
  vpc peer-link

interface port-channel101
  description vpc to FEX101
  switchport mode fex-fabric
  fex associate 101

interface port-channel103
  description vpc to FEX103
  switchport mode fex-fabric
  fex associate 103

interface port-channel105
  description vpc to FEX105
  switchport mode fex-fabric
  fex associate 105

interface port-channel107
  description vpc to FEX107
  switchport mode fex-fabric
  fex associate 107

interface port-channel109
  description vpc to FEX109
  switchport mode fex-fabric
  fex associate 109

interface port-channel111
  description vpc to FEX111
  switchport mode fex-fabric
```

```
  fex associate 111

interface port-channel113
  description vpc to FEX113
  switchport mode fex-fabric
  fex associate 113

interface port-channel198
  description *****To-UCS6120_01-E1/1-4*****
  switchport mode trunk
  spanning-tree port type network
  vpc 198

interface port-channel199
  description *****To-UCS6120_02-E1/1-4*****
  switchport mode trunk
  spanning-tree port type network
  vpc 199

interface port-channel201
  description *****To-N7K_01-E1/3-8*****
  switchport mode fabricpath

interface port-channel203
  description *****To-N7K_02-E1/3-8*****
  switchport mode fabricpath

interface Ethernet1/1
  description *****VPC+Peer-Link-To-N5K_02*****
  switchport mode fabricpath
  channel-group 100 mode active

interface Ethernet1/2
  description *****VPC+Peer-Link-To-N5K_02*****
  switchport mode fabricpath
  channel-group 100 mode active

interface Ethernet1/3
  description *****VPC+Peer-Link-To-N5K_02*****
  switchport mode fabricpath
  channel-group 100 mode active

interface Ethernet1/4
  description *****VPC+Peer-Link-To-N5K_02*****
  switchport mode fabricpath
  channel-group 100 mode active
```

```
interface Ethernet1/5
  description *****To-N7K_01-E1/3-8*****
  switchport mode fabricpath
  channel-group 201 mode active

interface Ethernet1/6
  description *****To-N7K_01-E1/3-8*****
  switchport mode fabricpath
  channel-group 201 mode active

interface Ethernet1/7
  description *****To-N7K_01-E1/3-8*****
  switchport mode fabricpath
  channel-group 201 mode active

interface Ethernet1/8
  description *****To-N7K_01-E1/3-8*****
  switchport mode fabricpath
  channel-group 201 mode active

interface Ethernet1/9
  description *****To-N7K_02-E1/3-8*****
  switchport mode fabricpath
  channel-group 203 mode active

interface Ethernet1/10
  description *****To-N7K_02-E1/3-8*****
  switchport mode fabricpath
  channel-group 203 mode active

interface Ethernet1/11
  description *****To-N7K_02-E1/3-8*****
  switchport mode fabricpath
  channel-group 203 mode active

interface Ethernet1/12
  description *****To-N7K_02-E1/3-8*****
  switchport mode fabricpath
  channel-group 203 mode active

interface Ethernet1/13
  description vpc to FEX101
  switchport mode fex-fabric
  fex associate 101
  channel-group 101

interface Ethernet1/14
```

```
  description vpc to FEX101
  switchport mode fex-fabric
  fex associate 101
  channel-group 101

interface Ethernet1/15
  description vpc to FEX101
  switchport mode fex-fabric
  fex associate 101
  channel-group 101

interface Ethernet1/16
  description vpc to FEX101
  switchport mode fex-fabric
  fex associate 101
  channel-group 101

interface Ethernet1/17
  description vpc to FEX103
  switchport mode fex-fabric
  fex associate 103
  channel-group 103

interface Ethernet1/18
  description vpc to FEX103
  switchport mode fex-fabric
  fex associate 103
  channel-group 103

interface Ethernet1/19
  description vpc to FEX103
  switchport mode fex-fabric
  fex associate 103
  channel-group 103

interface Ethernet1/20
  description vpc to FEX103
  switchport mode fex-fabric
  fex associate 103
  channel-group 103

interface Ethernet1/21
  description vpc to FEX105
  switchport mode fex-fabric
  fex associate 105
  channel-group 105
```

```
interface Ethernet1/22
  description vpc to FEX105
  switchport mode fex-fabric
  fex associate 105
  channel-group 105

interface Ethernet1/23
  description vpc to FEX105
  switchport mode fex-fabric
  fex associate 105
  channel-group 105

interface Ethernet1/24
  description vpc to FEX105
  switchport mode fex-fabric
  fex associate 105
  channel-group 105

interface Ethernet1/25
  description vpc to FEX107
  switchport mode fex-fabric
  fex associate 107
  channel-group 107

interface Ethernet1/26
  description vpc to FEX107
  switchport mode fex-fabric
  fex associate 107
  channel-group 107

interface Ethernet1/27
  description vpc to FEX107
  switchport mode fex-fabric
  fex associate 107
  channel-group 107

interface Ethernet1/28
  description vpc to FEX107
  switchport mode fex-fabric
  fex associate 107
  channel-group 107

interface Ethernet1/29
  description vpc to FEX109
  switchport mode fex-fabric
  fex associate 109
  channel-group 109
```

```
interface Ethernet1/30
  description vpc to FEX109
  switchport mode fex-fabric
  fex associate 109
  channel-group 109

interface Ethernet1/31
  description vpc to FEX109
  switchport mode fex-fabric
  fex associate 109
  channel-group 109

interface Ethernet1/32
  description vpc to FEX109
  switchport mode fex-fabric
  fex associate 109
  channel-group 109

interface Ethernet1/33
  description vpc to FEX111
  switchport mode fex-fabric
  fex associate 111
  channel-group 111

interface Ethernet1/34
  description vpc to FEX111
  switchport mode fex-fabric
  fex associate 111
  channel-group 111

interface Ethernet1/35
  description vpc to FEX111
  switchport mode fex-fabric
  fex associate 111
  channel-group 111

interface Ethernet1/36
  description vpc to FEX111
  switchport mode fex-fabric
  fex associate 111
  channel-group 111

interface Ethernet1/37
  description vpc to FEX113
  switchport mode fex-fabric
  fex associate 113
  channel-group 113
```

```
interface Ethernet1/38
  description vpc to FEX113
  switchport mode fex-fabric
  fex associate 113
  channel-group 113

interface Ethernet1/39
  description vpc to FEX113
  switchport mode fex-fabric
  fex associate 113
  channel-group 113

interface Ethernet1/40
  description vpc to FEX113
  switchport mode fex-fabric
  fex associate 113
  channel-group 113

interface Ethernet1/41

interface Ethernet1/48

interface Ethernet2/1
  switchport mode trunk

interface Ethernet2/5
  description *****To-UCS6120_02-E1/1-4*****
  switchport mode trunk
  spanning-tree port type network
  channel-group 198 mode active

interface Ethernet2/6
  description *****To-UCS6120_02-E1/1-4*****
  switchport mode trunk
  spanning-tree port type network
  channel-group 198 mode active

interface Ethernet2/7
  description *****To-UCS6120_02-E1/1-4*****
  switchport mode trunk
  spanning-tree port type network
  channel-group 198 mode active

interface Ethernet2/8
  description *****To-UCS6120_02-E1/1-4*****
  switchport mode trunk
  spanning-tree port type network
```

```
  channel-group 198 mode active

interface Ethernet2/9

interface Ethernet2/16

interface mgmt0
  ip address 192.168.10.253/24

interface Ethernet101/1/1

interface Ethernet101/1/2

interface Ethernet101/1/31

interface Ethernet101/1/32
line console
line vty
boot kickstart bootflash:/n5000-uk9-kickstart.5.2.1.N1.4.bin

boot system bootflash:/n5000-uk9.5.2.1.N1.4.bin
fabricpath domain default
fabricpath switch-id 101
```

2.5.2 Nexus 5000-2 的配置

以下是 Nexus 5000-2 的配置：

```
sh run
! Command: show running-config
! Time: Sun Mar 29 06:06:27 2009

version 5.2(1)N1(4)
install feature-set fabricpath
feature-set fabricpath
hostname N5K-02

feature privilege
feature telnet
cfs eth distribute
feature interface-vlan
feature lacp
feature vpc
feature lldp
feature fex

username admin password 5 $1$uP20glel$tM8IMqS6UB2.ubwADeqCE0  role network-admin
no password strength-check
```

```
banner motd #Nexus 5000 Switch
ip domain-lookup
class-map type qos class-fcoe
class-map type queuing class-fcoe
  match qos-group 1
class-map type queuing class-all-flood
  match qos-group 2
class-map type queuing class-ip-multicast
  match qos-group 2
class-map type network-qos class-fcoe
  match qos-group 1
class-map type network-qos class-all-flood
  match qos-group 2
class-map type network-qos class-ip-multicast
  match qos-group 2
fex 102
  pinning max-links 1
  description "FEX0102"
fex 104
  pinning max-links 1
  description "EX0104"
fex 106
  pinning max-links 1
  description "EX0106"
fex 108
  pinning max-links 1
  description "EX0108"
fex 110
  pinning max-links 1
  description "EX0110"
fex 112
  pinning max-links 1
  description "EX0112"
fex 114
  pinning max-links 1
  description "EX0114"
snmp-server user admin network-admin auth md5 0x15a89f7692289c89597b5e4f5a541b71
priv 0x15a89f7692289c89597b5e4f5a541b71 localizedkey

vrf context management
vlan 1
vlan 10
  name Network-Device-Mgmt
  mode fabricpath
vlan 100
  name server1
  mode fabricpath
```

```
vlan 112
  name iLO
  mode fabricpath
vpc domain 10
  role priority 120
  peer-keepalive destination 192.168.10.253 source 192.168.10.254
  peer-gateway
  auto-recovery
  fabricpath switch-id 1100
port-profile default max-ports 512

interface Vlan1

interface Vlan10

interface port-channel100
  description ***** VPC + Peer-Link-To-N5K_01 *****
  switchport mode fabricpath
  vpc peer-link

interface port-channel102
  description vpc to FEX102
  switchport mode fex-fabric
  fex associate 102

interface port-channel104
  description vpc to FEX104
  switchport mode fex-fabric
  fex associate 104

interface port-channel106
  description vpc to FEX106
  switchport mode fex-fabric
  fex associate 106

interface port-channel108
  description vpc to FEX108
  switchport mode fex-fabric
  fex associate 108

interface port-channel110
  description vpc to FEX110
  switchport mode fex-fabric
  fex associate 110

interface port-channel112
  description vpc to FEX112
```

```
  switchport mode fex-fabric
  fex associate 112

interface port-channel114
  description vpc to FEX114
  switchport mode fex-fabric
  fex associate 114

interface port-channel202
  description *****To-N7K_01-E1/9-12*****
  switchport mode fabricpath

interface port-channel204
  description *****To-N7K_02-E3/9-12*****
  switchport mode fabricpath

interface Ethernet1/1
  description *****VPC+Peer-Link-To-N5K_01*****
  switchport mode fabricpath
  speed auto
  channel-group 100 mode active

interface Ethernet1/2
  description *****VPC+Peer-Link-To-N5K_01*****
  switchport mode fabricpath
  speed auto
  channel-group 100 mode active

interface Ethernet1/3
  description *****VPC+Peer-Link-To-N5K_01*****
  switchport mode fabricpath
  speed auto
  channel-group 100 mode active

interface Ethernet1/4
  description *****VPC+Peer-Link-To-N5K_01*****
  switchport mode fabricpath
  speed auto
  channel-group 100 mode active

interface Ethernet1/5
  description *****To-N7K_01-E1/3-8*****
  switchport mode fabricpath
  channel-group 202 mode active

interface Ethernet1/6
  description *****To-N7K_01-E1/3-8*****
```

```
  switchport mode fabricpath
  channel-group 202 mode active

interface Ethernet1/7
  description ***** To-N7K_01-E1/3-8 *****
  switchport mode fabricpath
  channel-group 202 mode active

interface Ethernet1/8
  description ***** To-N7K_01-E1/3-8 *****
  switchport mode fabricpath
  channel-group 202 mode active

interface Ethernet1/9
  description ***** To-N7K_02-E3/9-12 *****
  switchport mode fabricpath
  channel-group 204 mode active

interface Ethernet1/10
  description ***** To-N7K_02-E3/9-12 *****
  switchport mode fabricpath
  channel-group 204 mode active

interface Ethernet1/11
  description ***** To-N7K_02-E3/9-12 *****
  switchport mode fabricpath
  channel-group 204 mode active

interface Ethernet1/12
  description ***** To-N7K_02-E3/9-12 *****
  switchport mode fabricpath
  channel-group 204 mode active

interface Ethernet1/13
  description vpc to FEX102
  switchport mode fex-fabric
  fex associate 102
  channel-group 102

interface Ethernet1/14
  description vpc to FEX102
  switchport mode fex-fabric
  fex associate 102
  channel-group 102

interface Ethernet1/15
  description vpc to FEX102
```

```
  switchport mode fex-fabric
  fex associate 102
  channel-group 102

interface Ethernet1/16
  description vpc to FEX102
  switchport mode fex-fabric
  fex associate 102
  channel-group 102

interface Ethernet1/17
  description vpc to FEX104
  switchport mode fex-fabric
  fex associate 104
  channel-group 104

interface Ethernet1/18
  description vpc to FEX104
  switchport mode fex-fabric
  fex associate 104
  channel-group 104

interface Ethernet1/19
  description vpc to FEX104
  switchport mode fex-fabric
  fex associate 104
  channel-group 104

interface Ethernet1/20
  description vpc to FEX104
  switchport mode fex-fabric
  fex associate 104
  channel-group 104

interface Ethernet1/21
  description vpc to FEX106
  switchport mode fex-fabric
  fex associate 106
  channel-group 106

interface Ethernet1/22
  description vpc to FEX106
  switchport mode fex-fabric
  fex associate 106
  channel-group 106

interface Ethernet1/23
```

```
  description vpc to FEX106
  switchport mode fex-fabric
  fex associate 106
  channel-group 106

interface Ethernet1/24
  description vpc to FEX106
  switchport mode fex-fabric
  fex associate 106
  channel-group 106

interface Ethernet1/25
  description vpc to FEX108
  switchport mode fex-fabric
  fex associate 108
  channel-group 108

interface Ethernet1/26
  description vpc to FEX108
  switchport mode fex-fabric
  fex associate 108
  channel-group 108

interface Ethernet1/27
  description vpc to FEX108
  switchport mode fex-fabric
  fex associate 108
  channel-group 108

interface Ethernet1/28
  description vpc to FEX108
  switchport mode fex-fabric
  fex associate 108
  channel-group 108

interface Ethernet1/29
  description vpc to FEX110
  switchport mode fex-fabric
  fex associate 110
  channel-group 110

interface Ethernet1/30
  description vpc to FEX110
  switchport mode fex-fabric
  fex associate 110
  channel-group 110

interface Ethernet1/31
```

```
  description vpc to FEX110
  switchport mode fex-fabric
  fex associate 110
  channel-group 110

interface Ethernet1/32
  description vpc to FEX110
  switchport mode fex-fabric
  fex associate 110
  channel-group 110

interface Ethernet1/33
  description vpc to FEX112
  switchport mode fex-fabric
  fex associate 112
  channel-group 112

interface Ethernet1/34
  description vpc to FEX112
  switchport mode fex-fabric
  fex associate 112
  channel-group 112

interface Ethernet1/35
  description vpc to FEX112
  switchport mode fex-fabric
  fex associate 112
  channel-group 112

interface Ethernet1/36
  description vpc to FEX112
  switchport mode fex-fabric
  fex associate 112
  channel-group 112

interface Ethernet1/37
  description vpc to FEX114
  switchport mode fex-fabric
  fex associate 114
  channel-group 114

interface Ethernet1/38
  description vpc to FEX114
  switchport mode fex-fabric
  fex associate 114
  channel-group 114

interface Ethernet1/39
```

```
    description vpc to FEX114
    switchport mode fex-fabric
    fex associate 114
    channel-group 114

  interface Ethernet1/40
    description vpc to FEX114
    switchport mode fex-fabric
    fex associate 114
    channel-group 114

  interface Ethernet1/41

  interface Ethernet1/48

  interface Ethernet2/1

  interface Ethernet2/2
  …
  interface Ethernet2/16
  interface mgmt0
    ip address 192.168.10.254/24

  interface Ethernet102/1/1

  interface Ethernet102/1/32
  line console
  line vty
  boot kickstart bootflash:/n5000-uk9-kickstart.5.2.1.N1.4.bin

  boot system bootflash:/n5000-uk9.5.2.1.N1.4.bin
  fabricpath domain default
  fabricpath switch-id 102
```

2.5.3 N7K-02-1 的配置

以下是 N7K-02-1 的配置：

```
N7K-01# sh run
! Command: show running-config
! Time: Wed Nov  5 17:38:25 2014

version 6.2(2a)
hostname N7K-01
no system admin-vdc
install feature-set fabricpath
vdc N7K-01 id 1
  limit-resource module-type f2e
  allow feature-set fabricpath
  cpu-share 5
```

```
  allocate interface Ethernet3/1-48
  limit-resource vlan minimum 16 maximum 4094
  limit-resource monitor-session minimum 0 maximum 2
  limit-resource monitor-session-erspan-dst minimum 0 maximum 23
  limit-resource vrf minimum 2 maximum 4096
  limit-resource port-channel minimum 0 maximum 768
  limit-resource u4route-mem minimum 96 maximum 96
  limit-resource u6route-mem minimum 24 maximum 24
  limit-resource m4route-mem minimum 58 maximum 58
  limit-resource m6route-mem minimum 8 maximum 8
  limit-resource monitor-session-inband-src minimum 0 maximum 1

  limit-resource anycast_bundleid minimum 0 maximum 16
  limit-resource monitor-session-mx-exception-src minimum 0 maximum 1
  limit-resource monitor-session-extended minimum 0 maximum 12
feature-set fabricpath

feature privilege
feature telnet
cfs eth distribute
feature ospf
feature interface-vlan
feature hsrp
feature lacp
feature vpc

username admin password 5 $1$Ger4U2T0$s1d0FAX7pQ3zkG3T197e7/  role network-admin
no password strength-check
ip domain-lookup
copp profile strict
snmp-server user admin network-admin auth md5 0xdde65d6090179de9e2976e654dbd3153
priv 0xdde65d6090179de9e2976e654dbd3153 localizedkey
rmon event 1 log trap public description FATAL(1) owner PMON@FATAL
rmon event 2 log trap public description CRITICAL(2) owner PMON@CRITICAL
rmon event 3 log trap public description ERROR(3) owner PMON@ERROR
rmon event 4 log trap public description WARNING(4) owner PMON@WARNING
rmon event 5 log trap public description INFORMATION(5) owner PMON@INFO

vlan 1,10,100,112
fabricpath topology 1
  member vlan 1,10,100,112
vlan 10
  mode fabricpath
  name Network-Deivce-Mgmt
vlan 100
  mode fabricpath
  name server1
```

```
vlan 112
  mode fabricpath
  name iLO

vrf context management
  ip route 0.0.0.0/0 10.1.1.1
fabricpath switch-id 201
vpc domain 1
  role priority 110
  peer-keepalive destination 192.168.10.252 source 192.168.10.251
  peer-gateway

  auto-recovery
  fabricpath switch-id 1000
  fabricpath multicast load-balance
  ipv6 nd synchronize
  ip arp synchronize

interface Vlan1
  no ip redirects

interface Vlan10
  no shutdown
  no ip redirects
  ip address 10.237.128.130/25
  no ipv6 redirects
  ip router ospf 100 area 0.0.0.0
  hsrp 10
    preempt
    priority 110
    ip 10.237.128.129

interface Vlan100
  no shutdown
  no ip redirects
  ip address 10.237.8.2/23
  no ipv6 redirects
  ip router ospf 100 area 0.0.0.0
  hsrp 100
    preempt
    priority 110
    ip 10.237.8.1

interface Vlan112
  no shutdown
  no ip redirects
  ip address 10.237.4.2/25
```

```
  no ipv6 redirects
  ip router ospf 100 area 0.0.0.0
  hsrp 112
    preempt
    priority 110
    ip 10.237.4.1

interface port-channel31
  description ***** To-ASR9000 *****
  ip address 10.237.2.18/30
  ip router ospf 100 area 0.0.0.0

interface port-channel200
  description ***** To-N7K_02-e3/1-4 vpc + link *****
  switchport
  switchport mode fabricpath
  medium p2p
  vpc peer-link
  fabricpath topology-member 1

interface port-channel201
  description ***** FP-To-N5K_01-E1/1-4 *****
  switchport
  switchport mode fabricpath
  medium p2p

interface port-channel202
  description ***** FP-To-N5K_02-E1/3-8 *****
  switchport
  switchport mode fabricpath
  medium p2p

interface Ethernet3/1
  description ***** To-N7K_02-e3/1-4 *****
  switchport
  switchport mode fabricpath
  medium p2p
  channel-group 200 mode active
  no shutdown

interface Ethernet3/2
  description ***** To-N7K_02-e3/1-4 *****
  switchport
  switchport mode fabricpath
  medium p2p
  channel-group 200 mode active
  no shutdown
```

```
interface Ethernet3/3
  description *****To-N7K_02-e3/1-4*****
  switchport
  switchport mode fabricpath
  medium p2p
  channel-group 200 mode active
  no shutdown

interface Ethernet3/4
  description *****To-N7K_02-e3/1-4*****
  switchport
  switchport mode fabricpath
  medium p2p
  channel-group 200 mode active
  no shutdown

interface Ethernet3/5
  description *****To-N5K_01-E1/1-4*****
  switchport
  switchport mode fabricpath
  medium p2p
  channel-group 201 mode active
  no shutdown

interface Ethernet3/6
  description *****To-N5K_01-E1/1-4*****
  switchport
  switchport mode fabricpath
  medium p2p
  channel-group 201 mode active
  no shutdown

interface Ethernet3/7
  description *****To-N5K_01-E1/1-4*****
  switchport
  switchport mode fabricpath
  medium p2p
  channel-group 201 mode active
  no shutdown

interface Ethernet3/8
  description *****To-N5K_01-E1/1-4*****
  switchport
  switchport mode fabricpath
  medium p2p
  channel-group 201 mode active
  no shutdown
```

```
interface Ethernet3/9
  description ***** To-N5K_02-E1/3-8 *****
  switchport
  switchport mode fabricpath
  medium p2p
  channel-group 202 mode active
  no shutdown

interface Ethernet3/10
  description ***** To-N5K_02-E1/3-8 *****
  switchport
  switchport mode fabricpath
  medium p2p
  channel-group 202 mode active
  no shutdown

interface Ethernet3/11
  description ***** To-N5K_02-E1/3-8 *****
  switchport
  switchport mode fabricpath
  medium p2p
  channel-group 202 mode active
  no shutdown

interface Ethernet3/12
  description ***** To-N5K_02-E1/3-8 *****
  switchport
  switchport mode fabricpath
  medium p2p
  channel-group 202 mode active
  no shutdown

interface Ethernet3/13
  no shutdown

interface Ethernet3/40
  no shutdown

interface Ethernet3/41
description ***** To-ASR9000 *****
  channel-group 32 mode active
  no shutdown

interface Ethernet3/42
description ***** To-ASR9000 *****
  channel-group 32 mode active
  no shutdown
```

```
interface Ethernet3/43
description *****To-ASR9000*****
  channel-group 32 mode active
  no shutdown

interface Ethernet3/44
description *****To-ASR9000*****
  channel-group 32 mode active
  no shutdown

interface Ethernet3/45
  channel-group 31 mode active
  no shutdown

interface Ethernet3/46
  channel-group 31 mode active
  no shutdown

interface Ethernet3/47
  channel-group 31 mode active
  no shutdown

interface Ethernet3/48
  channel-group 31 mode active
  no shutdown

interface mgmt0
  vrf member management
  ip address 192.168.10.251/24

interface loopback0
  description local-loopback
  ip address 10.237.3.3/32
  ip router ospf 100 area 0.0.0.0
line console
line vty
boot kickstart bootflash:/n7000-s2-kickstart.6.2.2a.bin sup-1
boot system bootflash:/n7000-s2-dk9.6.2.2a.bin sup-1
router ospf 100
  router-id 10.237.3.3
fabricpath domain default
  maximum-paths 128
  root-priority 110
no system auto-upgrade epld
```

2.5.4 N7K-02-2 的配置

以下是 N7K-02-2 的设备配置：

```
N7K-02# sh run
! Command: show running-config
! Time: Thu Nov  6 17:05:49 2014

version 6.2(2a)
hostname N7K-02
no system admin-vdc
install feature-set fabricpath
vdc N7K-02 id 1
  limit-resource module-type f2e
  allow feature-set fabricpath
  cpu-share 5
  allocate interface Ethernet3/1-48
  limit-resource vlan minimum 16 maximum 4094
  limit-resource monitor-session minimum 0 maximum 2
  limit-resource monitor-session-erspan-dst minimum 0 maximum 23
  limit-resource vrf minimum 2 maximum 4096
  limit-resource port-channel minimum 0 maximum 768
  limit-resource u4route-mem minimum 96 maximum 96
  limit-resource u6route-mem minimum 24 maximum 24
  limit-resource m4route-mem minimum 58 maximum 58
  limit-resource m6route-mem minimum 8 maximum 8
  limit-resource monitor-session-inband-src minimum 0 maximum 1

  limit-resource anycast_bundleid minimum 0 maximum 16
  limit-resource monitor-session-mx-exception-src minimum 0 maximum 1
  limit-resource monitor-session-extended minimum 0 maximum 12
feature-set fabricpath

feature privilege
feature telnet
feature vrrp
cfs eth distribute
feature ospf
feature interface-vlan
feature hsrp
feature lacp
feature vpc

username admin password 5 $1$JzbMa5Au$h9//Kl82x3hksIeukt5tM1  role network-admin
no password strength-check
ip domain-lookup
copp profile strict
snmp-server user admin network-admin auth md5 0xd2865b9752f6ea9fe0a6d7d0bbfdb7e0
priv 0xd2865b9752f6ea9fe0a6d7d0bbfdb7e0 localizedkey
rmon event 1 log trap public description FATAL(1) owner PMON@FATAL
rmon event 2 log trap public description CRITICAL(2) owner PMON@CRITICAL
```

```
rmon event 3 log trap public description ERROR(3) owner PMON@ERROR
rmon event 4 log trap public description WARNING(4) owner PMON@WARNING
rmon event 5 log trap public description INFORMATION(5) owner PMON@INFO

vlan 1,10,100,112
fabricpath topology 1
  member vlan 1,10,100,112
vlan 10
  mode fabricpath
  name Network-Device-Mgmt
vlan 100
  mode fabricpath
vlan 112
  mode fabricpath
  name iLO

vrf context management
  ip route 0.0.0.0/0 10.1.1.1
fabricpath switch-id 202
vpc domain 1
  role priority 120
  peer-keepalive destination 192.168.10.251 source 192.168.10.252
  peer-gateway

  auto-recovery
  fabricpath switch-id 1000
  fabricpath multicast load-balance
  ipv6 nd synchronize
  ip arp synchronize

interface Vlan1
  no ip redirects

interface Vlan10
  no shutdown
  no ip redirects
  ip address 10.237.128.131/25
  no ipv6 redirects
  ip router ospf 100 area 0.0.0.0
  hsrp 10
    preempt
    ip 10.237.128.129

interface Vlan100
  no shutdown
  no ip redirects
  ip address 10.237.8.3/23
```

```
  no ipv6 redirects
  ip router ospf 100 area 0.0.0.0
  hsrp 100
    preempt
    ip 10.237.8.1

interface Vlan112
  no shutdown
  no ip redirects
  ip address 10.237.4.3/25
  no ipv6 redirects
  ip router ospf 100 area 0.0.0.0
  hsrp 112
    preempt
    ip 10.237.4.1

interface port-channel32
  description ***** To-ASR9000 *****
  ip address 10.237.2.22/30
  ip router ospf 100 area 0.0.0.0

interface port-channel200
  description ***** To-N7K_01-e3/1-4 vpc + link *****
  switchport
  switchport mode fabricpath
  medium p2p
  vpc peer-link
  fabricpath topology-member 1

interface port-channel203
  description ***** FP-To-N5K_01-E1/9-12 *****
  switchport
  switchport mode fabricpath
  medium p2p

interface port-channel204
  description ***** FP-To-N5K_02-E1/9-12 *****
  switchport
  switchport mode fabricpath
  medium p2p

interface Ethernet3/1
  description ***** To-N7K_01-e3/1-4 *****
  switchport
  switchport mode fabricpath
  medium p2p
  channel-group 200 mode active
```

```
  no shutdown

interface Ethernet3/2
  description ***** To-N7K_01-e3/1-4 *****
  switchport
  switchport mode fabricpath
  medium p2p
  channel-group 200 mode active
  no shutdown

interface Ethernet3/3
  description ***** To-N7K_01-e3/1-4 *****
  switchport
  switchport mode fabricpath
  medium p2p
  channel-group 200 mode active
  no shutdown

interface Ethernet3/4
  description ***** To-N7K_01-e3/1-4 *****
  switchport
  switchport mode fabricpath
  medium p2p
  channel-group 200 mode active
  no shutdown

interface Ethernet3/5
  description ***** To-N5K_01-E1/9-12 *****
  switchport
  switchport mode fabricpath
  medium p2p
  channel-group 203 mode active
  no shutdown

interface Ethernet3/6
  description ***** To-N5K_01-E1/9-12 *****
  switchport
  switchport mode fabricpath
  medium p2p
  channel-group 203 mode active
  no shutdown

interface Ethernet3/7
  description ***** To-N5K_01-E1/9-12 *****
  switchport
  switchport mode fabricpath
  medium p2p
```

```
    channel-group 203 mode active
    no shutdown

  interface Ethernet3/8
    description *****To-N5K_01-E1/9-12*****
    switchport
    switchport mode fabricpath
    medium p2p
    channel-group 203 mode active
    no shutdown

  interface Ethernet3/9
    description *****To-N5K_02-E1/9-12*****
    switchport
    switchport mode fabricpath
    medium p2p
    channel-group 204 mode active
    no shutdown

  interface Ethernet3/10
    description *****To-N5K_02-E1/9-12*****
    switchport
    switchport mode fabricpath
    medium p2p
    channel-group 204 mode active
    no shutdown

  interface Ethernet3/11
    description *****To-N5K_02-E1/9-12*****
    switchport
    switchport mode fabricpath
    medium p2p
    channel-group 204 mode active
    no shutdown

  interface Ethernet3/12
    description *****To-N5K_02-E1/9-12*****
    switchport
    switchport mode fabricpath
    medium p2p
    channel-group 204 mode active
    no shutdown

  interface Ethernet3/13
    no shutdown

  interface Ethernet3/40
```

```
  no shutdown

interface Ethernet3/41
  description *****To-ASR9000*****
  channel-group 32 mode active
  no shutdown

interface Ethernet3/42
  description *****To-ASR9000*****
  channel-group 32 mode active
  no shutdown

interface Ethernet3/43
  description *****To-ASR9000*****
  channel-group 32 mode active
  no shutdown

interface Ethernet3/44
  description *****To-ASR9000*****
  channel-group 32 mode active
  no shutdown

interface Ethernet3/45
  description *****To-ASR9000*****
  channel-group 32 mode active
  no shutdown

interface Ethernet3/46
  description *****To-ASR9000*****
  channel-group 32 mode active
  no shutdown

interface Ethernet3/47
  description *****To-ASR9000*****
  channel-group 32 mode active
  no shutdown

interface Ethernet3/48
  description *****To-ASR9000*****
  channel-group 32 mode active
  no shutdown

interface mgmt0
  vrf member management
  ip address 192.168.10.252/24

interface loopback0
  description local-loopback
  ip address 10.237.3.4/32
  ip router ospf 100 area 0.0.0.0
```

```
line console
line vty
boot kickstart bootflash:/n7000-s2-kickstart.6.2.2a.bin sup-1
boot system bootflash:/n7000-s2-dk9.6.2.2a.bin sup-1
router ospf 100
  router-id 10.237.3.4

fabricpath domain default
  maximum-paths 128
  root-priority 100
no system auto-upgrade epld
```

2.6 小　结

思科 Nexus 交换机是目前在数据中心最有竞争力的产品，其领先的架构、强劲的交换能力和灵活的扩展能力都是业界最领先的。思科最新推出的是 Nexus 9000 系列。Nexus 交换机要使用 VN-tag 和 VN-LINK 才能发挥其强大的性能，目前大多数数据中心还没有建立统一的网络和计算系统，因而 Nexus 的性能还未得到很好的发挥，相信随着业务的变化和 Trill 标准的完善，Nexus 会有更加出色的表现。

第3章 数据中心存储交换机

3.1 MDS 9000 系列介绍

3.1.1 Cisco MDS 9100 系列

Cisco MDS 9100 系列通过将思科智能化网络带入中小型 SAN 和数据中心边缘应用等领域，提升了光纤通道交换机的标准。Cisco MDS 9100 系列可以通过一个小巧的 1RU 机型，在成本、性能和企业级功能之间实现完美的平衡。Cisco MDS 9100 系列提供了 20 端口和 40 端口两种配置，因而可以提供多种存储环境所需要的端口密度。通过提供业界领先的可扩展性、可用性和管理功能，Cisco MDS 9100 系列能够以很低的总成本（Total Cost of Ownership, TCO）部署高性能的 SAN。通过在一个经济有效、外型小巧的交换平台上添加一组范围广泛的智能化功能，Cisco MDS 9100 系列可以满足中小型存储环境对于成本、性能、管理性和连通性的要求，并提供与 Cisco MDS 9500 多层控制器和 Cisco MDS 9216 多层光纤通道交换机的全面兼容，从而可以在大型数据中心的部署中实现透明和端到端的存储服务供应。图 3-1 到图 3-4 显示的是 MDS 9100 系列不同型号的存储交换机。

图 3-1 Cisco MDS 9120 20 端口智能矩阵交换机

图 3-2 Cisco MDS 9140 40 端口智能矩阵交换机

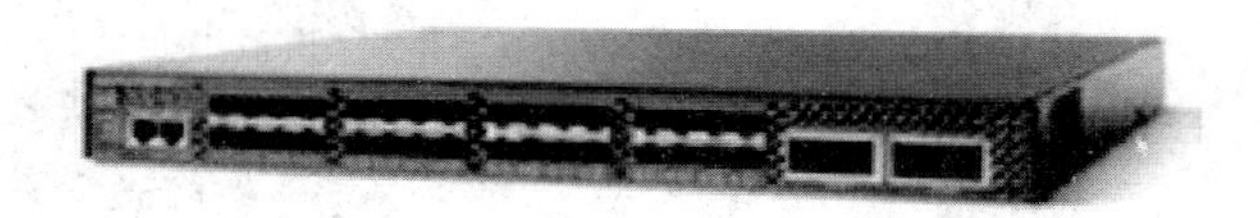

图 3-3 Cisco MDS 9134 多层矩阵交换机

Cisco MDS 9134 是使用 10G BASE 铜缆 CX4 堆叠在一起的两个 Cisco MDS 9134 交换机，如图 3-4 所示。

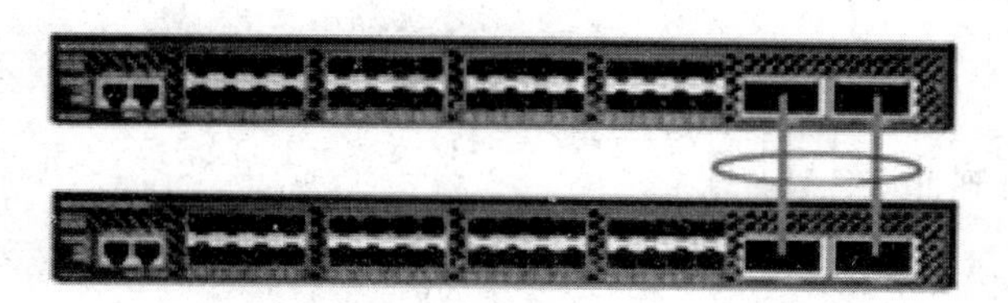

图 3-4 Cisco MDS 9134

3.1.2 Cisco MDS 9200 系列

Cisco MDS 9216 多层光纤通道交换机为光纤通道交换机市场带来了新的功能和投资保护能力。Cisco MDS 9216 的系统架构与 Cisco MDS 9500 系列相同，它将多层的智能和一个模块化的机型紧密地结合在一起，从而成为了业界最智能、最灵活的光纤通道交换机。Cisco MDS 9216具有 16 个 1/2 Gbit/s 自检测光纤通道端口，而它的扩展槽使用户可以添加 Cisco MDS 9000 系列的任何模块，因而它最多可以支持 48 个光纤通道端口。随着存储网络的进一步发展，用户可以将 Cisco MDS 9000 系列模块从 Cisco MDS 9216 光纤通道交换机中卸载下来，移植到 Cisco MDS 9500 系列多层控制器中，从而获得平稳的移植、通用的部件和出色的投资保护。Cisco MDS 9200 系列存储交换机如图 3-5 所示。

图 3-5 Cisco MDS 9200 系列

3.2 MDS 9000 基本配置

MDS 存储交换机在网络中的位置如图 3-6 所示。

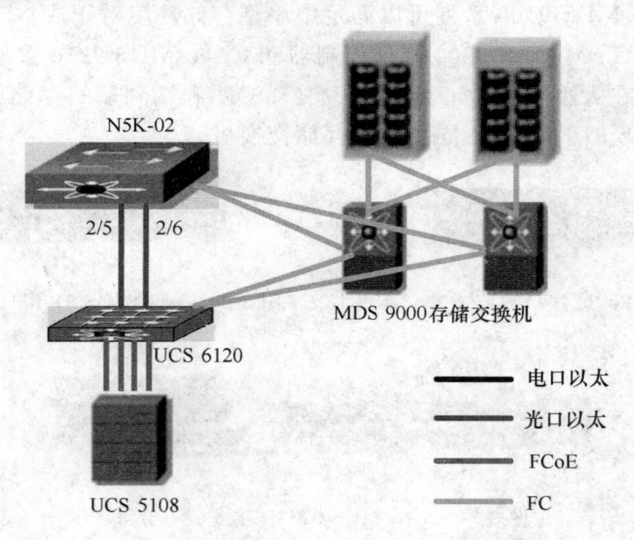

图 3-6 MDS 存储交换机在网络中的位置

思科 MDS 9000 的配置跟思科路由器命令相似，这里我们仅给出满足最低要求的配置命令，主要配置步骤：

① 创建 VSAN；

② 添加相关端口到此 VSAN；

③ 创建 ZONE；

④ 添加 ZONE 成员；

⑤ 创建 ZONESET；

⑥ 添加 ZONE 到 ZONESET 中；

⑦ 激活 ZONESET。

下面以 MDS 9134 为例，阐述思科存储交换机的基本配置。

3.2.1 命令行初始化配置

(1) 通过 Console 接入 MDS 9134，第一次加电，CLI 配置如下：

```
--- Basic System Configuration Dialog ---
This setup utility will guide you through the basic configuration of the system. Setup configures only enough connectivity for management of the system.
Press Enter in case you want to skip any dialog. Use ctrl-c at anytime to skip away remaining dialogs.
Would you like to enter the basic configuration dialog (yes/no): y
Enter the password for "admin" : admin
Create another login account (yes/no) [n]: n
Configure SNMPv3 Management parameters (yes/no) [y]: y
SNMPv3 user name [admin]: admin
SNMPv3 user authentication password : CISCO
The same password will be used for SNMPv3 privacy as well.
Configure read-only SNMP community string (yes/no) [n]: <Enter>
Configure read-write SNMP community string (yes/no) [n]: <Enter>
Enter the switch name: CISCO-MDS                                     ---- MDS 9000 的主机名
Continue with Out-of-band (mgmt0) management configuration? (yes/no) [y]: y
Mgmt0 IP address : 192.168.10.100  Mgmt0 IP netmask : 255.255.255.0
Continue with In-band (vsan1) management configuration? (yes/no) [n]: <Enter>
Enable the ip routing capabilities? (yes/no) [y]: y
Configure static route? (yes/no) [y]: n
Configure the default network? (yes/no) [y]: n
Configure the default gateway? (yes/no) [y]: <Enter>
  IPv4 address of the default gateway : 172.16.2.1
  Configure advanced IP options? (yes/no) [n]:
  Enable the telnet service? (yes/no) [n]: y
  Enable the ssh service? (yes/no) [y]:
  Type of ssh key you would like to generate (dsa/rsa) : dsa
 Configure clock? (yes/no) [n]:
 Configure timezone? (yes/no) [n]:
 Configure summertime? (yes/no) [n]:
 Configure the ntp server? (yes/no) [n]:
 Configure default switchport interface state (shut/noshut) [shut]: noshut
 Configure default switchport trunk mode (on/off/auto) [on]: off
 Configure default switchport port mode F (yes/no) [n]:
 Configure default zone policy (permit/deny) [deny]:
 Enable full zoneset distribution? (yes/no) [n]:
 Configure default zone mode (basic/enhanced) [basic]:
The following configuration will be applied:
```

```
  password strength-check
  switchname MDS9000P-FCSW1
  interface mgmt0
    ip address 172.16.2.4 255.255.255.0
    no shutdown
  ip default-gateway 172.16.2.1
  telnet server enable
  ssh key dsa  force
  ssh server enable
  no system default switchport shutdown
  system default switchport trunk mode off
  no system default zone default-zone permit
  no system default zone distribute full
  no system default zone mode enhanced

Would you like to edit the configuration? (yes/no) [n]:

Use this configuration and save it? (yes/no) [y]: y

[########################################] 100%
```

(2) 完成初始化后进行登录,用户名为 admin。

```
User Access Verification
MDS9000-FCSW1 login: admin
Password:
Cisco Nexus Operating System (NX-OS) Software
TAC support: http://www.cisco.com/tac
Copyright (c) 2002-2008, Cisco Systems, Inc. All rights reserved.
The copyrights to certain works contained in this software are
owned by other third parties and used and distributed under
license. Certain components of this software are licensed under
the GNU General Public License (GPL) version 2.0 or the GNU
Lesser General Public License (LGPL) Version 2.1. A copy of each
such license is available at
http://www.opensource.org/licenses/gpl-2.0.php and
http://www.opensource.org/licenses/lgpl-2.1.php

MDS9000-FCSW1#
```

(3) 配置实例:

```
MDS9000-FCSW1# conft t
! 进入配置模式
Enter configuration commands, one per line.  End with CNTL/Z.
MDS9000-FCSW1(config)# vsan database
! 进入 VSAN 配置模式
MDS9000-FCSW1(config-vsan-db)# vsan 10 name P570
```

```
! 创建 VSAN,10 是自定义的 VSAN ID
MDS9000-FCSW1(config-vsan-db)# vsan 10 interface fc1/3
! 添加接口到新建的 VSAN 中
Traffic on fc1/3 may be impacted. Do you want to continue? (y/n) y
MDS9000-FCSW1(config-vsan-db)# vsan 10 interface fc1/4
Traffic on fc1/4 may be impacted. Do you want to continue? (y/n) y
MDS9000-FCSW1(config-vsan-db)# exit
MDS9000-FCSW1(config)# zone name zone1 vsan 10
! 创建 ZONE
MDS9000-FCSW1(config-zone)# member interface fc1/3
! 添加 ZONE 成员
MDS9000-FCSW1(config-zone)# member interface fc1/7
MDS9000-FCSW1(config-zone)# exit
MDS9000-FCSW1(config)# zone name zone2 vsan 10
MDS9000-FCSW1(config-zone)# member interface fc1/8
MDS9000-FCSW1(config-zone)# member interface fc1/4
MDS9000-FCSW1(config-zone)# exit
MDS9000-FCSW1(config)# zoneset name P570 vsan 10
! 创建 ZONESET
MDS9000-FCSW1(config-zoneset)# member zone1
! 添加 ZONE 到 ZONESET 中
MDS9000-FCSW1(config-zoneset)# member zone2
MDS9000-FCSW1(config-zoneset)# exit
MDS9000-FCSW1(config)#zoneset activate name P570 vsan 10
! 激活 ZONESET
```

3.2.2 图形界面配置

(1) MDS 9134 还可进行图形化配置,通过随机光盘安装 Cisco 的 FabricManager 软件。

(2) 运行"X:\Program Files\Cisco Systems\MDS 9000\bin\FabricManagerSA. bat"文件,出现的登录界面(默认 pwd 为 password)如图 3-7 所示,其显示了图形化登录 MDS 9000 的画面。

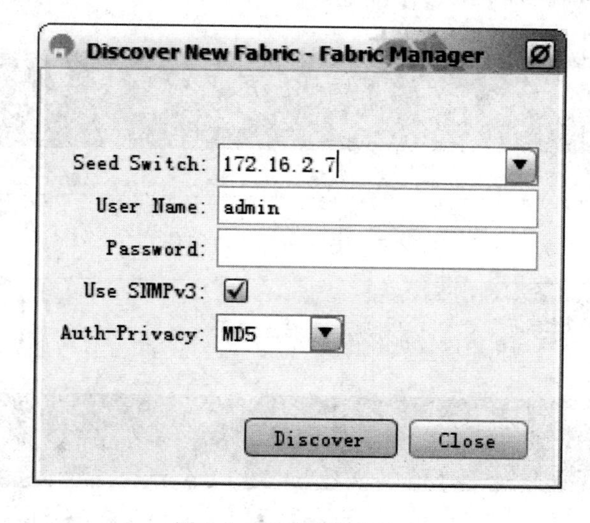

图 3-7　登录 MDS 9000

(3) 选择对应的光纤交换机后，单击“Discover”按钮进入，出现如图 3-8 所示的界面。

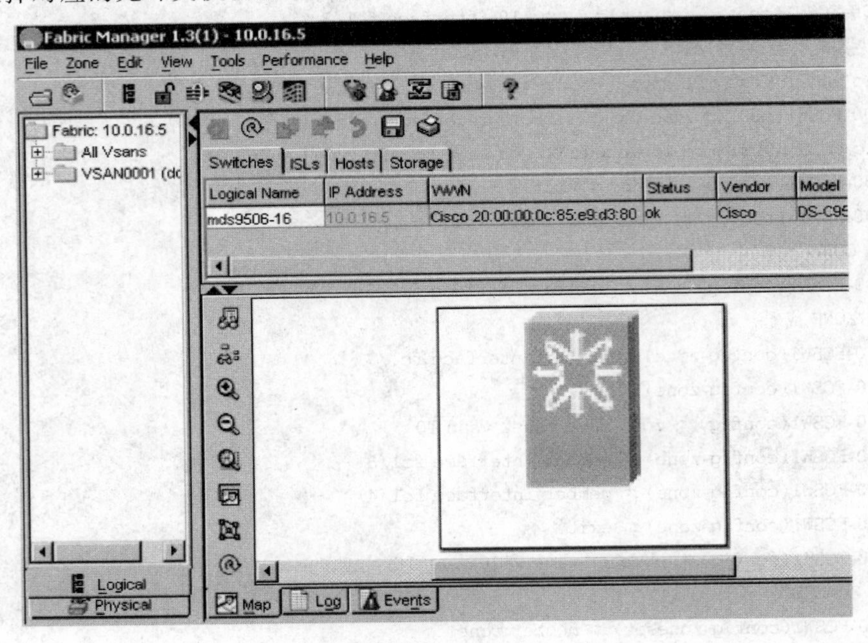

图 3-8 存储交换机配置图

注意:MDS 系统会自己绘出相关拓扑，但前提是光纤交换机、主机和存储之间的光纤正常连接，光纤交换机上的对应端口处于 UP 状态(初始化的时候系统会让你选择所有端口默认状态是 shut 还是 no shut)。如果 MDS 交换机刚启动，则需等待一段时间，否则可能出现拓扑图不全的问题。

(4) 双击拓扑图中的交换机图标，可以查看交换机端口情况(如图 3-9 所示)。

在“VSAN ALL”中可以选择某个 VSAN 中的端口，如图 3-10 所示。

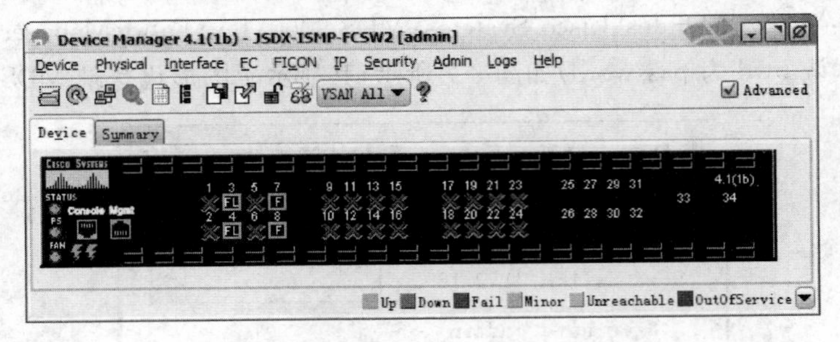

图 3-9 展开的交换机图标

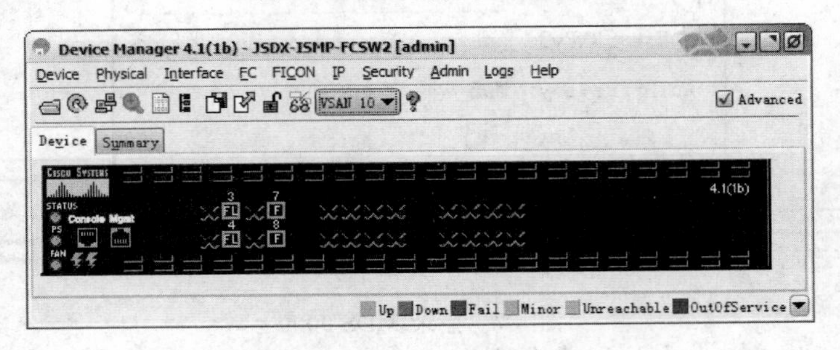

图 3-10 选择 VSAN 接口

双击某个端口，可以查看该端口的详细情况，如图 3-11 所示。

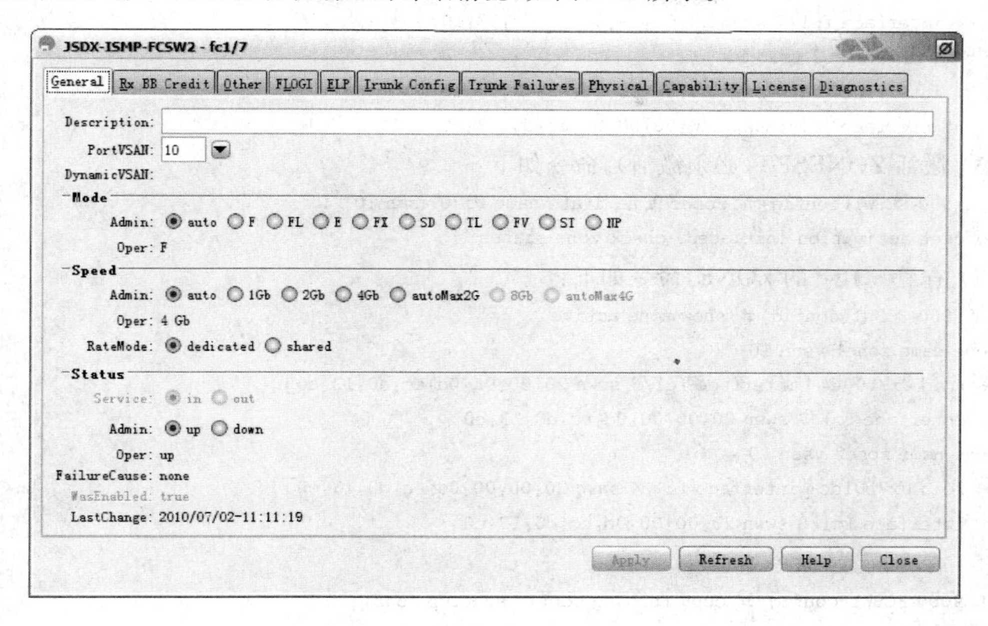

图 3-11　端口详细情况

（5）配置好之后，点击某个 ZONE，可以在右侧看到对应的端口情况和拓扑中的对应线路，如图 3-12 所示。

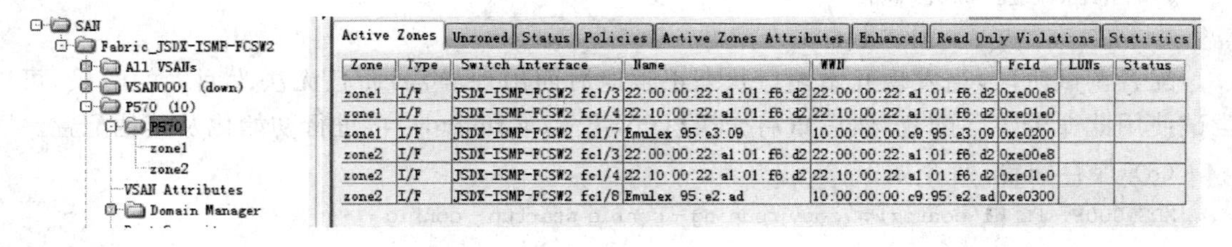

Zone	Type	Switch Interface	Name	WWN	FcId	LUNs	Status
zone1	I/F	JSDX-ISMP-FCSW2 fc1/3	22:00:00:22:a1:01:f6:d2	22:00:00:22:a1:01:f6:d2	0xe00e8		
zone1	I/F	JSDX-ISMP-FCSW2 fc1/4	22:10:00:22:a1:01:f6:d2	22:10:00:22:a1:01:f6:d2	0xe01e0		
zone1	I/F	JSDX-ISMP-FCSW2 fc1/7	Emulex 95:e3:09	10:00:00:00:c9:95:e3:09	0xe0200		
zone2	I/F	JSDX-ISMP-FCSW2 fc1/3	22:00:00:22:a1:01:f6:d2	22:00:00:22:a1:01:f6:d2	0xe00e8		
zone2	I/F	JSDX-ISMP-FCSW2 fc1/4	22:10:00:22:a1:01:f6:d2	22:10:00:22:a1:01:f6:d2	0xe01e0		
zone2	I/F	JSDX-ISMP-FCSW2 fc1/8	Emulex 95:e2:ad	10:00:00:00:c9:95:e2:ad	0xe0300		

图 3-12　ZONE 与端口对应关系

3.2.3　常用检查命令

（1）查看 ZONE 配置情况，命令如下：

```
MDS9000-FCSW1(config)# show zone
zone name zone1 vsan 10
   interface fc1/3 swwn 20:00:00:0d:ec:d0:13:c0
   interface fc1/7 swwn 20:00:00:0d:ec:d0:13:c0
zone name zone2 vsan 10
   interface fc1/4 swwn 20:00:00:0d:ec:d0:13:c0
   interface fc1/8 swwn 20:00:00:0d:ec:d0:13:c0
MDS9000-FCSW1(config)# show zoneset
```

（2）查看 ZONESET 配置情况，命令如下：

```
MDS9000-FCSW1(config)# show zoneset
zoneset name P570 vsan 10
  zone name zone1 vsan 10
```

```
    interface fc1/3 swwn 20:00:00:0d:ec:d0:13:c0
    interface fc1/7 swwn 20:00:00:0d:ec:d0:13:c0
  zone name zone2 vsan 10
    interface fc1/4 swwn 20:00:00:0d:ec:d0:13:c0
    interface fc1/8 swwn 20:00:00:0d:ec:d0:13:c0
```

(3) 激活 ZONESET(必须激活),命令如下:

```
MDS9000-FCSW1(config)# zoneset activate name P570 vsan 10
Zoneset activation initiated. check zone status
```

(4) 查看 active 的 ZONE,命令如下:

```
MDS9000-FCSW1(config)# show zone active
zone name zone1 vsan 10
 * fcid 0x7b00e4 [interface fc1/3 swwn 20:00:00:0d:ec:d0:13:c0]
   interface fc1/7 swwn 20:00:00:0d:ec:d0:13:c0
zone name zone2 vsan 10
 * fcid 0x7b01dc [interface fc1/4 swwn 20:00:00:0d:ec:d0:13:c0]
   interface fc1/8 swwn 20:00:00:0d:ec:d0:13:c0
```

(5) 保存配置,命令如下:

```
MDS9000-FCSW1(config)# copy running-config startup-config
[########################################] 100%
```

(6) 查看系统当前配置,命令如下:

```
MDS9000-FCSW1(config)# show runn
```

(7) 清除配置,命令如下:

```
MDS9000 - FCSW1# write erease
```

配置调试阶段,当发现开始端口状态正常但后面出现异常,而且无法改回去的时候,可以尝试使用此命令清除配置,然后重启交换机(N7K-02# reload),再进行初始化及其他配置。

(8) 保存配置,命令如下:

```
MDS9000P - FCSW1(config)# copy running - config startup - config
```

3.3 配置案例

以下是 MDS 9000 的配置案例:

```
version 5.1(1b)
role name default-role
  description This is a system defined role and applies to all users.
  rule 5 permit show feature environment
  rule 4 permit show feature hardware
  rule 3 permit show feature module
  rule 2 permit show feature snmp
  rule 1 permit show feature system
username admin password 5 $1$6tiqPmaH$exkTt0fqezo4tLGiDLhYx1  role network-admin
feature telnet
ssh key dsa
ip domain-lookup
```

```
aaa group server radius radius
snmp-server user admin network-admin auth md5 0x829d954ae056812e01f62004532c538a priv 0x829d954ae056812e01f62004532c538a local izedkey
snmp-server enable traps license
snmp-server enable traps entity fru
vsan database
  vsan 10 name "P570"
fcdomain fcid database
  vsan 1 wwn 20:03:00:0d:ec:d0:13:c0 fcid 0x790000 area dynamic
  vsan 1 wwn 20:04:00:0d:ec:d0:13:c0 fcid 0x790100 area dynamic
  vsan 1 wwn 10:00:00:00:c9:95:da:53 fcid 0x790200 dynamic
  vsan 1 wwn 10:00:00:00:c9:95:d5:da fcid 0x790300 dynamic
  vsan 10 wwn 20:03:00:0d:ec:d0:13:c0 fcid 0x7b0000 area dynamic
  vsan 10 wwn 20:04:00:0d:ec:d0:13:c0 fcid 0x7b0100 area dynamic
  vsan 10 wwn 10:00:00:00:c9:95:da:53 fcid 0x7b0200 dynamic
  vsan 10 wwn 10:00:00:00:c9:95:d5:da fcid 0x7b0300 dynamic
vsan database
  vsan 10 interface fc1/3
  vsan 10 interface fc1/4
  vsan 10 interface fc1/7
  vsan 10 interface fc1/8
interface fc1/1
  port-license acquire
interface fc1/2
  port-license acquire
...
interface fc1/34
interface mgmt0
  ip address 172.16.2.4 255.255.255.0
system default switchport trunk mode off
ip default-gateway 172.16.2.1
switchname MDS9000P-FCSW1
boot kickstart bootflash:/m9100-s2ek9-kickstart-mz.5.1.1b.bin
boot system bootflash:/m9100-s2ek9-mz.5.1.1b.bin
interface fc1/1
interface fc1/2
...
interface fc1/33
interface fc1/34
! Full Zone Database Section for vsan 10
zone name zone1 vsan 10
    member interface fc1/3 swwn 20:00:00:0d:ec:d0:13:c0
    member interface fc1/7 swwn 20:00:00:0d:ec:d0:13:c0
zone name zone2 vsan 10
    member interface fc1/4 swwn 20:00:00:0d:ec:d0:13:c0
    member interface fc1/8 swwn 20:00:00:0d:ec:d0:13:c0
zoneset name P570 vsan 10
```

```
    member zone1
    member zone2
zoneset activate name P570 vsan 10
no system default switchport shutdown
```

3.4 小 结

思科存储交换机和思科路由器命令相似，这就降低了配置的复杂度，存储交换机主要的接口是 FC，但是演进的趋势是采用 FCoE，通过阵列与 Nexus 互联也是数据中心存储的发展趋势，思科已经在 Nexus 5000 和 7000 系列交换机上通过集成 FCoE 模块实现了一体化的目标。

第 4 章　ASR 9000 路由器配置

4.1　ASR 9000 系列产品介绍

Aggregation Service Router 9000（ASR 9000）是思科针对下一代 IP 网络（IP NGN）转型而设计的超大容量运营商级别的边缘路由器平台，以帮助运营商应对“Zettabyte 时代”到来之后激增的网络容量需求，以及以太网服务和视频业务为流量主体的服务提供，目前大多部署在城域网的边缘，作为城域网的互联路由器，在数据中心可以定位为对外出口的路由器或者是数据中心互联路由器。防火墙在数据中心的位置拓扑如图 4-1 所示。

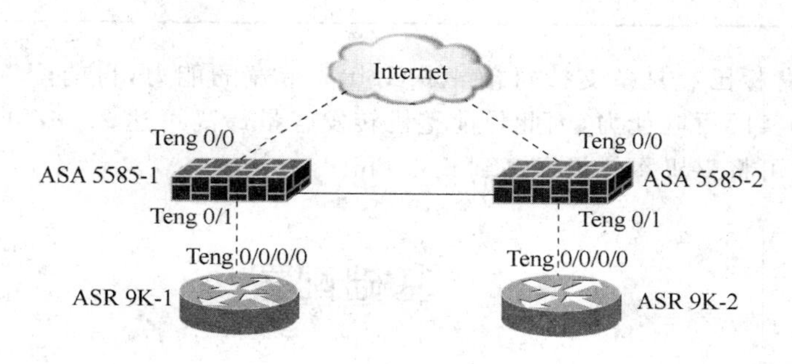

图 4-1　数据中心防火墙位置拓扑

ASR 9000 产品的特点包括：超大容量、极高密度的高速以太网接口，可持续的系统运行能力，不间断的视频体验和更低的碳排放量。思科 ASR 9000 采用了创新的分布式操作系统 IOS-XR 软件，该软件此前一直运行在思科高端集群路由器平台 CRS-1 上，可实现路由器系统的永续运行。

思科 ASR 9000 产品的推出标志着运营商已经做好迎接 IP NGN 转型的准备，运营商可以通过部署 ASR 9000，极大程度地提高网络边缘的高速率以太网端口密度、系统转发容量和新型业务的承载能力，更好地应对未来在有线、固网宽带以及移动网络中快速增长的业务需求。

思科 ASR 9000 系列型号（如表 4-1 所示）从低端到高端包括：Cisco ASR 9001、Cisco ASR 9006、Cisco ASR 9010、Cisco ASR 9904、Cisco ASR 9912、Cisco ASR 9922。

思科 ASR 9000 是一个全分布式的路由器平台，主要的功能组件包括：路由控制引擎、交换矩阵、线路线卡、可按需配置的电源模块（AC/DC）、可变速冗余风扇、机框。

ASR 9010 是一种竖插槽的 10 槽机框，宽 5.27 厘米，高 14.47 厘米，长 11.02 厘米，在标准机柜中可以放入 2 个 10 槽的 ASR 9000，每个 10 槽机框可插入 8 个线路板卡和 2 个路由控制引擎,支持前进风的散热方式。冗余风扇位置,水平插入在线路板卡的下方。

表 4-1　ASR 9000 系列设备图片

	ASR 9001	ASR 9904	ASR 9006	ASR 9010	ASR 9912	ASR 9922
产品图片						
尺　寸	2RU	6RU	10RU	21RU	30RU	43RU
板卡数量	2	4	6	10	12	22

ASR 9000 的背板已经具备支持每槽 400 Gbit/s 带宽的能力,目前配置的交换矩阵支持每槽 180 Gbit/s 的实际吞吐能力,因此目前整机转发容量最高可达 2.88 Tbit/s,未来只需通过升级交换矩阵就可将整机吞吐量提高到 6.4 Tbit/s。

4.2　基础配置

4.2.1　IOS XR 用户操作界面

(1) 用户登录格式,显示如下:

```
User Access Verification
Username: iosxr
Password: password
RP/0/RSP0/CPU0:router#
```

(2) 命令提示符含义解释,命令如下:

```
RP/0/RSP0/CPU0:router#
```

- RP:表示路由器引擎。
- 0:Rack 机架编号,可以多机箱集群也可以是 ASR 9K 独立工作。
- RSP0:Slot 引擎编号,RSP0 或 RSP1。
- CPU0:Module 管理接口。
- router#:路由器主机名称。

(3) 在执行模式下输入 config 进入全局配置模式,与 IOS 一样:

```
RP/0/RSP0/CPU0:router#
RP/0/RSP0/CPU0:router#configure
RP/0/RSP0/CPU0:router(config)#
```

(4) 在特权模式下输入 admin 进入管理员配置模式,与 IOS 不一样,admin 模式的作用是对系统功能做升级、修改的操作：

```
RP/0/RSP0/CPU0:router#
RP/0/RSP0/CPU0:router#admin
RP/0/RSP0/CPU0:router(admin-config)#
```

（5）与 IOS 不同的是，刚刚输入的命令并不会马上生效，生效需要输入命令 commit：

```
RP/0/RSP0/CPU0:router(config)commit
```

（6）CLI 配置流程常用命令，图 4-2 为 CLI 配置流程。

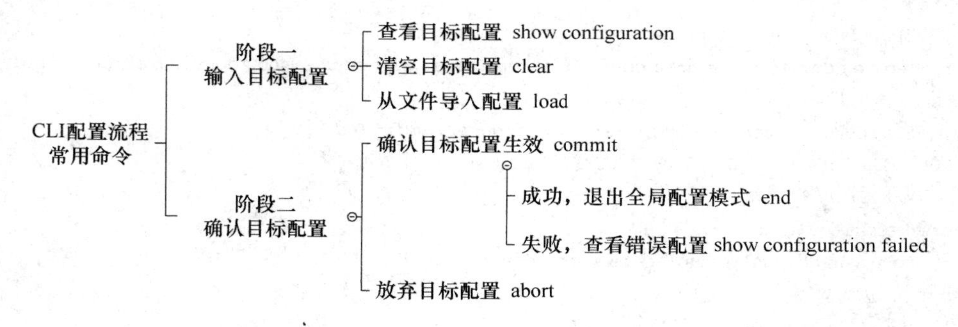

图 4-2 CLI 配置流程

（7）查看目标配置，命令如下：

```
RP/0/RSP0/CPU0:router(config)# show configuration ?
Building configuration…
interface Gi0/3/0/1
ipv4 address 10.1.1.1 255.0.0.0
shutdown
end
```

（8）清空目标配置，命令如下：

```
RP/0/RSP0/CPU0:router(config)# clear
RP/0/RSP0/CPU0:router(config)# commit
RP/0/RSP0/CPU0:router(config)# show configuration
Building configuration…
end
```

（9）从文件导入输入配置文件，命令如下：

```
磁盘文件路径 disk1/test.cfg
RP/0/RSP1/CPU0:router(config)# load disk1:/test.cfg
! 导入配置文件
Loading.
77 bytes parsed in 1 sec (76)bytes/sec
```

（10）确认目标配置生效，退出全局配置模式，命令如下：

```
RP/0/RSP0/CPU0:router(config)# commit
RP/0/0/0/0:Aug  6 09:26:13.781 : % LIBTARCFG-6-COMMIT Configuration committed by user 'cisco'.
Use 'show configuration commit changes 1000000101' to view the changes.
RP/0/RSP0/CPU0:router(config)# exit
RP/0/RSP0/CPU0:router#
```

（11）直接退出全局配置模式，命令如下：

```
RP/0/RSP0/CPU0:router(config-if)# end
Uncommitted changes found, commit them before exiting (yes/no/cancel)? [cancel]:
```

(12) 出现错误的情形，显示如下：

```
RP/0/RSP0/CPU0:router# configure
RP/0/RSP0/CPU0:router(config)# interface Gi 0/3/0/2
RP/0/RSP0/CPU0:router(config-if)# ipv4 address 10.1.1.1/8
RP/0/RSP0/CPU0:router(config-if)# no shutdown
RP/0/RSP0/CPU0:router(config-if)# exit
RP/0/RSP0/CPU0:router(config)#
```

确认目标配置失败：

```
RP/0/RSP0/CPU0:router(config)# commit
% Failed to commit one or more configuration items. Please use 'show configuration failed' to view the errors
RP/0/RSP0/CPU0:router(config)# show configuration failed
!! CONFIGURATION FAILED DUE TO SEMANTIC ERRORS
interface Gi 0/3/0/2
  ipv4 address 10.1.1.1/8
!!% the ipv4 address is duplicate with interfae Gi 0/3/0/1
! 提示地址冲突
RP/0/RSP0/CPU0:router(config)#
```

(13) 路由器配置的变更管理，命令如下：

每次成功的 commit 操作，路由器生成一项历史记录，分配一个 commit ID 标识：

```
RP/0/RSP0/CPU0:router# configure
RP/0/RSP0/CPU0:router(config)# interface Gi 0/3/0/1
RP/0/RSP0/CPU0:router(config-if)# ipv4 address 10.1.1.1/8
RP/0/RSP0/CPU0:router(config-if)# no shutdown
RP/0/RSP0/CPU0:router(config-if)# exit
RP/0/RSP0/CPU0:router(config)#
RP/0/RSP0/CPU0:router(config)# commit
RP/0/0/0:Aug  6 09:26:13.781 : % LIBTARCFG-6-COMMIT Configuration committed by user 'cisco'. Use 'show configuration commit changes 1000000219 ' to view the changes.
RP/0/RSP0/CPU0:router(config)#
```

(14) 查看配置文件 running-config 的变更记录，命令如下：

```
RP/0/RSP1/CPU0:router# show configuration commit list
SNo. Label/ID    User    Line        Client   Time Stamp
~~   ~~~~~       ~~      ~~          ~~~      ~~~~~~
1    1000000219  cisco   vty0        CLI      12:27:50 UTC Wed Mar 22 2013
2    1000000218  cisco   vty1        CLI      11:43:31 UTC Mon Mar 20 2013
3    1000000217  cisco   con0_RSP0_C CLI      17:44:29 UTC Wed Mar 15 2013
```

(15) 查看某次变更情况：

```
RP/0/RSP1/CPU0:router# show configuration commit changes 1000000219
Building configuration…
interface Gi 0/3/0/1
  ipv4 address 10.1.1.1/8
  no shutdown
```

(16) 预览配置，命令如下：

```
RP/0/RSP1/CPU0:router# show configuration rollback changes to 1000000219
Building configuration...
interface Gi 0/3/0/1
  no ipv4 address 10.1.1.1/8
```

```
  shutdown
!
end
```

(17) 还原变化,命令如下:

```
RP/0/RSP1/CPU0:router# rollback configuration to 1000000219
Loading Rollback Changes.
Loaded Rollback Changes in 1 sec
Committing.
2 items committed in 1 sec (1)items/sec
Updating.
Updated Commit database in 1 sec
Configuration successfully rolled back to '1000000219'
```

(18) CLI 配置变更管理小结。

查看历史记录,命令如下:

```
show configuration commit list
```

查看历史记录内容,命令如下:

```
show configuration commit changes <Commit ID>
```

预览配置还原,命令如下:

```
show configuration rollback changes to <Commit ID>
```

实施配置还原,命令如下:

```
rollback configuration to <Commit ID>
```

4.2.2 板卡的顺序

以 ASR 9922 为例,考虑到是多机箱结构,思科 ASR 9922 的板卡命名规则如下。

如 0/1/2/3,则 0 代表机箱号,1 代表板卡号,2 代表子卡号,3 代表端口号。

机箱上面一排接口卡从左到右分别是板卡 0～9,机箱下面一排从左到右分别是包板卡 10～19,需要注意的是 10～19 卡的端口号是从下往上的顺序。引擎在机箱的中部,交换矩阵也在机箱中部,共可插 7 块。ASR 9922 机箱和板片顺序如图 4-3 所示。

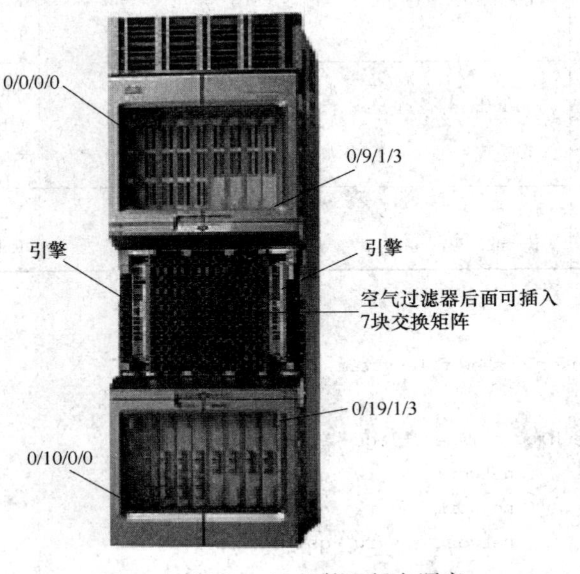

图 4-3 ASR 9922 机箱和板卡顺序

4.2.3 IOS XR 文件存储系统

表 4-2 展示了引擎 RSP 的存储设备(注意,随着版本的变化可能有所出入)。

表 4-2 引擎 RSP 里的存储设备

存储设备	介质	数量	容量	特点
Harddisk	SAS	1	70 GB	硬盘,吞吐高
eUSB	NAND	2	2 GB	闪存、擦写速度快、功耗低、适合存储文件
C. Flash(removable)	Compact Flash	1	1 GB	CF 卡
Flash	NOR	2	64 MB	闪存、随机存取,以存储代码为主
NVRAM		1	512 KB	存储配置文件

表 4-3 RSP 存储设备分区

存储设备	分区	容量	用途
Harddisk	/harddisk: /harddiska: /harddiskb:	70 GB (total)	dumps/ etc. Primary for kernel dumps
eUSB	/disk0: /disk0a: /disk1: /disk1a:	1.6 GB 0.4 GB 1.6 GB 0.4 GB	IOS-XR Packages Golden Disk or Mirror Disk
C. Flash (removable)	/compactflash:	1 GB	General Purpose Storage
Flash 1	/bootflash:	43 MB	MONLIB MBI images ROMMON/Firmware
Flash 2	/configflash:	28 MB	MONLIB OBFL data Secondary for kernel dumps
NVRAM	/nvram: /nvram-raw:		Variables, logs Reboot history; crashinfo, PCDS

显示文件系统:

```
RP/0/RSP0/CPU0:router# show filesystem
File Systems:

     Size(b)  Free(b)      Type  Flags  Prefixes
           -        -   network     rw  qsm/dev/fs/tftp:
           -        -   network     rw  qsm/dev/fs/rcp:
           -        -   network     rw  qsm/dev/fs/ftp:
          ...
```

```
  1644150784    1404023296   flash-disk  rw  disk1:
  1644150784    1404023296   flash-disk  rw  disk0:
 35645292544   35628865024     harddisk  rw  harddisk:
  1022427136    1022418944   flash-disk  rw  compactflash:
  8075067392    8073070080     harddisk  rw  harddiska:
  7840202752    7838258688     harddisk  rw  harddiskb:
   411041792     410914816   flash-disk  rw  disk1a:
   411041792     410903552   flash-disk  rw  disk0a:
      224256        165888        nvram  rw  nvram:
      290816        290816        nvram  rw  nvram-raw:
    44695552      23889200        flash  rw  bootflash:
    28966912      28862044               rw  configflash:
```

4.2.4 软件管理

版本号格式:A.F.M。

- A:Architecture Release 支持新硬件,如 4.0.0 开始支持 OC48/192 SPA。
- F:Feature Release 增加新功能或特性,大约每 8 个月发布一次。

例如,3.3.0 支持 L3VPN,3.4.0 支持 L2VPN。

- M:Maintain Release 功能改进,Bug 修复,大约每 4 个月发布一次。

(1) IOS XR 软件包的组成以 4.1.1 版本为例,分为 8 个软件功能包。

- asr9k-mini-p.pie-4.1.1(MBI):内核与操作系统。
- asr9k-mcast-p.pie-4.1.1(mcast):组播。
- asr9k-mpls-p.pie-4.1.1(mpls):MPLS。
- asr9k-video-p.pie-4.1.1(video):视频。
- asr9k-optic-p.pie-4.1.1(optic):IPoDWDM。
- asr9k-mgbl-p.pie-4.1.1(mgbl):管理。
- asr9k-k9sec-p.pie-4.1.1(k9sec):安全。
- asr9k-doc-p.pie-4.1.1(doc):联机文档。

(2) 查看正在运行的软件包。

```
RP/0/RSP0/CPU0:router# show install active summary
Default Profile:
  SDRs:
    Owner
  Active Packages:
    disk0:asr9k-mini-p.pie-4.1.1
    disk0:asr9k-mcast-p.pie-4.1.1
    disk0:asr9k-mpls-p.pie-4.1.1
    disk0:asr9k-video-p.pie-4.1.1
    disk0:asr9k-optic-p.pie-4.1.1
    disk0:asr9k-mgbl-p.pie-4.1.1
    disk0:asr9k-doc-p.pie-4.1.1
```

ASR 9000 功能软件的缺省只有核心模块,其他的软件模块需要单独安装,例如,缺省不支持 MPLS,这也是 ASR 跟 IOS 的区别之一,从软件上看,不需要的功能模块就可以不装,如果需要的话,需要安装相应的功能模块。例如,对 MPLS 的支持,如果装的是 disk0:asr9k-mpls-px-4.2.3 ,那么内核模块为 disk0:asr9k-mini-px-4.2.3 是必须的。

下面是笔者安装的国内第一台 IOS-XR 软件构成:

```
RP/0/RP0/CPU0:(admin)#sh install active summ
Default Profile:
  SDRs:
    Owner
  Active Packages:
    disk0:asr9k-px-4.2.3.CSCud29892-1.0.0
    disk0:asr9k-px-4.2.3.CSCuc84257-1.0.0
    disk0:asr9k-px-4.2.3.CSCuc59492-1.0.0
    disk0:asr9k-mini-px-4.2.3
    disk0:asr9k-mpls-px-4.2.3
    disk0:asr9k-px-4.2.3.CSCud07536-1.0.0
    disk0:asr9k-px-4.2.3.CSCuc47831-1.0.0
    disk0:asr9k-fpd-px-4.2.3
```

4.2.5 软件包安装与卸载

需要在 admin 模式下执行的命令如下所示。

1. IOS XR 软件包的安装

(1) 检查文件系统和磁盘空间:

```
router(admin)#cfs check
showfilesystem | in "Free|disk0:"
```

升级 4.2.3 至少要 700M 空余空间,到 Cisco 网站下载大版本号的 IOS XR,解包得到 pie,如下:

```
RP/0/RSP0/CPU0:router(admin)#
```

(2) 把 pie 添加到设备:

```
Install add source tftp://61.236.123.11 asr9k-mini-p.pie-4.2.3 asr9k-mpls-p.pie-4.2.3 sync
```

(3) 激活 pie:

```
install activate disk0:*4.2.3* sync
```

此步骤完成后,ASR 9010 会重启。

(4) 安装确认:

```
install commit
```

(5) 检查 Firmware 是否需要升级(最后一项是 yes 的要升级):

```
Show hw-module fpd location all
```

(6) 升级 Firmware 的命令(要安装 asr9k-upgrade-p.pie-4.2.3):

```
upgradehw-module fpd alllocation all
```

需要重启,可以和 SMU 一起重启。

(7) 升级 SMU:

```
Install-add-source-tftp://61.236.123.11asr9k-p-4.2.3.CSCuc84253.pie asr9k-p-4.2.3.CSCud07536.pie asr9k-p-4.2.3.CSCud29892.pie sync
```

(8) 激活升级包：

```
install activate disk0:*4.2.3* sync
```

此步会重启。

(9) 升级确认(必须)：

```
install commit
```

(10) 升级验证：

```
show install active summary
```

2. IOS XR 软件包的卸载

(1) 进入管理员模式，命令如下：

```
RP/0/RSP0/CPU0:router# admin
```

(2) 去激活软件包，命令如下：

```
install deactivate disk0:asr9k-mpls-4.2.3
```

(3) 查看不激活，命令如下：

```
Show install inactive summary
```

(4) 命令确认，命令如下：

```
install commit
```

(5) 删除软件包文件，命令如下：

```
install remove inactive
```

4.2.6 IOS XR 接口配置

(1) 物理端口名称表示。

常见物理端口类型

```
interface MgmtEth              0/RSP0/CPU0/0
! 管理口
interface GigabitEthernet      0/1/0/1
! 千兆口
interface TenGigE              0/2/0/2
! 万兆口
interface POS                  0/3/1/3
! POS 口
```

(2) 端口名称的定义。

如 TenGigE　0 / 1 / 2 / 3，解释如下：

- TenGigE：物理端口类型；
- 0：Rack 机架编号；
- 1：Slot 插槽编号；
- 2：Module 子插槽；
- 3：Port 端口编号。

(3) 以太网端口的配置选项。

- 配置模式：configure
- 端口路径：interface [GigabitEthernet | TenGigE] interface-path-id
- IPv4 地址：ipv4 address ip address mask

- 以太网流控：flow-control {bidirectional | egress | ingress}
- 最大帧长：mtu bytes
- 指定 MAC 地址：mac-address value1. value2. value3
- 启用端口：no shutdown
- 确认配置：commit
- 退出配置模式：end
- 查看端口信息：show interfaces [GigabitEthernet | TenGigE] interface-path-id

（4）子接口与 VLAN 配置，命令如下：

```
RP/0/RSP0/CPU0:router# show run interface GigabitEthernet0/5/0/0
interface GigabitEthernet0/5/0/0
  no shutdown
!
interface GigabitEthernet0/5/0/0.1
! 配置子接口
  encapsulation dot1q 10
! 802.1q 封装
  ipv4 address 10.3.1.1/24
!
interface GigabitEthernet0/5/0/0.2
  encapsulation dot1q 20
  ipv4 address 10.3.2.1/24
!
```

（5）查看接口运行状态，命令如下：

```
RP/0/RSP0/CPU0:router# show interfaces GigabitEthernet 0/5/0/0
```

（6）链路捆绑（Link Bundling）配置，命令如下：

```
RP/0/RSP0/CPU0:router# show run interface bundle 1

interface Bundle-Ether 1
! Bundle 逻辑接口
  ipv4 address 1.2.3.4/24
!
interface TenGigE 0/3/0/0
! 物理端口加入 Bundle
  bundle id 1 mode active
  no shutdown
!
interface TenGigE 0/3/0/1
! 物理端口加入 Bundle
  bundle id 1 mode active
  no shutdown
!
```

（7）查看 Bundle 接口配置命令：

```
RP/0/RSP0/CPU0:router# show bundle Bundle-Ether 1
RP/0/RSP0/CPU0:router# show lacp bundle Bundle-Ether 1
```

4.2.7 常用路由协议配置

ASR 9000路由器支持表4-4的所有路由协议。

表4-4 路由协议分类

按协议分类	域内路由协议 IGP				域间路由协议 BGP
按算法分类	距离矢量路由算法		链路状态路由算法		Path Vector
IPv4 Class	RIP	IGRP			
IPv4 Classless	RIPv2	EIGRP	OSPFv2	ISIS	BGP-4
IPv6	RIPng	EIGRP for IPv6	OSPFv3	ISIS for IPv6	BGP-4 Multiprotocol Extensions

(1) 配置静态路由命令如下：

```
router static
address-family ipv6 unicast
! IPv6 单播
  2001:0DB8::/32 2001:0DB8:3000::1
!
vrf vrf1
  address-family ipv4 multicast
! IPv4 组播
   11.0.0.0/16 TenGigE0/2/0/0
!
```

(2) OSPF基本配置，命令如下：

```
RP/0/RSP0/CPU0:router# show run router ospf
router ospf ISP-OSPFv2
! 启动 OSPFv2 进程，名称为字符或数字

router-id  1.2.3.4
! 路由器标识，在路由域内唯一
area 0
  interface GigabitEthernet 0/3/0/0
! 接口运行 OSPF
  cost 10
  !
  Interface Loopback 0
    Passive enable
  !
!
RP/0/RSP0/CPU0:router# show run router ospf
router ospf ISP-OSPFv2
  router-id  1.2.3.4
  redistribute connected
! 重分发路由
  redistribute static
area 1
  hello-interval 10
```

```
! hello 包发送间隔
  dead-interval 40
! hello 包的最长等待时间
  interface GigabitEthernet 0/3/0/1
   authentication message-digest
! 邻居连接认证,MD5 算法
   message-digest-key 1 md5 clear Password
  !
  interface Loopback0
   passive enable
  !
```

(3) ISIS 基本配置,命令如下:

```
RP/0/RSP0/CPU0:router# show run router isis
router isis TEST
! 启动 ISIS 进程,名称为字符或数字
  is-type level-2-only
! 骨干区域
  net 49.0192.0168.0001.0001.00
  address-family ipv4 unicast
  interface TenGigE0/7/0/0
! 接口运行 ISIS
  address-family ipv4 unicast
   !
!
interface loopback0
  address-family ipv4 unicast
   !
!
```

(4) BGP 基本配置,命令格式如下:

```
RP/0/RSP0/CPU0:router# show run router bgp
router bgp 100
! 启动 BGP 进程
  bgp router-id 192.0.0.1
  address-family ipv4 unicast
! 指定 IPv4 地址簇
   redistribute ospf isp
! 重分发 OSPF 路由信息
   !
  neighbor 10.2.2.2
! EBGP 邻居
   remote-as 200
! EBGP 邻居自治域编号
   address-family ipv4 unicast
     next-hop-self
! 设置通告路由的下一跳属性
     route-policy pass-all in
! 对接收的路由应用路由策略
```

```
    route-policy pass-all out
! 对通告的路由应用路由策略
  !
!
route-policy pass-all
! 定义路由策略
   pass
end-policy
```

例 1 修改某些路由的属性(整个 prefix-set 要一起改掉)。

```
prefix-set Prefix
  10.0.2.0/24,
  10.0.3.0/24 ge 28,
  10.0.4.0/24 ge 26 le 30,
end-set
route-policy prefer
  if destination in Prefix then
   set local-preference 200
  endif
end-policy

router bgp 100
  neighbor 10.2.2.2
  address-family ipv4 unicast
    policy prefer in
```

例 2 屏蔽 AS-Path 包含某 AS 的路由。

```
as-path-set ignore_path
  ios-regex '_11_',
end-set
route-policy ignore_path_as
  if as-path in ignore_path then
    drop
  else
    pass
  endif
end-policy

router bgp 100
  neighbor 10.2.2.2
  address-family ipv4 unicast
     policy ignore_path_as in
```

(5) VRRP 的配置。

虚拟路由器同样支持冗余协议 VRRP,实例如图 4-4 所示。

终端网关 IP:10.0.0.4。

应用场景:主备冗余,消除单个设备故障的影响;负载分担,形成多个虚拟路由组,以对应不同流量;VRRP 可以最大在 3 台网关上实现。

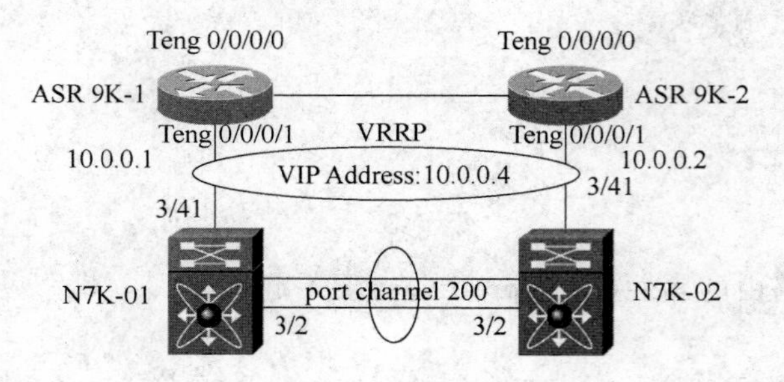

图 4-4　VRRP 应用示意图

VRRP 配置:

```
RP/0/RSP0/CPU0:ASR9K-1# show run
interface GigabitEthernet0/0/0/1
ipv4 address 10.0.0.1

router vrrp
! 进入 VRRP 配置模式
interface GigabitEthernet0/0/0/1
address-family ipv4
vrrp 1
! VRRP 组号
    ipv4 10.0.0.4
    ! Vitrual IP 地址
track interface Gig 0/0/0/1 30
! 若所跟踪的端口 down,优先级减少 30,优先级变成 90,低于缺省的备用链路 1 000,来确保备用端口变成主用端口
  !
priority 120
! 优先级 1~255,默认 100,值越大优先级越高
```

在 RouterB 上做类似的配置,优先级使用缺省的 100 即可。

查看 VRRP 运行状态:

```
RP/0/0/CPU0:ASR9K-1#sh vrrp
Sat Nov 15 12:29:57.689 UTC
IPv4 Virtual Routers:
                    A indicates IP address owner
                    | P indicates configured to preempt
                    | |
Interface   vrIDPrio A P State     Master addrVRouteraddr
Gi0/0/0/0     1  120   P Master    local            10.0.0.4
RP/0/0/CPU0:ASR9K-1#
```

配置管理接口(配备双引擎,每个引擎有 2 个管理口,所以双引擎有 4 个管理口)

```
interface MgmtEth0/RSP0/CPU0/0
  ipv4 address 192.168.0.1 255.255.255.0
!
```

```
interface MgmtEth0/RSP0/CPU0/1
  ipv4 address 192.168.10.1 255.255.255.0
!
interface MgmtEth0/RSP1/CPU0/0
  ipv4 address 192.168.0.2 255.255.255.0
!
interface MgmtEth0/RSP1/CPU0/1
  ipv4 address 192.168.10.2 255.255.255.0
```

配置虚管理地址，始终与主引擎同一子网的物理管理接口相关联：

```
ipv4 address virtual address 192.168.0.3/24
```

无论主引擎是哪个，通过虚管理地址都能接入到路由器。

4.2.8 NetFlow 配置

NetFlow 配置的基本步骤：

① 创建一个 exporter map;

② 创建一个 monitor map;

③ 创建一个 sampler map;

④ 在接口里调用 monitor map 和 sampler map。

(1) 创建一个 exporter map：

```
flow exporter-map FEM1
version v9
  options interface-table
  template data timeout 600
!
dscp 10
transport udp 50040
source Loopback0
destination 21.1.1.1
!
```

(2) 创建一个 monitor map：

```
flow monitor-map FMM1
  record ipv4
  exporter FEM1
  cache entries 10000
  cache timeout active 5
  cache timeout inactive 5
!
```

(3) 创建一个 sampler map：

```
sampler-map FSM1
  random 1 out-of 5000
!
```

(4) 在接口里调用 monitor map 和 sampler map：

```
interface TenGigE0/0/1/1
  ipv4 address 22.9.9.1 255.255.255.252
  flow ipv4 monitor FMM1 sampler FSM1 ingress
```

4.2.9 远程控制访问

(1) 先定义 access-class 名称，命令如下：

```
line default
access-class ingress LOGIN-LIMIT
```

(2) 再定义可访问的源地址，命令如下：

```
ipv4 access-list LOGIN-LIMIT
20 permit ipv4 host 161.235.206.2 any
! 只允许源地址 161.235.206.2 访问该设备，缺省拒绝 any
```

4.3 高级配置

4.3.1 按时间执行的命令

时间策略的控制需要较复杂的命令组合，基本的思路是先把命令写成不同文件存在设备的某个存储设备中，在不同时间会对不同脚本文件进行调用执行，下面是定义一个时间段(00:00 到 08:00)执行的命令实例。

在 ASR 9000 上输入：

```
aaa authorization eventmanager default local
event manager environment _cron_entry1 00 00 ***
event manager environment _cron_entry2 00 08 ***
event manager directory user policy disk0:
event manager policy acl-1.tcl username CISCO persist-time 28800 type user
event manager policy acl-2.tcl username CISCO persist-time 57600 type user
```

下面是需要上传到设备 DISK0 的两个文件。

(1) acl-1.tcl 文件的配置

```
::cisco::eem::event_register_timer cron name crontimer1 cron_entry $_cron_entry1
#::cisco::eem::event_register_none
namespace import ::cisco::eem::*
namespace import ::cisco::lib::*
if {[catch {cli_open} result]} {
    action_syslog priority info msg "CLI Open Failed: $result"
    exit
}
set t_acl "ipv4 access-list TEST
  10 permit ipv4 222.72.75.0/24 183.61.190.0/24 nexthop1 ipv4 21.198.0.60
  20 permit ipv4 122.172.75.0/24 211.151.181.0/24 nexthop1 ipv4 21.198.0.60
  !"
array set cli1 $result
action_syslog priority info msg "Starting ACl script"
cli_exec $cli1(fd) "conf t
                    $t_acl
                    commit"
```

```
action_syslog priority info msg "ACl ScriptEnd   "
if {[catch {cli_close $cli1(fd) $cli1(tty_id)} result]} {
    action_syslog priority info msg $result
}
```

(2) acl-2. tcl 文件的配置

```
# List of event manager environment variables
# _cron_entry - cron value
#

::cisco::eem::event_register_timer cron name crontimer2 cron_entry $_cron_entry2
#::cisco::eem::event_register_none

namespace import ::cisco::eem::*
namespace import ::cisco::lib::*
if {[catch {cli_open} result]} {
    action_syslog priority info msg "CLI Open Failed: $result"
    exit
}
set t_acl "ipv4 access-list TEST
  no 10 permit ipv4 222.72.75.0/24 183.61.190.0/24 nexthop1 ipv4 21.198.0.60
  no 20 permit ipv4 122.172.75.0/24 211.151.181.0/24 nexthop1 ipv4 21.198.0.60
!"
array set cli1 $result
action_syslog priority info msg "Starting ACl script"
cli_exec $cli1(fd) "conf t
                    $t_acl
                   commit"
action_syslog priority info msg "ACl ScriptEnd"
if {[catch {cli_close $cli1(fd) $cli1(tty_id)} result]} {
    action_syslog priority info msg $result
}
```

4.3.2 配置 BGP 路由

ASR 系列采用 IOS-XR 软件，命令格式与 IOS 有所不同，在编写 BGP 相关属性的时候尤其要注意，举例如下。

```
prefix-set 9394to65532deny
! 定义 prefix-set，名为 9394to65532deny
  10.0.0.0/8 ge 8 le 32,
  61.234.206.0/24 ge 24 le 32,
  36.192.176.0/20 ge 20 le 32
end-set

prefix-set 9394to65532permit
! 定义 prefix-set，名为 9394to65532permit
  0.0.0.0/0 le 32
end-set
```

```
route-policy 9394to65532
! 定义一个 route-policy,名为 9394to65532
if destination in 9394to65532deny then
drop
elseif destination in 9394to65532permit then
pass
endif
end-policy
!
community-setwangnei
! 定义名为 wangnei 的团体
  9394:50100,
  9394:50700
end-set
community-setmianfei
! 定义名为 mianfei 的团体
  9394:135,
  9394:136,
  9394:137
end-set

as-path-set210
! 定义名为 210 的 as-path 属性
ios-regex '_1394 $ ',
! 以 AS1393 开头的路由
ios-regex '^1394_65235 $ ',
! 包含 AS1394 和 65235 的路由
ios-regex '^1394_9531 $ '
! 包含 AS1394 和 9531 的路由
end-set

prefix-setredirect_to_chinanet
! 定义名为 redirect_to_chinanet 的 prefix
  121.56.0.0/15 ge 15 le 32,
  222.73.0.0/16
end-set

prefix-setredirect_to_cnc
! 定义名为 redirect_to_cnc 的 prefix
  221.217.0.0/18 ge 18 le 32
end-set

route-policy bgp2
! 定义一个名为 bgp2 的路由策略
apply 9394to65532
! 调用 9394to65532 策略
if community matches-any wangnei then
```

```
！检查匹配 wangnei 的策略
done
elseif community matches-any mianfei then
！检查匹配 mianfei 的策略
done
elseif as-path in 210 then
！检查满足 as-path in 210 的策略
done
elseif destination in redirect_to_chinanet then
set next-hop 222.43.3.12
setlocal-preference 150
！去往 chinanet 的路由设置下一跳为 222.43.3.12,并且把 local-preference 改成 150
elseif destination in redirect_to_cnc then
set next-hop 222.43.3.12
set local-preference 150
！去往 cnc 的路由设置下一跳为 222.43.3.12 并且把 local-preference 设置为 150
elseif destination in any then
set next-hop 222.43.3.12
！缺省的下一跳都送往 222.43.3.12
endif
end-policy

routerbgp 5523
！配置 BGP 路由和邻居
nsr
bgp router-id 22.143.3.11
bgp cluster-id 22.143.3.11
address-family ipv4 unicast
network 36.192.176.0/21
network 36.192.185.0/21
!

neighbor 21.98.0.160
remote-as 9394
ebgp-multihop 3
description Connect to NE5000
update-source Loopback0
address-family ipv4 unicast
route-policy bgp2 in
！入方向的流量匹配 BGP2 的策略,则执行相应动作
```

4.3.3 MTU 值的设置

新的设备上线后,有些业务正常,有些业务不正常,这是因为不同应用软件对延时的敏感性不同所造成的。

OSI 第三层网络层协议的包头是在第二层帧头之上的,也就是说在封装二层帧头的时候,是将数据内容和三层包头全部作为数据封装在里面的,对于二层来说,之前的数据最大是多

少，是由 MTU 来决定的，正常情况下 MTU 就是第三层数据和包头的最大尺寸，这时无须分段就能传输，如果数据包比 MTU 大，就得分段后传输。

MPLS 的标签是在二层帧头之后的，所以二层帧头将标签的大小和三层包的内容累加到一起作为数据封装，如果三层包的所有内容正好和 MTU 一样大，在此基础上加上 MPLS 标签，就肯定比额定的 MTU 要大，所以这时 MPLS 的标签数据是会被分段后传输的，如果不想被分段，就得更新 MTU 的大小。

在有 MPLS 的网络环境下，MTU 需要特别的设置，所有 IPv4 地址有一个甚至多个对应的标签，但这并不意味着可以随意加几个字节，由于标签占用 4 字节，所以在 MPLS 的网络中增加的包数量是 4 乘以 N 的整数倍。如果遇到的项目是 2 层标签，最低加 8 个字节，即 1 508。

ASR 9922 的 MTU 缺省值是 1 514，对 IEEE 802.1Q 的数据包建议值为 1 518，对 QinQ 的数据包 MTU 建议值为 1 522。

4.3.4　MPLS VPN 的设置

图 4-5 是 MPLS 实验需要的拓扑。

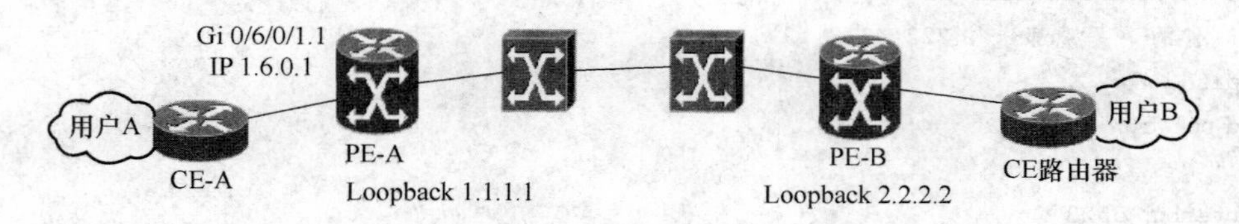

图 4-5　MPLS VPN 实验拓扑

1. 配置 MPLS VPN——建立 VRF

(1) 定义 VRF

```
vrf vpnA
address-family ipv4 unicast
  import route-target
   100:1
  !
  export route-target
   100:1
  !
!
```

(2) 接口关联 VRF

```
interface GigabitEthernet0/6/0/1.1
  vrf vpnA
  ipv4 address 1.6.0.1 255.255.255.252
  encapsulation dot1q 1
```

2. 配置 MPLS VPN-PE 连接 CE 的路由协议

(1) 静态路由：

```
router static
  vrf vpnA
  address-family ipv4 unicast
```

```
  192.168.1.0/24 1.6.0.2
!
```

或 OSPF 路由：

```
router ospf VPNA
  vrf vpnA
   router-id 1.6.0.1
   redistribute bgp 100
   area 0
     interface GigabitEthernet0/6/0/0.1
     !
  !
```

（2）配置 MPLS VPN：

PE-A：

```
router bgp 100
bgp router-id 192.0.0.1
address-family vpnv4 unicast
!
neighbor 2.2.2.2
  remote-as 100
  update-source Loopback0
  address-family vpnv4 unicast
!
!
vrf vpnA
    rd 100:20
    address-family ipv4 unicast
      redistribute ospf VPNA
    !
```

PE-B：

```
router bgp 100
bgp router-id 192.0.0.2
address-family vpnv4 unicast
!
neighbor 1.1.1.1
  remote-as 100
  update-source Loopback0
  address-family vpnv4 unicast
!
!
vrf vpnA
    rd 100:20
    address-family ipv4 unicast
      redistribute ospf VPNA
    !
```

查看设立的 VRF，命令如下：

```
RP/0/RP0/CPU0:FJFZ-ASR# sh vrf all
```

查看 VRF 路由，命令如下：

```
RP/0/RP0/CPU0:FJFZ-ASR# sh route vrf all
```

4.4 常用命令

4.4.1 查看路由器工作状态

查看系统版本，命令如下：

```
show version
```

查看线卡工作状态，命令如下：

```
show platform
```

主备用引擎冗余状态，命令如下：

```
show redundancy
```

监控温度及指示灯状态，命令如下：

```
show environment
```

（1）查看系统版本状态

```
RP/0/RSP0/CPU0:router# show version
Cisco IOS XR Software, Version 5.0.0[Default]
ROM: System Bootstrap, Version 1.4(20100216:021454) [ASR9K ROMMON],

router uptime is 2 weeks, 6 days, 10 hours, 30 minutes
System image file is "bootflash:disk0/asr9k-os-mbi-5.0.0/mbiasr9k-rp.vm"
cisco ASR9K Series (MPC8641D) processor with 4194304K bytes of memory.
MPC8641D processor at 1333MHz, Revision 2.2
2 Management Ethernet
12 TenGigE
40 GigabitEthernet
219k bytes of non-volatile configuration memory.
975M bytes of compact flash card.
33994M bytes of hard disk.
```

（2）查看线卡工作状态

```
RP/0/RSP0/CPU0:router(admin)# show platform
Node            Type                State           Config State
---------------------------------------------------
0/RSP0/CPU0     A9K-RSP-4G(Active)  IOS XR RUN      PWR,NSHUT,MON
0/FT0/SP        FAN TRAY            READY
0/FT1/SP        FAN TRAY            READY
0/1/CPU0        A9K-40GE-B          IOS XR RUN      PWR,NSHUT,MON
0/4/CPU0        A9K-8T/4-B          IOS XR RUN      PWR,NSHUT,MON
0/6/CPU0        A9K-4T-B            IOS XR RUN      PWR,NSHUT,MON
0/PM0/SP        A9K-3KW-AC          READY           PWR,NSHUT,MON
0/PM1/SP        A9K-3KW-AC          READY           PWR,NSHUT,MON
0/PM2/SP        A9K-3KW-AC          READY           PWR,NSHUT,MON
```

（3）查看主备用引擎状态

```
RP/0/RSP0/CPU0:router# show redundancy
Redundancy information for node 0/RSP0/CPU0:
==================================
Node 0/RSP0/CPU0 is in ACTIVE role
Node 0/RSP0/CPU1 is NSR ready
Reload and boot info
----------------------
A9K-RSP-4G reloaded Mon May 10 15:53:34 2010: 2 weeks, 6 days, 10 hours, 44 minutes ago
Active node booted Mon May 10 15:53:34 2010: 2 weeks, 6 days, 10 hours, 44 minutes ago
Active node reload "Cause: pID node reload is required by install operation"
```

（4）监控环境温度

```
RP/0/RSP0/CPU0:router# show environment temperatures
R/S/I   Modules         Inlet           Hotspot
                        Temperature     Temperature
                        (deg C)         (deg C) 0/1/*
        host            36.1            43.1            0/RSP0/*
        host            30.8            41.5            0/4/*
        host            35.4            45.5            0/6/*
```

（5）时钟设置

```
RP/0/RP0/CPU0:ios#clock set 11:19:20 23 november 2013
```

4.4.2 常用检修命令

（1）查看已经执行的命令。

查看已经递交的命令列表：

```
show configuration commit list
! 先查看递交过的命令
```

查看序号为1000000808递交的命令的内容：

```
show configuration commit changes 1000000808
```

（2）静态路由数量的限制。

静态路由超过3 000条时Console上会提醒出错，好在有改正的方法，在静态路由中设置如下：

```
maximum path ipv4 100000
```

ASR 9922版本5.2.3最大支持的静态路由数量为15万条。

（3）有路由，traceroute为什么没有下一跳？

一般出现这种情况的原因是下一跳设备是防火墙，但是作者的这个项目中，不是因为下一跳设备是防火墙，ASR 9922接到NE80E再连接到地市BR，在ASR 9922上显示有路由，trace却连下一跳都不可达。

由于省内核心设备都启用MPLS标签交换，最后发现NE80E（ME60支持的标签转发数量更少）上居然是部分路由没有标签转发表，最后厂家的解释是这是由该设备标签转发表满引起的故障，清除标签转发表暂时解决问题，命令如下所示。

```
PTBR1#traceroute X.X.X.X
Type escape sequence to abort.
```

```
Tracing the route to X.X.X.X

  1   *   *   *
  2   *   *   *
  3   *   *   *
```

（4）按照 BGP 相应属性查看路由。

查看该 BGP 属性下的路由，命令如下：

```
sh bgp community 9394:136
sh bgp regexp _9394 $
sh bgp community 9394:135
```

查看端口策略条目数，命令如下：

```
sh access-lists cttfj hardware ingress location 0/12/cpu0
```

（5）查看当前的软件 PIE。

```
show install act summary
```

（6）其他常用命令。

让某个子卡业务下线，命令如下：

```
hw-module service offline location 0/10/0
```

给某个子卡下电，命令如下：

```
RP/0/RP0/CPU0:(config) # hw-module subslot 0/0/0 shutdown unpowered
```

设置时间，不可以在 admin 下执行，命令如下：

```
RP/0/RP0/CPU0:ios # clock set 11:19:20 23 november 2012
```

查看单个引擎情况，命令如下：

```
RP/0/RP0/CPU0:ios(admin) # show platform summ location 0/RP1/CPU0
```

查看所有板卡状态，命令如下：

```
RP/0/RP0/CPU0:ios(admin) # show platform summ
```

查看 OSPF 路由，命令如下：

```
RP/0/RP0/CPU0:ios # sh ospf   summ
```

查看 BGP 路由，命令如下：

```
RP/0/RP0/CPU0:ios # sh bgp summ
```

检修 OSPF 邻居，命令如下：

```
RP/0/RP0/CPU0:ios # debug ospf 1 adj
```

修改每次显示的长度，命令如下：

```
RP/0/RP0/CPU0:ios # terminal length 40
```

必须在 config 模式下，不是 admin 模式下才能有效，命令如下：

```
RP/0/RP0/CPU0:ios(config) # no hw-module subslot 0/0/1 shutdown
```

想要定义某个账号有效，必须放在管理组里，命令如下：

```
RP/0/RP0/CPU0:ios(config) # username ciscocisco
RP/0/RP0/CPU0:ios(config-un) # group root-system
RP/0/RP0/CPU0:ios(config-un) # password ciscocisco
RP/0/RP0/CPU0:ios(config-un) # commit
! 提交 commit 后命令才生效
```

在 ASK 9000 上查看 VPN 路由，命令如下：

```
RP/0/RP0/CPU0:ios #sh bgp vpnv4 unicast
```

查看接口物理收发功率，命令如下：

```
RP/0/RP0/CPU0:ios#sh controllers tenGigE 0/2/1/2 phy
```

查看 BGP 路由，命令如下：

```
RP/0/RP0/CPU0:ios#sh bgp
```

查看 BGP 邻居是否建立，命令如下：

```
RP/0/RP0/CPU0:ios#sh bgp summ
```

查看 OSPF 邻居是否建立，命令如下：

```
RP/0/RP0/CPU0:ios#sh ospf nei
```

查看总体路由条目数，命令如下：

```
RP/0/RP0/CPU0:ios#sh route summary
```

如何删除一条 BGP 邻居，命令如下：

```
进到 router bgp 655532
no neighbor 1.1.1.1
commit
```

备份配置文件到电脑 TFTP，命令如下：

```
copy disk0a:/usr/asr9k-config tftp:
! 10.1.1.100/asr9k-config
```

不中断 BGP 邻居关系的情况下刷新路由表，命令如下：

```
clear bgp ipv4 unicast 10.1.1.200 soft
```

配置 ASR9K 的管理口参与路由进程，命令如下：

```
rp mgm forwarding
```

查看静态路由表里匹配所有 0.0.0.0 项的条目，命令如下：

```
sh run router static | include 0.0.0.0
```

查看标签转发表，命令如下：

```
RP/0/RP0/CPU0:FJFZ-ASR#sh mpls forwarding
```

查看 32 位标签转发表，命令如下：

```
RP/0/RP0/CPU0:FJFZ-ASR#sh mpls forwarding | include /32
```

查看某条路由的标签转发情况，命令如下：

```
RP/0/RP0/CPU0:FJFZ-ASR#sh mpls forwarding prefix XX.XX.XX.XX/XX
```

修改一个 prefix-set，as-path-set 或 community-set，需要先做一个完整的脚本，然后一下全部替换，命令如下：

```
prefix-set SET-NAME
  221.173.5.0/24,
  222.43.0.0/16 ge 16 le 32,
  122.90.0.0/16 ge 16 le 32,
  36.192.176.0/21 ge 21 le 32
end-set
```

查看路由转发表，命令如下：

```
RP/0/RP0/CPU0:XM#sh cef x.x.x.x
```

建立一个用户的账号,可分 4 步来进行。

第一步,建立权限。

```
taskgroup priv1
  task read bgp
  task read rib
  task read ipv4
  task read sysmgr
  task read system
  task read logging
  task read network
  task read fault-mgr
  task read interface
  task read basic-services
```

第二步,建立一个用户组。

```
usergroup priv1
  taskgroup priv1
  taskgroup operator
```

第三步,把一个账号放到权限组里。

```
username ciscocisco
  group priv1
```

第四步,新建一个账号。

```
Username ciscocisco password ciscocisco
```

4.5 实际配置举例

为了减少篇幅,作者删掉了一些不必要的配置。

```
RP/0/RP0/CPU0:SP--ASR# sh run
Building configuration…
IOS XR Configuration 4.2.3
Last configuration change at Fri Jan 25 21:31:12 2013 by CORE
!
hostname SP--ASR
logging buffered 4096000
logging 61.233.13.254 vrf default
logging source-interface Loopback0
telnet vrf default ipv4 server max-servers 15
taskgroup priv1
  task read bgp
  task read rib
  task read ipv4
  task read sysmgr
  task read system
  task read logging
```

```
   task read network
   task read fault-mgr
   task read interface
   task read basic-services
 !
 usergroup priv1
 ! 配置账号组
   taskgroup priv1
 ! 授予组相应权限
   taskgroup operator
 ! 授予组相应权限
 !
 username CORE
   group netadmin
   group operator
   group sysadmin
   group root-system
   group serviceadmin
   group cisco-support
   password 7 15340118100A7669306374
 !
 username ciscocisco
 ! 对一个账号配置组,该账号相应获得该组权限
   group priv1
   password 7 05080F1C22434D000A0618
 !
 cdp
 vrf cww
 ! 配置 MPLS VRF
   address-family ipv4 unicast
     import route-target
       5523:900
       5523:901
     !
     export route-target
       5523:900
     !
   !
 vrf ip-pbx
   address-family ipv4 unicast
     import route-target
       5523:0
         5523:6000
     !
     export route-target
       5523:8
```

```
    5523:6000
  !
!
line console
  exec-timeout 0 0
  length 0
!
line default
  timestamp disable
  exec-timeout 15 0
!
vty-pool default 0 15
snmp-server ifindex persist
! 配置网管需要的参数
snmp-server trap link ietf
snmp-server engineID local 123456789
snmp-server host 22.143.23.2 traps solarwinds@CORE
snmp-server host 22.143.23.4 traps solarwinds@CORE
snmp-server host 22.143.26.9 traps solarwinds@CORE
snmp-server host 22.143.26.10 traps solarwinds@CORE
snmp-server community solarwinds@CORE RO
snmp-server community solarwinds@CORE50160 RW
snmp-server traps snmp
snmp-server traps config
snmp-server traps entity
snmp-server traps syslog
snmp-server correlator buffer-size 4096
snmp-server trap-source Loopback0
!
ipv4 access-list PORT-POLICY
  100 permit ipv4 any 110.125.0.0 0.0.253.255
  105 permit ipv4 any 10.0.0.0 0.253.253.255
  110 permit ipv4 any 22.143.0.0 0.0.253.255
…
! 这里略去类似的配置

5000 permit ipv4 any any
! 注意这句必须添加,缺省拒绝 any
ipv4 access-list TEST
! 配置 ACL 列表
  20 permit ipv4 host 61.235.206.2 any
  30 permit ipv4 host 61.235.206.8 any
  200 deny ipv4 any any
!
!
ipv4 access-list PORT-POLICY-001
  1620 permit ipv4 host 22.143.62.58 any nexthop1 ipv4 22.9.15.37
```

```
  5000 permit ipv4 any any
!
flow exporter-map FEM1
! 配置 NetFlow
  version v9
  options interface-table
    template data timeout 600
!
  dscp 10
  transport udp 8881
  source Loopback0
  destination 22.143.23.3
!
flow exporter-map FEM2
! 配置 NetFlow
  version v9
    options interface-table
    template data timeout 600
  !
  dscp 10
  transport udp 50040
  source Loopback0
  destination 218.203.150.66
!
flow monitor-map FMM2
! 配置 NetFlow
  record ipv4
  exporter FEM1
  exporter FEM2
  cache entries 900000
  cache timeout active 5
  cache timeout inactive 5
!
sampler-map FSM1
  random 1 out-of 5000
!
interface Bundle-Ether1
  description Connect to DSAS9306
  mtu 1540
  bundle maximum-active links 2
!
interface Bundle-Ether1.1
  ipv4 address 22.9.18.141 255.255.255.252
  flow ipv4 monitor FMM2 sampler FSM1 ingress
! 在接口调用,用作 NetFlow
  encapsulation dot1q 8
  ipv4 access-group PORT-POLICY ingress
```

```
!
!
interface Bundle-Ether1.1200
  description to-DSAoa
  vrf PORT-POLICY-oa
  ipv4 address 192.168.1.5 255.255.255.0
  encapsulation dot1q 1200
!
interface Bundle-Ether1.1201
  description to fjoa-ma5200f
  vrf PORT-POLICY-oa
  ipv4 address 192.168.101.1 255.255.255.0
  encapsulation dot1q 1201
!
interface Bundle-Ether1.1211
  description cww
  vrf cww
  ipv4 address 172.16.33.1 255.255.255.224
  encapsulation dot1q 1211
!
interface Loopback0
  ipv4 address 22.143.3.11 255.255.255.255
!

interface MgmtEth0/RP0/CPU0/0
  ipv4 address 10.0.0.120 255.255.255.0
!
interface MgmtEth0/RP0/CPU0/1
  ipv4 address 172.16.2.5 255.255.255.252
!
interface MgmtEth0/RP1/CPU0/0
  shutdown
!
interface MgmtEth0/RP1/CPU0/1
  shutdown

!
interface TenGigE0/0/0/0
  description connect-to-NE5000 1/1/0
  mtu 1540
  ipv4 address 22.9.9.2 255.255.255.252
  flow ipv4 monitor FMM2 sampler FSM1 ingress
!
interface TenGigE0/0/0/1
  description Connect to DSAS9306
  bundle id 1 mode on
  bundle port-priority 10
```

```
!
interface TenGigE0/0/0/2
  description to_jq9306_G1/1/0/0
  bundle id 2 mode on
!
interface TenGigE0/0/0/3
  description to SP-10G-02-yichang
  ipv4 address 22.9.15.145 255.255.255.252
  flow ipv4 monitor FMM2 sampler FSM1 ingress
  ipv4 access-group PORT-POLICY ingress
!
interface TenGigE0/0/1/0
  description to SP_03-hudong
  ipv4 address 211.138.159.149 255.255.255.252
  flow ipv4 monitor FMM2 sampler FSM1 ingress
  transport-mode wan
  ipv4 access-group PORT-POLICY ingress
!
interface TenGigE0/0/1/1
  description to 7609-3/0/0
  ipv4 address 22.143.3.222 255.255.255.252
  flow ipv4 monitor FMM2 sampler FSM1 ingress
!
interface TenGigE0/0/1/2
  description connect to DSB-ASR9K Teng 0/0/1/2
  mtu 1540
  ipv4 address 22.9.18.85 255.255.255.252
!
interface TenGigE0/0/1/3
  description connect to DSA7609 Teng 2/3
  ipv4 address 22.9.18.230 255.255.255.252
!
!
prefix-set any
  0.0.0.0/0 le 32
end-set
!
prefix-set PERMIT-deny
  221.173.5.0/23,
  0.0.0.0/0 le 32
end-set
!
prefix-set 1314to5523deny
  22.143.0.0/16 ge 16 le 32,

  12.143.0.0/16 ge 16 le 32
end-set
```

```
!
prefix-set redirect_to_cnc
  221.213.0.0/18 ge 18 le 32
end-set
!
prefix-set PERMIT-permit
  22.143.0.0/16 ge 16 le 32,

  12.143.0.0/16 ge 16 le 32
end-set
!
prefix-set AS-PERMIT
  0.0.0.0/0 le 32
end-set
!
prefix-set OSPF-DENY
  0.0.0.0/0 le 32
end-set
!
prefix-set redirect_to_chinanet
  121.56.0.0/15 ge 15 le 32,
  113.108.65.0/19 ge 19 le 32,
  222.73.0.0/16
end-set
!
prefix-set OSPF-PER
  12.142.180.0/22,

  12.143.0.0/16 ge 16 le 32
end-set
!
as-path-set 21
  ios-regex '_1314 $',

  ios-regex '^1314_9831 $'
end-set
!
community-set MF
  1314:135,
  1314:136,
  1314:137
end-set
!
community-set WN
  1314:50100,

  1314:50600,
```

```
  1314:50700
 end-set
 !
 route-policy bgp2
 ! 配置 BGP 策略
   apply 1314to5523
   if community matches-any wangnei then
     done
   elseif community matches-any mianfei then
     done
   elseif as-path in 210 then
     done
   elseif destination in redirect_to_chinanet then
     set next-hop 22.143.3.12
     set local-preference 150
   elseif destination in redirect_to_cnc then
     set next-hop 22.143.3.12
     set local-preference 150
   elseif destination in any then
     set next-hop 22.143.3.12
   endif
 end-policy
 !
 route-policy pass-all
   pass
 end-policy
 !
 route-policy PERMIT
   if destination in PERMIT-permit then
     pass
   elseif destination in PERMIT-deny then
     drop
   endif
 end-policy
 !
 route-policy 1314to5523
   if destination in 1314to5523deny then
     drop
   elseif destination in AS-PERMIT then
     pass
   endif
 end-policy
 !
 route-policy NDC-SPT
   if community matches-any wangnei then
     pass
   elseif as-path in 210 then
```

```
    pass
  endif
end-policy
!
route-policy ospf-in-static
  if destination in OSPF-PER then
    done
  elseif destination in OSPF-DENY then
    drop
  endif
end-policy
!
router static
! 配置静态路由
  maximum path ipv4 100000
  address-family ipv4 unicast
    0.0.0.0/0 22.143.3.221 100
    0.0.0.0/0 TenGigE0/0/0/0 22.9.9.1 10
    0.0.0.0/0 TenGigE0/1/0/0 22.9.9.233 10
    0.0.0.0/0 TenGigE0/2/0/0 22.9.15.37 10
…
  中间的静态路由就删掉了,以免浪费篇幅
…
  119.188.73.147/32 11.158.10.1 description NDC-56
  119.188.73.148/32 11.158.10.1 description NDC-56
  119.188.73.149/32 11.158.10.1 description NDC-56
  119.188.73.150/32 11.158.10.1 description NDC-56
  222.191.228.101/32 11.158.10.1 description NDC-qiyi
!
vrf cww
  address-family ipv4 unicast
    0.0.0.0/0 172.16.36.1
  !
!
vrf PORT-POLICY-oa
  address-family ipv4 unicast
    0.0.0.0/0 192.168.101.2
  !
!
!
router ospf 1
! 配置 OSPF 路由和邻居
  nsr
  router-id 22.143.3.11
  default-information originate always metric-type 1
  redistribute static metric-type 1 route-policy ospf-in-static
  address-family ipv4 unicast
```

```
  area 0
    interface Bundle-Ether1.1
    !
    interface Bundle-Ether2.26
    !
    interface Bundle-Ether4
      cost 20
    !
    !
    interface TenGigE0/10/1/1
      cost 20
    !
    interface TenGigE0/10/1/2
      cost 30
    !
    interface TenGigE0/10/1/3
      cost 30

    !
    interface TenGigE0/12/1/3
      cost 20
    !
  !
!
router bgp 5523
! 配置 BGP 路由和邻居
  nsr
  bgp router-id 22.143.3.11
  bgp cluster-id 22.143.3.11
  address-family ipv4 unicast
    network 36.192.176.0/21
    network 36.192.185.0/21
    network 58.61.39.0/25
    network 22.143.192.0/18
    network 22.143.225.0/19
  !
  address-family vpnv4 unicast
  !
  neighbor-group CORE
    remote-as 5523
    update-source Loopback0
    address-family ipv4 unicast
      route-policy pass-all in
      route-reflector-client
      route-policy pass-all out
      next-hop-self
    !
```

```
  !
  neighbor 22.19.20.3
   remote-as 5523
   update-source Loopback0
   address-family ipv4 unicast
    route-policy pass-all in
    route-policy pass-all out
   !
   address-family vpnv4 unicast
    route-policy pass-all in
    route-reflector-client
    route-policy pass-all out
    next-hop-self
  !
 neighbor 22.90.1.33
  remote-as 5523
  description to-NDC
  update-source Loopback0
  address-family ipv4 unicast
   route-reflector-client
   route-policy NDC-TIETONG out
   next-hop-self

 neighbor 11.98.10.16
  remote-as 39
  ebgp-multihop 3
  description Connect to ZongbuNE5000
  update-source Loopback0
  address-family ipv4 unicast
   route-policy bgp2 in
   route-policy bgppermit out
  !
 !
 vrf cww
  rd 5523:500
  address-family ipv4 unicast
   network 0.0.0.0/0
   redistribute connected
   redistribute static
  !
 !
 vrf ip-pbx
  rd 5523:6000
  address-family ipv4 unicast
   redistribute connected
   redistribute static
  !
```

```
!
!
mpls ldp
! 将参与 MPLS 进程的接口加进来
  router-id 22.143.3.11
  nsr
  interface Bundle-Ether4
  !
  interface Bundle-Ether5
  !
  interface TenGigE0/0/1/2
  !
  interface TenGigE0/1/1/2
  !
  interface TenGigE0/2/1/0

  !
!
end
RP/0/RP0/CPU0:SP--ASR#
```

4.6 小　结

思科 ASR 9000 系列是思科在数据中心及核心网边缘的主打产品，其先进的架构、卓越的性能获得了市场很高的占有率，其设备软件为 IOS-XE，不足的是命令格式与 IOS 有所不同，也给使用者带来一定的不便。

第5章 思科自适应安全设备配置

5.1 ASA产品系列介绍

Cisco ASA 5500系列自适应安全设备是思科专门设计的安全解决方案的产品，将最高的安全性和出色的VPN服务与创新的可扩展服务架构有机地结合在一起。作为思科自防御网络的核心组件，Cisco ASA 5500系列能够提供主动威胁防御，在网络受到威胁之前就能及时阻挡攻击，控制网络行为和应用流量，并提供灵活的VPN连接。思科强大的多功能网络安全设备系列不但能部署在家庭办公室、分支机构、中小企业和大型企业网络的出口，提供广泛而深入的安全功能，而且可以部署在数据中心出口，笔者的案例版本是8.4.(7)。

思科防火墙型号从低端到高端有：ASA 5505，ASA 5510，ASA 5520，ASA 5540，ASA 5550，ASA 5580，主要参数介绍如表5-1所示。

表5-1 防火墙主要参数

型号 参数	ASA 5505	ASA 5510	ASA 5520	ASA 5540	ASA 5550
吞吐量	150 Mbit/s	300 Mbit/s	450 Mbit/s	650 Mbit/s	1.2 Gbit/s
3DES/AES VPN 吞吐率	100 Mbit/s	170 Mbit/s	225 Mbit/s	325 Mbit/s	425 Mbit/s
IPSec VPN	25	250	750	5 000	5 000
SSL VPN 缺省/最大	2/25	2/250	2/750	2/5 000	2/5 000
并发连接	10 000，25 000*	50 000，130 000*	280 000	400 000	650 000
新建连接数/秒	4 000	9 000	12 000	25 000	36 000
集成网络端口	8端口快速以太网交换机（包括2个PoE端口）	3个快速以太网端口，1个管理端口，5个快速以太网端口*	4个千兆以太网端口，1个快速以太网端口	4个千兆以太网端口，1个快速以太网端口	8个千兆以太网端口，4个SFP光纤端口，1个快速以太网端口
VLAN	3（无中继支持）/20（中继支持）	50/100	150	200	250
高可用性	不支持，无状态主用/备用和双ISP支持*	不支持，主用/主用和主用/备用*	主用/主用和主用/备用	主用/主用和主用/备用	主用/主用和主用/备用
最大内存	256 MB	256 MB	512 MB	1 024 MB	4 096 MB
最低内存	64 MB	64 MB	64 MB	64 MB	64 MB

注意："*"表示通过升级许可证提供的数量。

ASA 5580X 用在大型数据中心，对不同用户数量有着不同型号与之对应，详细参数如表 5-2 所示

表 5-2　ASA 5580X 的主要参数

参数 \ 型号	Cisco ASA 5585-X with SSP-10	Cisco ASA 5585-X with SSP-20	Cisco ASA 5585-X with SSP-40	Cisco ASA 5585-X with SSP-60
吞吐量	3 Gbit/s (multiprotocol),4 Gbit/s(max)	7 Gbit/s (multiprotocol),10 Gbit/s(max)	12 Gbit/s(multiprotocol),20 Gbit/s(max)	20 Gbit/s(multiprotocol),40 Gbit/s(max)
3DES/AES VPN 吞吐率	1 Gbit/s	2 Gbit/s	3 Gbit/s	5 Gbit/s
Site-to-site 及 IPSec VPN	5 000	10 000	10 000	10 000
SSL VPN 缺省/最大	2/5 000*	2/10 000*	2/10 000*	2/10 000*
并发连接	1 000 000	2 000 000	4 000 000	4 000 000
每秒新建连接数	65 000	140 000	240 000	350 000
集成网络端口	8-10/100/1 000, 2-10GE,SFP+ (withASA5585-Sec-Pllicense), 2-10/100/1 000 management +8-10/100/1 000, 2-10GE,SFP+, 2-10/100/1 000 management(with IPS SSP-10)	8-10/100/1 000, 2-10GE,SFP+ (withASA5585-Sec-Pllicense), 2-10/100/1 000 management +8-10/100/1 000, 2-10GE,SFP+, 2-10/100/1 000 management(with IPS SSP-20)	6-10/100/1 000, 4-10GE,SFP+, 2-10/100/1 000 management +6-10/100/1 000, 4-10GE,SFP+, 2-10/100/1 000 management(with IPSSSP-40)	6-10/100/1 000, 4-10GE, SFP+, 2-10/100/1 000 management +6-10/100/1 000, 4-10GE,SFP+, 2-10/100/1000 management (with IPSSSP-60)
VLAN	1 024	1 024	1 024	1 024
高可用性	A/A and A/S	A/A and A/S	A/A and A/S	A/A and A/S
最大内存	6 GB(SSP-10),12 GB (SSP-10 and IPS SSP-10)	12 GB(SSP-20),24 GB (SSP-20 and IPS SSP-20)	12 GB(SSP-40),36 GB (SSP-40 and IPS SSP-40)	24 GB(SSP-60),72 GB (SSP-60 and IPS SSP-60)
最低内存	2 GB	2 GB	2 GB	2 GB

注意："*"表示通过升级许可证提供的数量。

5.2 基础配置

5.2.1 防火墙模式

思科 ASA 防火墙主要有两种工作模式:透明模式和路由模式。缺省工作在路由模式,可以通过 show firewall 查看当前工作模式,透明模式应用不多,读者可以参看相关文档。

透明防火墙的基本配置如下所示。

启用透明防火墙模式:

```
firewall transparent
```

返回路由模式:

```
no firewall transparent
```

5.2.2 接口配置

思科自适应安全设备的接口配置如图 5-1 所示。

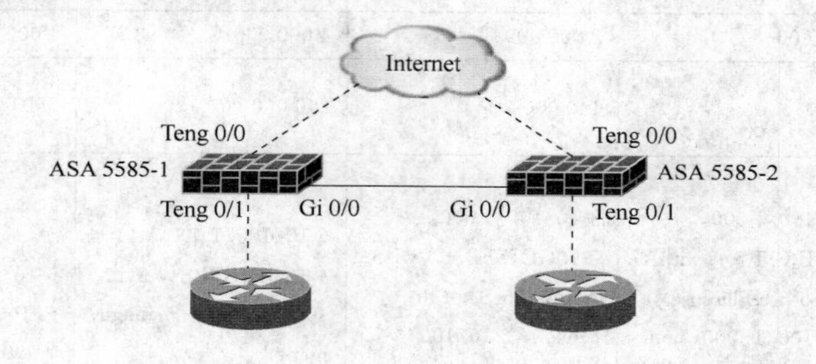

图 5-1 防火墙测试拓扑

外网口的配置(以版本软件 8.4(7)为例):

```
ASA 5585(config)# interface Teng 0/0
ASA 5585(config-if)# nameif outside
! 一般的情况将 E0/0 命为外网接口,而将 E0/1 命为内网接口。
```

配置接口安全级别,外网口缺省为零:

```
ASA 5585(config-if)# security-level 0
! 100 为最大,数字越高越靠近安全区域
```

配置 IP 地址:

```
ASA 5585(config-if)# ip address  x.x.x.x
```

关闭/激活接口:

```
ASA 5585(config-if)# shutdown/no shutdown
```

内网口配置:

```
ASA 5585(config)# interface Teng0/1
ASA 5585(config-if)# nameif inside
ASA 5585(config-if)# security-level 100
ASA 5585(config-if)# ip address  x.x.x.x
ASA 5585(config-if)# no shutdown
```

DMZ 区域的配置，通常将存储接到 DMZ 区域：

```
ASA 5585(config)# interface  Teng0/3
ASA 5585(config-if)# nameif inside
ASA 5585(config-if)# security-level 50
! 可以有多个区域，安全值介于 0～100
ASA 5585(config-if)# ip address  x.x.x.x
ASA 5585(config-if)# no shutdown
```

5.2.3 静态路由配置

在 inside 接口上创建一条到 192.168.3.0/24 网络走 192.168.10.1 的路由，ASA 会将到 192.168.3.0/24 网络的所有数据包转发给下一跳 192.168.10.1，命令如下：

```
ASA 5585(config)# route inside 192.168.3.0 255.255.255.0 192.168.10.1 100
```

创建一条外网默认路由，ASA 将所有互联网流量转发给 Internet 网关 219.139.50.1，后面的 100 表示 distance，命令如下：

```
ASA 5585(config)# route outside 0.0.0.0 0.0.0.0 219.139.50.1 100
```

如果有两条甚至多条出口，依据出口 distance 值，如果值相等则出去的数据包会分别发到两个不同出口做负载均衡，而 distance 值不同表示为冗余，或称为备份。

5.2.4 访问控制列表配置

表示允许外网访问内网为 192.168.11.11 主机的 SSH 端口，没有这个命令，后面的端口映射访问是不成功的，命令如下：

```
access-list outside_access extended permit tcp any host 192.168.11.11 eq ssh
```

5.2.5 网络地址转换配置

NAT 实现的方式有 3 种：动态 NAT 、静态 NAT、PAT。动态 NAT 指将内部网络私有 IP 地址转换为公有 IP 地址，IP 地址不确定，是给出的出口地址段，所有被授权访问 Internet 的私有 IP 地址可随机转换为任何指定合法 IP 地址。静态 NAT 指 IP 地址一对一的转换。PAT 指改变外出数据包的源端口并进行端口转换。内部所有网络均可以共享一个合法外部 IP 地址，实现对 Internet 的访问，从而可以最大限度节约 IP 地址资源。同时，又可以隐藏网络内部的所有主机，有效地避免来自自己 Internet 的攻击。做 NAT 时通常用 PAT。

动态 NA＋PAT 的配置：

```
object network Internal
 subnet 192.168.0.0 255.255.0.0
object network Internal
 nat (inside,outside) dynamic interface
```

静态地址转换配置：将外部地址 183.131.13.71 转换到内部服务器地址 192.168.11.11，注意这里没有对端口进行限制。

```
object network 192.168.11.11
 host 192.168.11.11
object network 192.168.11.11
 nat (inside,outside) static 183.131.13.71
```

要做一对一的端口和地址转换，以下例子表示将外网口的 TCP 端口 8443 映射到内网口的 SSH 端口。

```
object service tcp8443
! 定义外网口对应的端口
 service tcp source eq 8443
object service ssh
! 定义内网口对应的端口
 service tcp source eq ssh

object network SERVICE
 host 192.168.11.11
object network SERVICE
 nat (inside,outside) static interface service tcp ssh tcp8443
```

5.3 高级配置

5.3.1 防火墙工作状态调试

查看当前 ASA 配置，命令如下：

```
ASA 5585-1 # show running-config
```

查看 CPU 得用率（正常应该在 80%以下），命令如下：

```
show cpu usage
```

查看内存使用情况，命令如下：

```
ASA 5585-1 # show memory
```

检查当前连接和最大连接数，命令如下：

```
ASA 5585-1 # show conn count
```

查看端口状态，命令如下：

```
ASA 5585-1 # show interface interface_name
```

验证防火墙的连接性，命令如下：

```
ping ip_address
```

查看路由表，命令如下：

```
ASA 5585-1 # show route
```

ASA 防火墙 ACL 检查，命令如下：

```
ASA 5585-1 # show access-list
```

5.3.2 ASA 防火墙的冗余

ASA 防火墙可以配置成双活模式（Active-Active）或主备模式（Active-Standby），由于双活模式应用不多，这里不做介绍，下面主要介绍主备模式的配置。

配置状态交换端口：

```
interface GigabitEthernet0/0
 description LAN/STATE Failover Interface

failover
```

```
！启用主备模式
failover lan unit primary
！指明本防火墙为主用防火墙
failover lan interface folink GigabitEthernet0/0
！指明状态交换的物理端口名为 folink
failover link folink GigabitEthernet0/0
！指明状态信息交换的链路
failover interface ip folink 172.16.1.1 255.255.255.252 standby 172.16.1.2
！说明 folink 的主要地址和备用地址
```

在备用防火墙上配置：

```
interface GigabitEthernet0/0
 description LAN/STATE Failover Interface
failover
failover lan unit secondary
！指明本防火墙为备用
failover lan interface folink GigabitEthernet0/0 failover link folink GigabitEthernet0/0
failover interface ip folink 172.16.1.1 255.255.255.252 standby 172.16.1.2
```

如果 failover 连线口 G0/0 起来，则所有配置只需要在主用设备上配置，备用防火墙可以实时同步主防火墙的配置。

5.3.3 配置 ASDM 访问

ASDM 是一个基于 Web 浏览器的 Java 程序的图形化安全设备管理工具。ASDM 定位为配置工具，通过 ASDM 可以对安全设备进行配置和监控，ASDM 的配置功能十分强大，几乎可以实现命令行全部的操作，参见图 5-2。

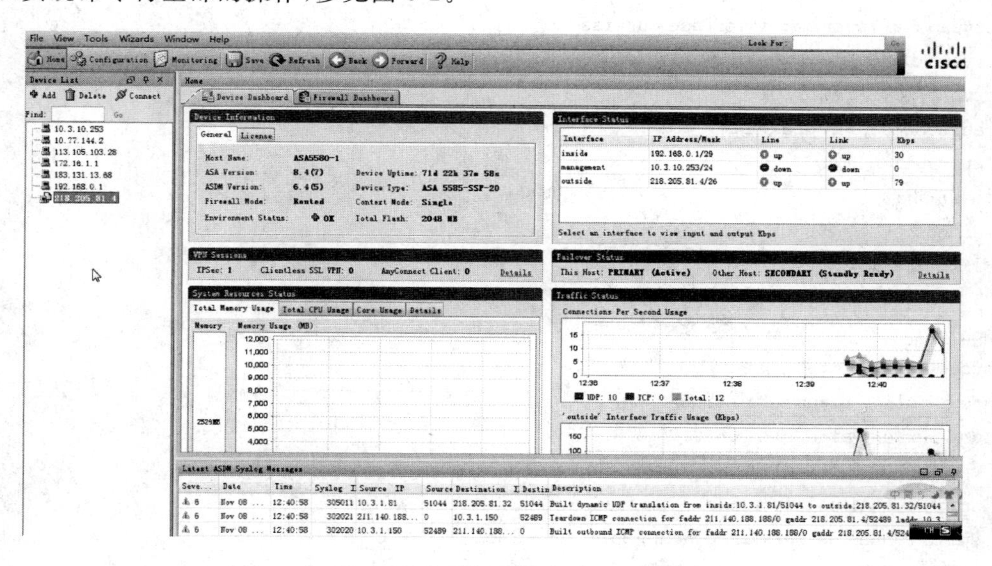

图 5-2 登录防火墙界面

首先，要确保 ASDM 软件在 flash 盘里。

```
ASA5580-1＃ dir
Directory of disk0:/
14     drwx     32768        15:40:26 Jul 17 2014     coredumpinfo
102    -rwx     38191104     15:42:08 Jul 17 2014    asa912-smp-k8.bin
```

```
103     -rwx    18097844    15:42:32 Jul 17 2014    asdm-713.bin
104     -rwx    31186944    19:56:32 Aug 27 2014    asa847-smp-k8.bin
2       drwx    32768       15:47:36 Jul 17 2014    log
5       drwx    32768       15:48:04 Jul 17 2014    crypto_archive
106     -rwx    16280544    19:57:36 Aug 27 2014    asdm-645.bin
107     -rwx    12998641    15:48:18 Jul 17 2014    csd_3.5.2008-k9.pkg
108     drwx    32768       15:48:20 Jul 17 2014    sdesktop
111     -rwx    4678691     15:48:22 Jul 17 2014    anyconnect-win-2.5.2014-k9.pkg
112     drwx    32768       21:01:14 Aug 27 2014    tmp
2007171072 bytes total (1870397440 bytes free)
ASA5580-1#
asdm image disk0:/asdm-645.bin
!请注意,一个防火墙可以有多个ASDM版本,但是同时只能运行一个版本
```

其次,开启 HTTPS 服务。

```
asa5580-1(config)# http server enable
```

最后,打开浏览器,在地址栏输入 https://interface ip address,即可打开 ASDM 页面。

5.3.4 配置 IPSEC VPN

如果需要用思科 VPN Client 客户端软件以 IPSEC 方式访问防火墙,则在防火墙上做如下配置。

```
crypto ipsec ikev1 transform-set rmvpn esp-3des esp-sha-hmac
crypto dynamic-map rmvpn 10 set ikev1 transform-set rmvpn
crypto dynamic-map rmvpn 10 set reverse-route
crypto map rmvpn-map 10 ipsec-isakmp dynamic rmvpn
crypto map rmvpn-map interface outside
crypto ikev1 enable outside
crypto ikev1 policy 10
 authentication pre-share
 encryption 3des
 hash sha
 group 2
 lifetime 86400

webvpn
group-policy rmvpn internal
group-policy rmvpn attributes
 dns-server value 202.106.0.20 8.8.8.8
 vpn-tunnel-protocol ikev1
 split-tunnel-policy tunnelspecified
 split-tunnel-network-list value SplitTunnelAcl
username CISCO password W0vbEX8WowujRxD/encrypted privilege 15
tunnel-group rmvpn type remote-access
tunnel-group rmvpn general-attributes
 address-pool vpn-pool
 default-group-policy rmvpn
tunnel-group rmvpn ipsec-attributes
 ikev1 pre-shared-key *****
```

!

在 VPN Client 客户端则需要配置防火墙的外网口地址和认证组以及登录账号就可以了，如图 5-3 所示。

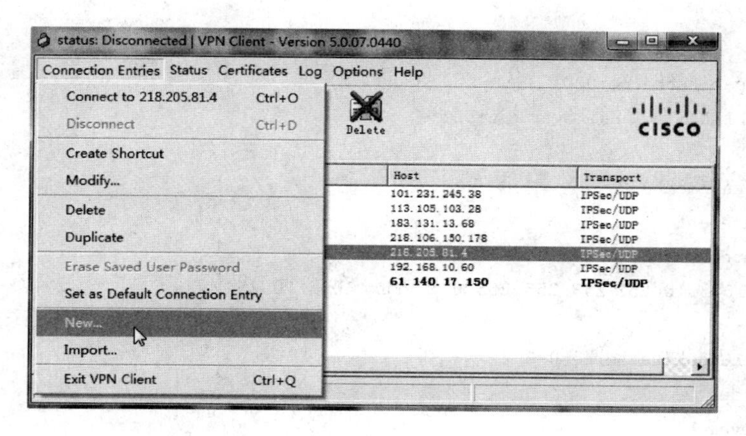

图 5-3　思科 IPSEC VPN Client 登录画面

选择地址后，进入配置画面，如图 5-4 所示。

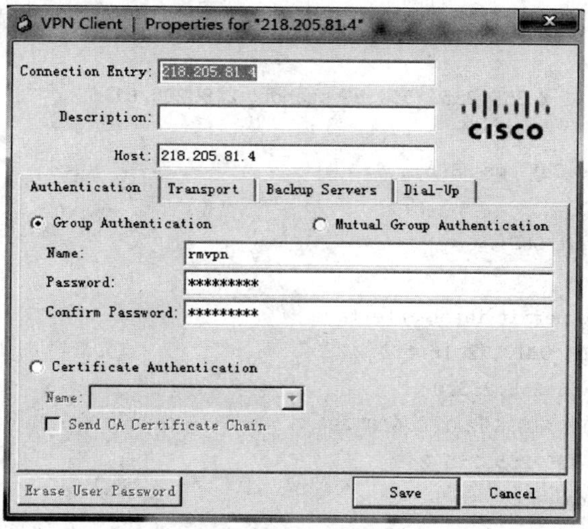

图 5-4　配置 VPN Client

5.4　实际案例

为了节省篇幅，只列出必要的配置，这个配置可以通过 TFTP 上传到 ASA 5585 防火墙上。

```
: Saved
: Written by wafer at 22:08:47.778 UTC Thu Aug 28 2014
!
ASA Version 8.4(7)
!
hostname ASA5580 - 1
```

```
domain-name cisco.com
enable password NJMXRSb4E0.1Zn5I encrypted
passwd 2KFQnbNIdI.2KYOU encrypted
names
!
interface GigabitEthernet0/0
 description LAN/STATE Failover Interface
!
interface Management0/0
 nameif management
 security-level 100
 ip address 10.3.10.253 255.255.255.0 standby 10.3.10.254
!
interface TenGigabitEthernet0/8
 nameif inside
 security-level 100
 ip address 192.168.0.1 255.255.255.248 standby 192.168.0.2
!
interface GigabitEthernet1/0
 nameif outside
 security-level 0
 ip address 218.205.81.4 255.255.255.192 standby 218.205.81.5
!
boot system disk0:/asa847-smp-k8.bin
ftp mode passive
dns server-group DefaultDNS
 domain-name cisco.com
same-security-traffic permit inter-interface
object network NETWORK_OBJ_172.16.1.0_24
 subnet 172.16.1.0 255.255.255.0
object network NETWORK_OBJ_192.168.0.0_29
 subnet 192.168.0.0 255.255.255.248
object network Internal
 subnet 10.3.0.0 255.255.0.0
object network rmvpn-address
 subnet 172.16.2.0 255.255.255.0
object network 10.3.1.56
 host 10.3.1.56
object service tcp8022
 service tcp source eq 8022
access-list inside extended permit icmp any any
access-list inside extended permit ip any any
access-list outside extended permit ip any any
access-list SplitTunnelAcl standard permit 172.16.0.0 255.255.255.0
access-list SplitTunnelAcl standard permit 10.3.0.0 255.255.0.0
access-list nonat extended permit ip 10.3.0.0 255.255.0.0 172.16.2.0 255.255.255.0
access-list 102 extended permit tcp any interface outside eq ssh
```

```
pager lines 24
logging enable
logging asdm informational
mtu management 1500
mtu inside 1500
mtu outside 1500
ip local pool vpn-pool 172.16.2.10-172.16.2.254 mask 255.255.255.0
failover
failover lan unit primary
failover lan interface folink GigabitEthernet0/0
failover link folink GigabitEthernet0/0
failover interface ip folink 172.16.1.1 255.255.255.252 standby 172.16.1.2
icmp unreachable rate-limit 1 burst-size 1
asdm image disk0:/asdm-645.bin
no asdm history enable
arp timeout 14400
no arp permit-nonconnected
nat (inside,outside) source static Internal Internal destination static rmvpn-address rmvpn-ad-
dress
!
object network Internal
 nat (inside,outside) dynamic interface
object network 10.3.1.56
 nat (inside,outside) static 218.205.81.6
access-group outside in interface outside
route outside 0.0.0.0 0.0.0.0 218.205.81.1 1
route inside 10.3.0.0 255.255.0.0 192.168.0.6 1
timeout xlate 3:00:00
timeout pat-xlate 0:00:30
timeout conn 1:00:00 half-closed 0:10:00 udp 0:02:00 icmp 0:00:02
timeout sunrpc 0:10:00 h323 0:05:00 h225 1:00:00 mgcp 0:05:00 mgcp-pat 0:05:00
timeout sip 0:30:00 sip_media 0:02:00 sip - invite 0:03:00 sip-disconnect 0:02:00
timeout sip-provisional-media 0:02:00 uauth 0:05:00 absolute
timeout tcp-proxy-reassembly 0:01:00
timeout floating-conn 0:00:00
dynamic-access-policy-record DfltAccessPolicy
user-identity default-domain LOCAL
aaa authentication enable console LOCAL
aaa authentication ssh console LOCAL
http server enable
http 0.0.0.0 0.0.0.0 management
http 0.0.0.0 0.0.0.0 inside
http 0.0.0.0 0.0.0.0 outside
no snmp-server location
no snmp-server contact
snmp-server community cmccrd_ops
snmp-server enable traps snmp authentication linkup linkdown coldstart warmstart
```

```
crypto ipsec ikev1 transform - set rmvpn esp-3des esp-sha-hmac
crypto dynamic-map rmvpn 10 set ikev1 transform-set rmvpn
crypto dynamic-map rmvpn 10 set reverse-route
crypto map rmvpn-map 10 ipsec - isakmp dynamic rmvpn
crypto map rmvpn-map interface outside
crypto ca trustpoint _SmartCallHome_ServerCA
 crl configure
crypto ca certificate chain _SmartCallHome_ServerCA
 certificate ca 6ecc7aa5a7032009b8cebcf4e952d491 ********

  quit
crypto ikev1 enable outside
crypto ikev1 policy 10
 authentication pre-share
 encryption 3des
 hash sha
 group 2
 lifetime 86400
telnet 0.0.0.0 0.0.0.0 inside
telnet timeout 5
ssh 0.0.0.0 0.0.0.0 inside
ssh 0.0.0.0 0.0.0.0 outside
ssh timeout 60
ssh version 2
ssh key-exchange group dh-group1-sha1
console timeout 0
!
tls-proxy maximum-session 1000
!
threat-detection basic-threat
threat-detection statistics access-list
no threat-detection statistics tcp-intercept
webvpn
group-policy rmvpn internal
group-policy rmvpn attributes
 dns-server value 202.106.0.20 8.8.8.8
 vpn-tunnel-protocol ikev1
 split-tunnel-policy tunnelspecified
 split-tunnel-network - list value SplitTunnelAcl
username wafer password W0vbEX8WowujRxD/ encrypted privilege 15
username cisco password TR2lDc8o7vYizX8v encrypted
tunnel-group rmvpn type remote-access
tunnel-group rmvpn general-attributes
 address-pool vpn-pool
 default-group-policy rmvpn
tunnel-group rmvpn ipsec-attributes
 ikev1 pre-shared-key Wafer1234
```

```
!
class-map inspection_default
 match default-inspection-traffic
!
!
policy-map type inspect dns preset_dns_map
 parameters
  message-length maximum client auto
  message-length maximum 512
policy-map global_policy
 class inspection_default
  inspect dns preset_dns_map
  inspect ftp
  inspect h323 h225
  inspect h323 ras
  inspect rsh
  inspect rtsp
  inspect esmtp
  inspect sqlnet
  inspect skinny
  inspect sunrpc
  inspect xdmcp
  inspect sip
  inspect netbios
  inspect tftp
  inspect ip-options
  inspect icmp
!
service-policy global_policy global
prompt hostname context
no call-home reporting anonymous
Cryptochecksum:c5a497399b97eefe968852d904724afb
: end
```

5.5 小　结

思科 ASA 防火墙以性能和稳定性著称，还可以通过专门的图形界面软件 ASDM 对防火墙进行配置，ASA 防火墙在数据中心可以根据需要配置成主备模式或双活模式，思科官方网站有更多的产品说明和配置介绍。

第6章 思科服务器 UCS 安装与配置

刀片服务器通过架顶的交换机将服务器直连数据中心，并在该统一计算系统内部实现自交换矩阵至虚拟服务器网卡的全链路层面的 FEX 贯通。网络层面可以识别并可控到每一台虚拟机网卡的流量，并进而在其上实现可自定义的各类功能，这便是思科的 UCS 统一计算系统。

6.1 UCS 产品系列介绍

6.1.1 UCS B 系列

Cisco UCS B 系列刀片服务器是思科统一计算系统的重要构建模块，为当今和未来的数据中心提供了灵活、可扩展的计算能力，同时能够帮助降低总体拥有成本(TCO)。

Cisco UCS B 系列刀片服务器构建于工业标准服务器技术基础之上，有以下特性：

① 多达两个英特尔至强系列 5500 多核处理器；

② 两个可选前置热插拔 SAS 硬盘；

③ 支持多达两个双端口扩展卡连接，可提供高达 40 Gbit/s 的冗余 I/O 吞吐率；

④ 工业标准 DDR3 内存；

⑤ 通过集成服务处理器实现远程管理，并可执行在 Cisco UCS Manager 软件中制定的策略；

⑥ 通过每台服务器前面板上的控制台端口使用本机 KVM(键盘、显示器和鼠标)；

⑦ 通过远程 KVM、安全外壳(SSH)协议、虚拟介质(vMedia)以及智能平台管理接口(IP-MI)协议实现带外管理。

Cisco UCS B 系列包括两款刀片服务器产品：Cisco UCS B200 M1 双插槽刀片服务器与 Cisco UCS B250 M1 双插槽内存扩展刀片服务器。Cisco UCS B200 M1 是一款半宽刀片服务器，拥有 12 个 DIMM 插槽，可支持高达 96 GB 的内存，同时还可支持一个扩展卡。Cisco UCS B250 M1 是一款全宽刀片服务器，拥有 48 个 DIMM 插槽，可支持高达 384 GB 的内存，同时还可支持两个扩展卡。

B 系列型号有：B22 M3、B200 M3、B230 M2、B260 M4、B420 M3、B440 M2、B460 M4 详细的设备参数可以参阅思科在线网站的文档。

图 6-1 是安装 B 系列服务器的 UCS 5108 机箱。

Cisco UCS 5108 刀片服务器机箱用于放置阵列扩展模块，多达 4 个电源和 8 台刀片服务器。作为系统大幅简化优势的一部分，刀片服务器机箱也可通过阵列互联进行管理，从而消除了另一个管理点。

刀片机箱能够支持最多 8 个半宽刀片服务器或 4 个全宽刀片服务器。该认证配置使用 8 台(每个机箱中 4 台)Cisco UCS B200 M1 刀片服务器，每台配有两枚 2.93 GHz 四核 Intel

Xeon 5500 系列处理器。每台刀片服务器均配置 24 GB 内存。通过使用 Cisco UCS B250 M1 内存扩展刀片服务器,最多可以提供 384 GB 的内存。

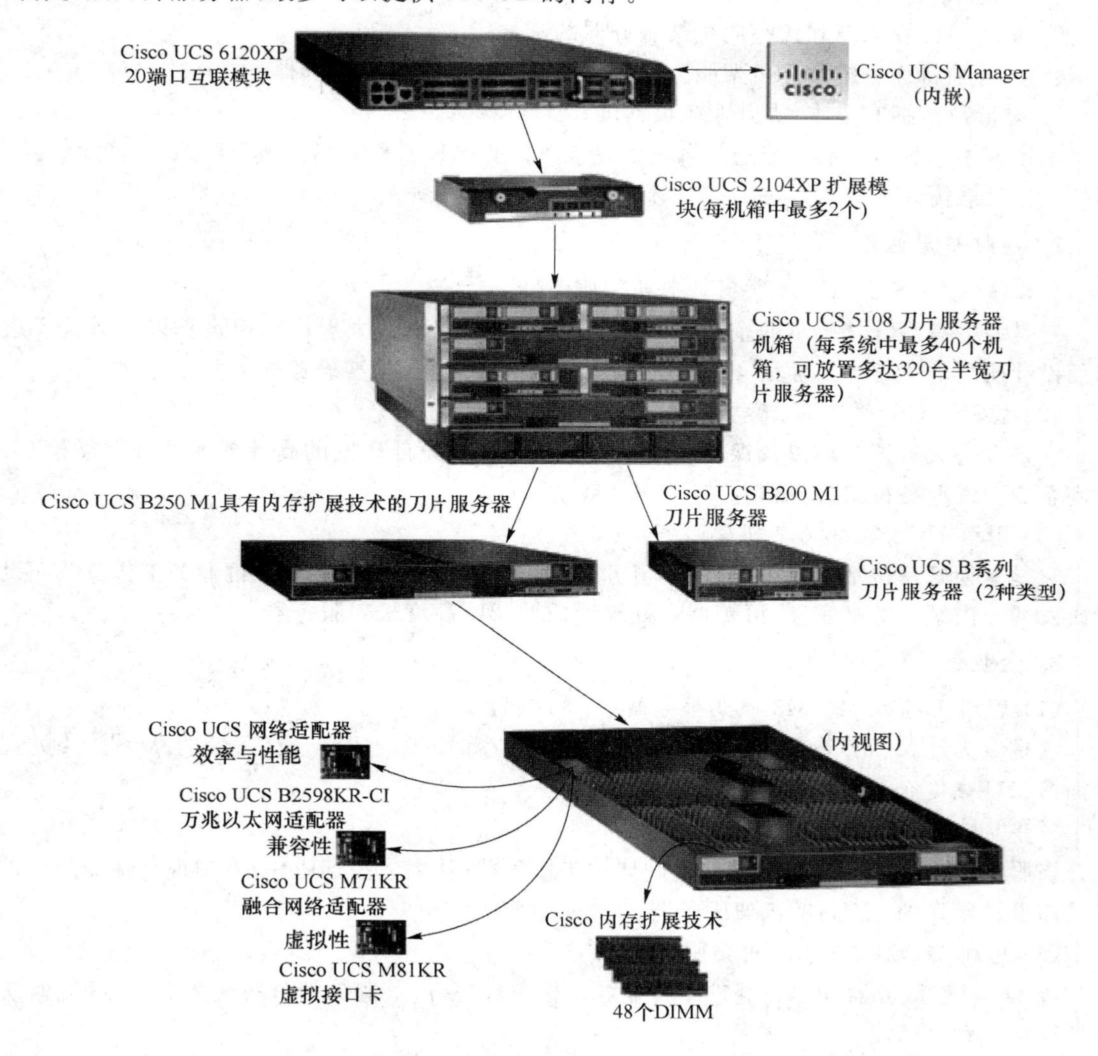

图 6-1　B 系列服务器的 UCS 5108

刀片服务器外形支持各种扩展格式的 Cisco UCS 网络适配器,包括用于提升效率和性能的万兆以太网适配器,用于为系统提供全面虚拟化支持的 Cisco UCS M81KR 虚拟接口卡,以及用于全面兼容现有以太网和光纤通道环境的 Cisco UCS M71KR 融合网络适配器集合。这些适配器在主机操作系统中将显示为 NIC(以太网接口卡)和 HBA(光纤通道主机总线适配器)。它们使统一阵列对于操作系统完全隐形,能够将来自 NIC 和 HBA 的流量传输到统一阵列之上。系统提供有 Emulex 或 QLogic HBA 芯片两种版本。认证配置使用 Cisco UCS M71KR-Q QLogic 融合网络适配器,通过连接至每个机箱阵列扩展模块,提供了 20 Gbit/s 的连接能力。

6.1.2　UCS C 系列

思科 UCS C 系列机架服务器具有以下优势:

① 与形状因数无关的入口点可直接切入思科 UCS 平台；

② 更简单、更快速地部署应用程序；

③ 是统一计算创新和优势向机架服务器的延伸；

④ 具有独特优势的常见机架服务器软件包为客户提供了更多选择；

⑤ 降低总体拥有成本，提升业务灵活度。

本系列中的每款产品都能通过各种处理能力、内存、I/O 和内部存储资源的组合来满足不同的工作负载需求。

1. 企业关键业务

(1) 思科 UCS C460 M4 关键业务机架服务器

它是专为有高计算要求的企业和处理关键业务工作负载、大规模模拟运算以及数据库应用而设计研发的 4 处理器插槽 4U 高度(机架单元)超高性能机架服务器。

(2) 思科 UCS C460 M2 高性能机架服务器

它是专为大数据应用以及裸机和虚拟化工作负载而设计研发的高性能 4 处理器插槽 4U 高度企业关键业务机架服务器。

(3) 思科 UCS C260 M2 机架服务器

它是专为管理包括存储服务、联机事务处理、数据仓库在内的企业关键业务工作负载而设计研发的业内领先的高密度、可扩展 2 处理器插槽、2U 高度机架服务器。

2. 企业级

(1) 思科 UCS C420 M3 机架服务器

它是专为计算、I/O、存储、以及内存密集型独立和虚拟应用而设计研发的高密度、4 处理器插槽、2U 高度企业级机架服务器。

(2) 思科 UCS C240 M3 机架服务器

该服务器为在大范围密集存储基础架构工作负载(从大数据到协作)下实现高性能及可扩展性而设计研发的 2U 高度机架服务器。

(3) 思科 UCS C220 M3 机架服务器

该 1U 机架服务器能够在处理多种业务工作负载(从网页服务到分布式数据库)的同时实现超高性能和高密度。

3. 向外扩展

(1) 思科 UCS C24 M3 机架服务器

该服务器为在大范围存储密集基础架构工作负载(从 IT 和网页基础架构到大数据)下实现高经济性和内部可扩展性而设计研发的 2 处理器插槽、2U 高度机架服务器。

(2) 思科 UCS C22 M3 机架服务器

该服务器为在大范围向外扩展工作负载(从 IT 和网页基础架构到分布式数据库)下实现高经济性和密度优化特性而设计研发的 2 处理器插槽、1U 高度机架服务器。

6.1.3 UCS E 系列

Cisco UCS E-Series Servers 与 ISR 系列路由器整合一起，设备定位用于分支机构，设备外观如图 6-2 所示。UCS E 系列需要客户安装需要的操作系统，像 Windows，Linux 或 VMware vSphere，跟 B 或 C 系列 UCS 一样，可以通过自带的 CIMC 安装操作系统，如何通过 CI-

MC 安装操作系统，后面章节会进一步介绍。

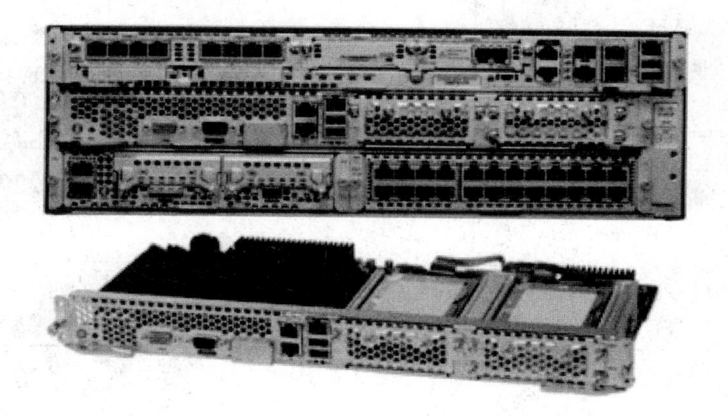

图 6-2　带有思科 UCS E 系列的 3945 ISR 路由器

思科 UCS E 系列参数说明如表 6-1 所示。

表 6-1　思科 UCS E 系列参数

特　征	Cisco UCS E140S M1 单宽服务器	Cisco UCS E140D M1 全宽服务器	Cisco UCS E160D M1 全宽服务器
外　形	Single-wide server module	Double-wide server module	Double-wide server module
CPU	Intel Xeon processor E3-1105C V1	Intel Xeon processor E3-2428L	Intel Xeon processor E3-2418L
CPU 核数	4 cores	4 cores	6 cores
DIMM 插槽	2 slots	3 slots	3 slots
内　存	8 to 16 GB Supports DDR3 1333 MHz very-low profile (VLP) unbuffered dual in-line memory module (UDIMM) at 1.5 volts (V) and 4 and 8 GB	8 to 48 GB Supports DDR3 1333 MHz registered DIMM (RDIMM) at 1.35V and 4, 8 and 16 GB	8 to 48 GB Supports DDR3 1333 MHz RDIMM at 1.35V and 4, 8 and 16 GB
RAID	RAID 0 and 1	RAID 0, 1 and 5	RAID 0, 1 and 5
存储类型	SATA, SAS, SSD and SED	SATA, SAS, SSD and SED	SATA, SAS, SSD and SED
硬　盘	SAS 10 000-rpm, SATA 7 200-rpm and SAS SSD drives Supports 2 drives	SAS 10 000-rpm, SATA 7 200-rpm and SAS SSD drives Supports 3 drives	SAS 10 000-rpm, SATA 7 200-rpm and SAS SSD drives Supports 3 drives
存储容量	200 GB to 2 TB	200 GB to 3 TB	200 GB to 3 TB
内部网络接口	2 Gigabit Ethernet interfaces	2 Gigabit Ethernet interfaces	2 Gigabit Ethernet interfaces

外部接口	1 USB Connector 1 RJ-45 Gigabit Ethernet connector 1 management port 1 keyboard, video and mouse port (supports VGA, 1 USB connector and 1 DB9 serial connector)	2 USB connectors 2 RJ-45 Gigabit Ethernet connectors 1 management port 1 VGA port 1 DB9 serial connector	2 USB connectors 2 RJ-45 Gigabit Ethernet connectors 1 management port 1 VGA port 1 DB9 serial connector
支持的路由器平台	Cisco 2911, 2921, 2951, 3925, 3925e, 3945, 3945e and 4451-X ISRs	Cisco 2921, 2951, 3925, 3925e, 3945, 3945e and 4451-X ISRs	Cisco 3925, 3925e, 3945, 3945e and 4451-X ISRs

6.2 通过 CIMC 安装 UCS 操作系统

CIMC(Cisco Integrated Management Controller)用于思科 UCS 服务器 C 系列机架式服务管理工具,该管理工具支持 HTTP 和 SSH 访问。使用 CIMC 可以对服务器进行以下几个方面的管理:① 开机、关机、重启服务器操作;② 打开或关闭服务器指示灯;③ 设置服务器的启动顺序以及其他 BIOS 选项设置;④ 查看服务器的性能、告警及服务器的状态。

配置 CIMC 的要求:① IE 浏览器;② 根据 CIMC Firmware 需要的 Java 版本,测试 Java 版本为 1.6.0_43,CIMC Firmware 版本为 1.4(3p);③ Adobe Flash Player 10.0 10.0 以上。

注意:CIMC 只要电源线接上,且电源线上已经供电,CIMC 就会工作。

检查当前使用 Java 版本如图 6-3 所示。

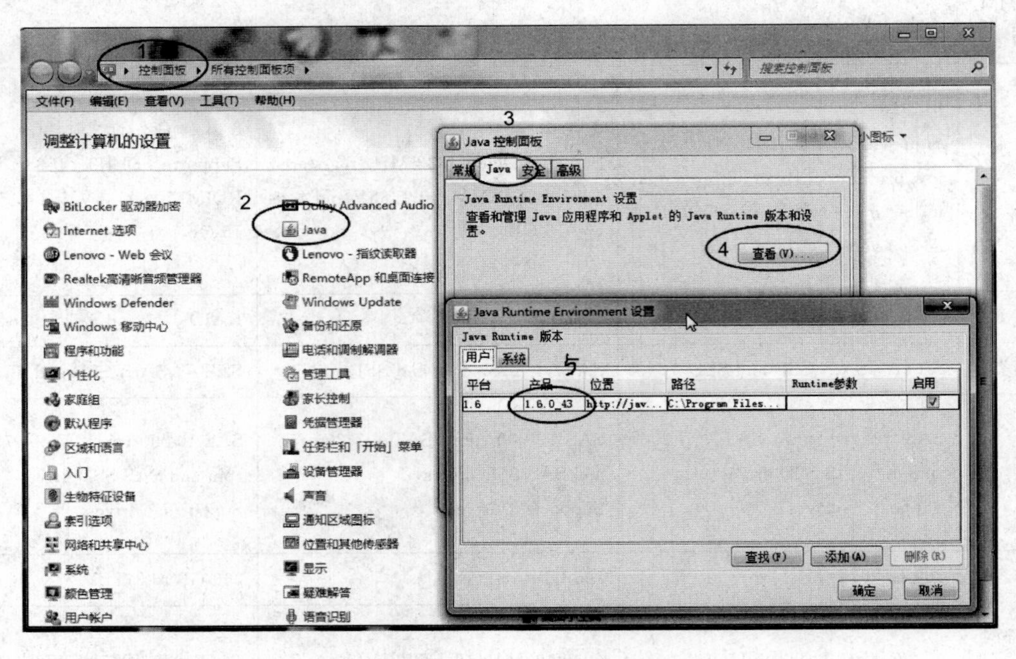

图 6-3 个人电脑 Java 版本的检查

修改 CIMC 登录账号，地址和账号可以在最初接显示器和键盘时设置，Web 登录画面如图 6-4 所示。

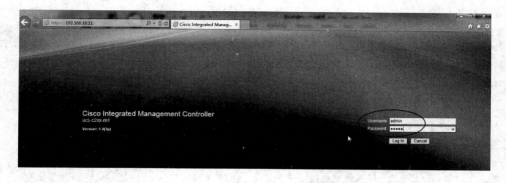

图 6-4　Web 登录画面

登录后的显示画面如图 6-5 所示。

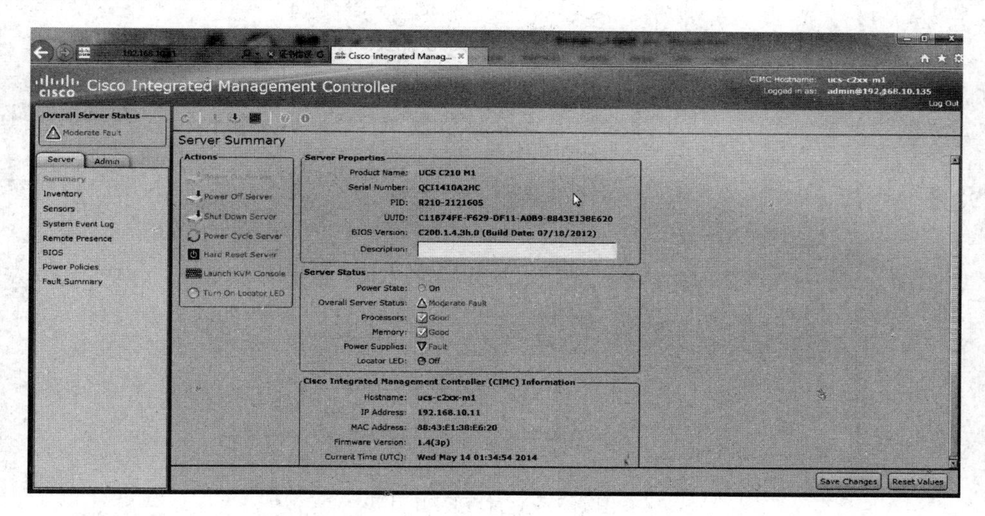

图 6-5　CIMC 登录后画面

检查运行的 Firmware 版本，如图 6-6 所示。

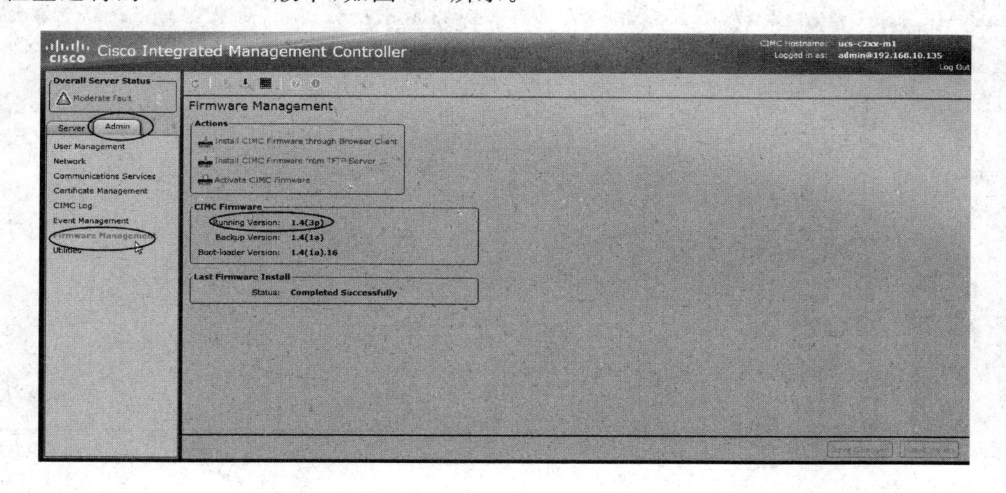

图 6-6　检查 Firmware 版本

如果通过 Web 上的 Launch KVM Console 打开 KVM 出现问题，可以通过升级 Firmware 版本来解决，升级的方式有两种，一种是直接通过 Web 界面，另一种是通过 TFTP，分别如图 6-7 和图 6-8 所示。

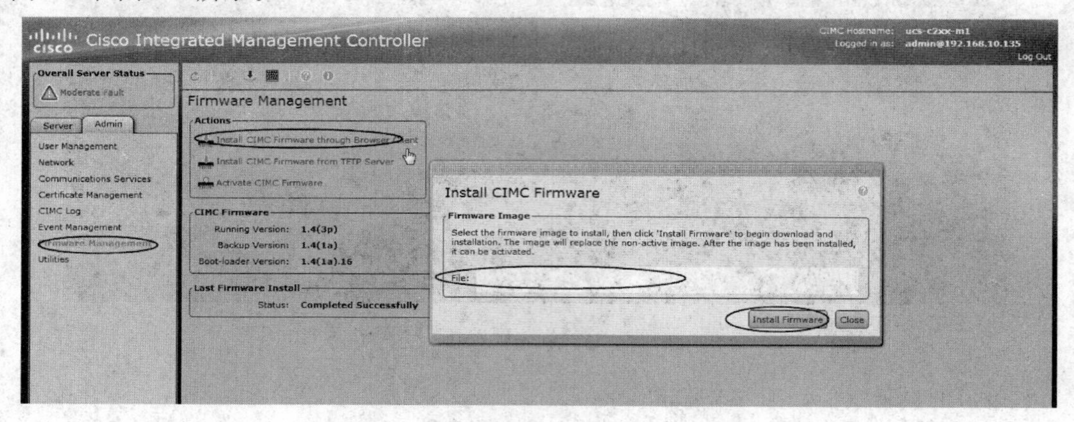

图 6-7　通过 Web 升级

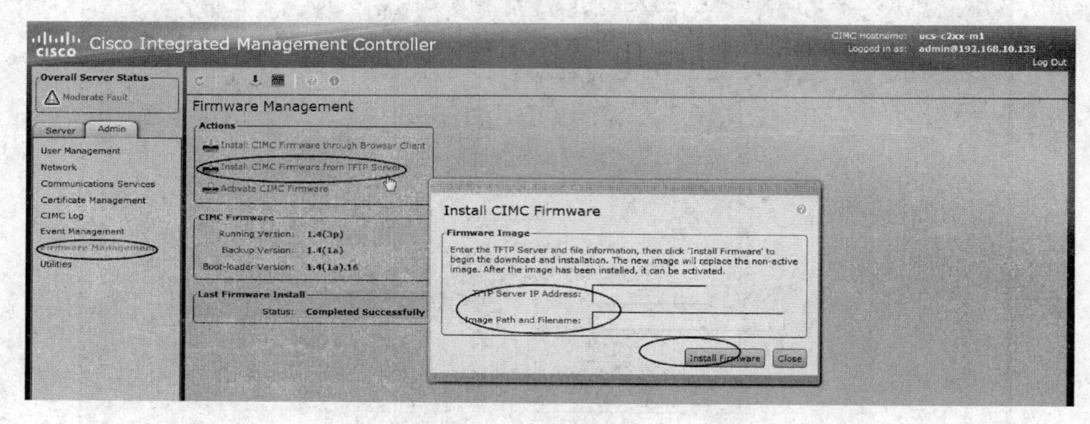

图 6-8　通过 TFTP 升级

点击“Launch KVM Console”进入 KVM 控制台，如图 6-9 所示。

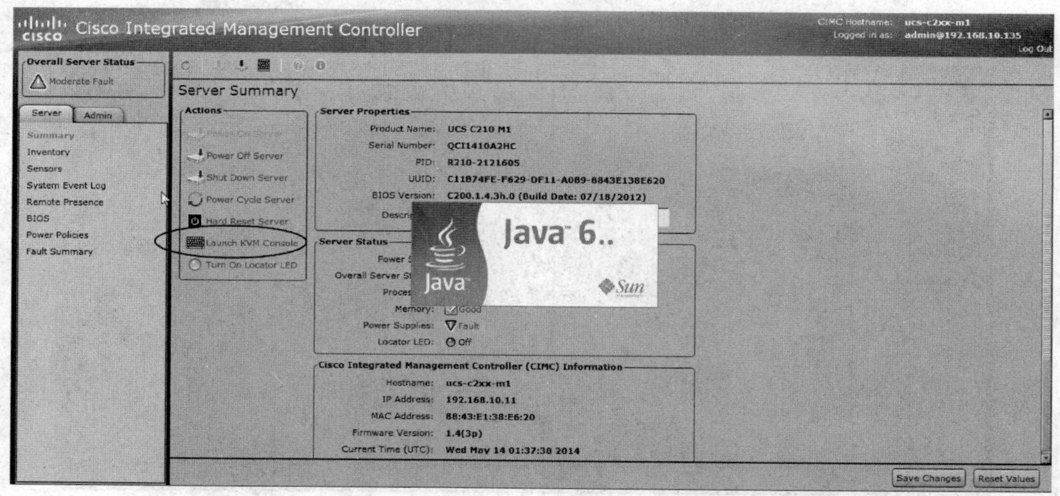

图 6-9　进入 KVM 控制台

进入 KVM 之前要进行 Java 验证，如图 6-10 所示。

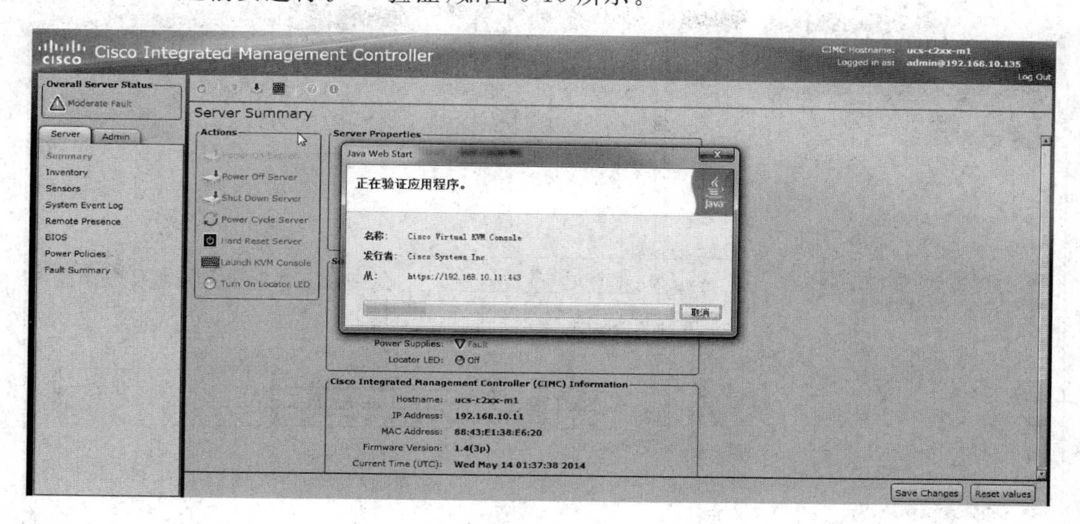

图 6-10　Java 验证过程

验证通过后，可以通过 Web 进入控制台，就跟直接连显示器键盘是一样的，如图 6-11 所示。

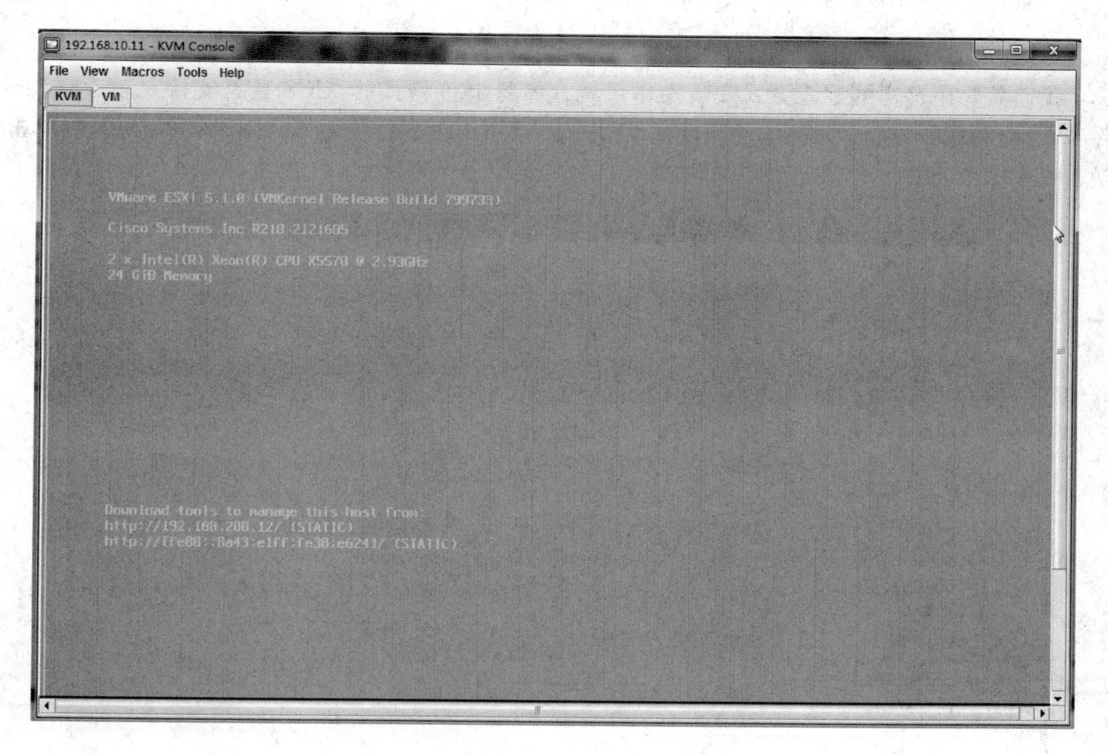

图 6-11　Web 进入控制台

这里特别要注意的是 Java 的版本不同往往导致不能进入控制台，Java 的版本请根据 CIMC的需要安装，在控制面板中可以选择需要的版本(前提是已经安装了多个版本)，步骤为在控制面板上双击“Java”，在打开的对话框单击“Java”再单击“查看”，然后启用不同的 Java 版本，如图 6-12 所示。

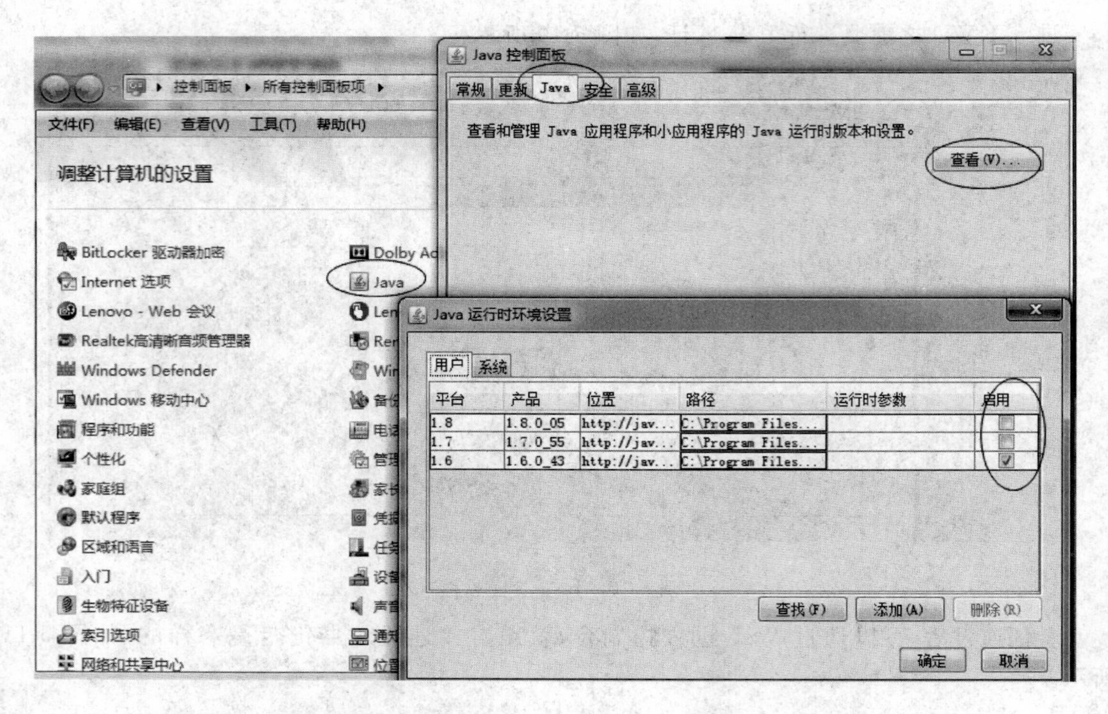

图 6-12　选择个人电脑上 Java 生效的版本

我们需要用这样的远程方式安装操作系统，OS 是 ISO 格式的，UCS 是否支持安装 VMware 平台，请查阅思科在线文档，作者测试的 UCS 是 C210 M1 支持安装 ESXi，单击“VM”，再单击“Add image”，如图 6-13 所示。

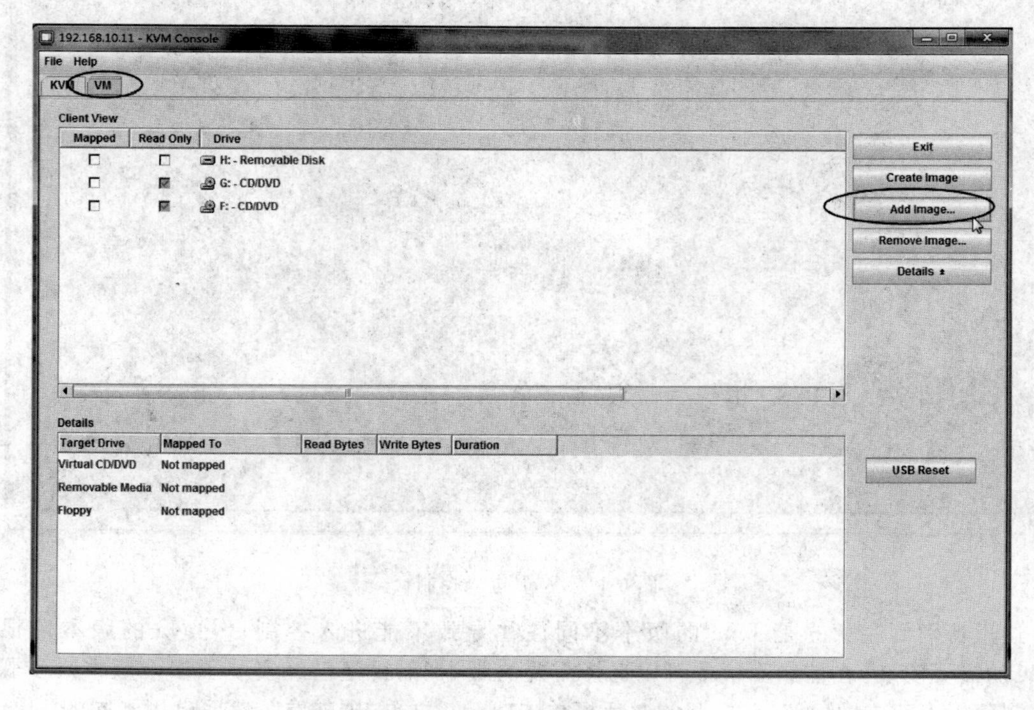

图 6-13　选择开始安装 ESXi

此时，会出现 ISO 文件路径选择提示，找到电脑上存放 ISO 文件的相应路径，如图 6-14 所示。

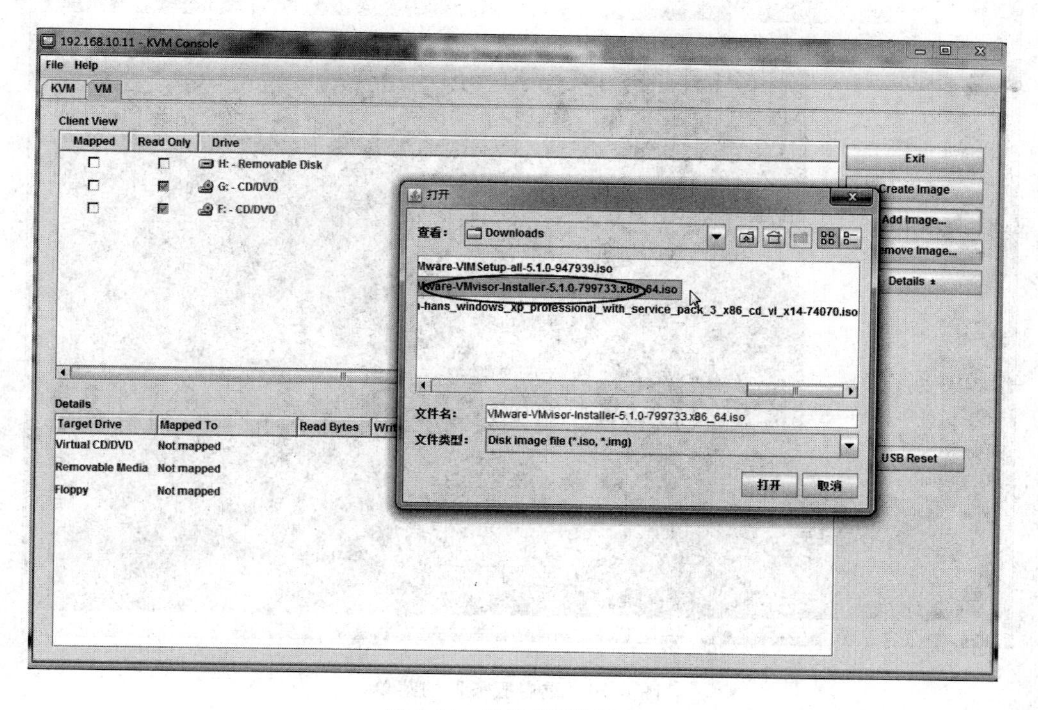

图 6-14 选择 ESXi 版本

注意选中该文件，以便在服务器重启的时候装载指定的 OS，这里安装的是 ESXi 6.1，如图 6-15 所示。

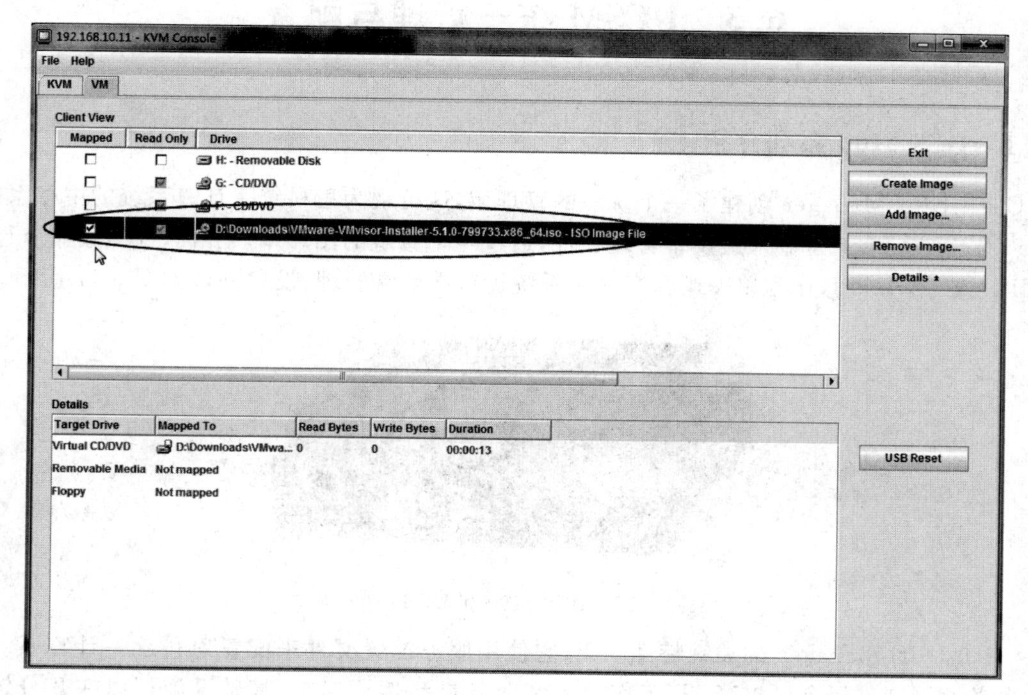

图 6-15 选中 ESXi 版本

跟随提示，选中引导的虚拟光盘，重启服务器进入安装过程，如图 6-16 所示。

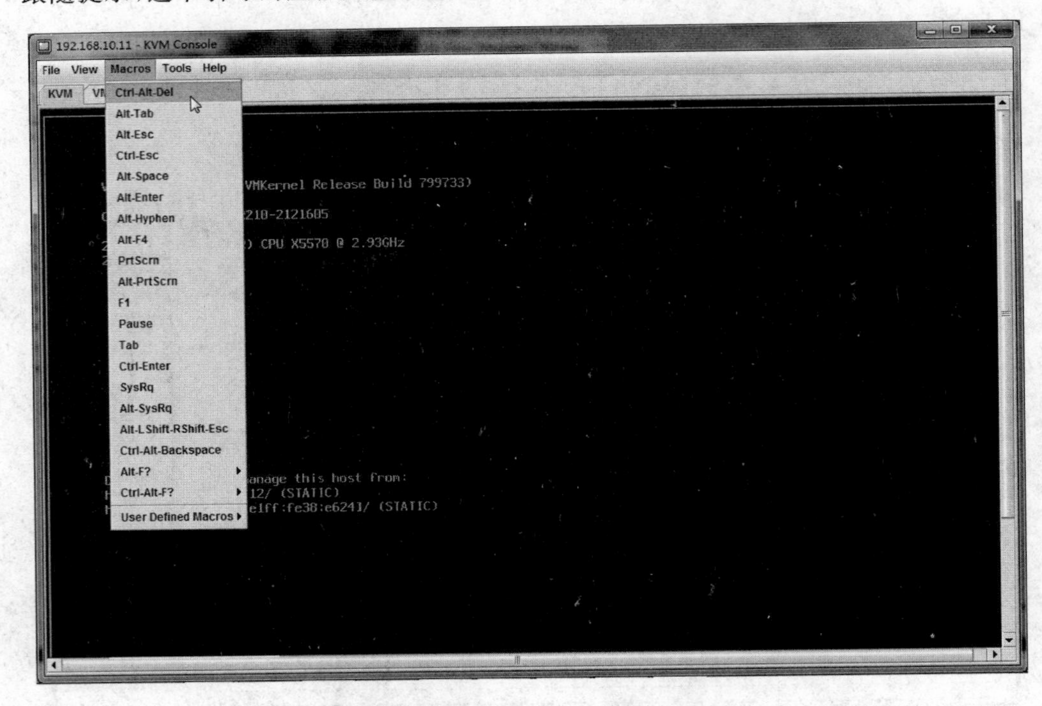

图 6-16　重启服务器的菜单

跟随向导，就会完成 ESXi 操作系统的安装。

6.3　UCSM 统一管理与配置

6.3.1　UCS 6100 系列介绍

Cisco UCS Manager 创建了一个统一的管理域，这可视为思科统一计算系统的中枢系统。Cisco UCS Manager 是嵌入式设备管理软件，通过一个直观的图形用户界面(GUI)、命令行界面(CLI)或 XMLAPI，作为单一逻辑实体对系统进行端到端管理，设备外观如图 6-17 所示。

图 6-17　UCS 6100 和 UCS 6200

Cisco UCS Manager 的关键特性在于，它使用服务配置文件来配置思科统一计算系统资源。服务配置文件概念可提高 IT 工作效率和业务灵活性。现在，基础设施可以在几分钟内配置完成，而不必再花费数天时间，从而使 IT 人员的关注重点能够从维护转向战略计划工作。

Cisco UCS Manager 使用服务配置文件来配置服务器及其 I/O 连接。服务配置文件由服务器、网络和存储管理员创建，并存储在 Cisco UCS 6100/6200 系列阵列互联交换机（Fabric Interconnect）。在当今的数据中心，服务器很难部署和改变使用目的，因为这通常要花费几天甚至几周的时间来实施。这一问题的出现是因为服务器、网络和存储团队需要仔细的人工协调，来确保其所有设备都能实现互操作。服务配置文件允许将思科统一计算系统中的服务器视作"裸计算能力"，在应用工作负载中进行分配和重新分配，从而能够更加动态、高效地使用当今数据中心内的服务器处理能力。

服务配置文件由服务器及服务器所需的相关局域网和存储域网络（SAN）连接组成。当为一台服务器部署了一个服务配置文件之后，Cisco UCS Manager 便可自动配置该服务器、适配器、扩展模块和互联阵列，以匹配该服务配置文件中所规定的配置。设备配置的自动化可减少配置服务器、网络接口卡（NIC）、主机总线适配器（HBAs）和局域网与 SAN 交换机等所需的人工步骤。人工步骤的减少可进一步降低人为错误的概率，改进一致性，缩短服务器部署时间。

服务配置文件可使虚拟化环境和非虚拟化环境同时受益。工作负载可能需要在服务器之间进行转移，以变更指定给一项工作负载的硬件资源或者使服务器离线进行维护或升级。服务配置文件可用于提高非虚拟化服务器的灵活性。它们还可与虚拟化集群一起使用，轻松增添新资源，来补充现有虚拟机的灵活性。同时，服务配置文件还可用于支持思科服务器的 VM-FEX 功能，以运行支持 VM-FEX 的虚拟机管理程序。

Cisco UCS Manager 安装在一对 Cisco UCS 6100/6200 系列互联阵列之上，使用主/被动集群配置来实现高可用性。该管理器不仅要参与服务器配置工作，还要参与设备发现、资产管理、配置、诊断、监控、故障检测、审核及统计数据收集等工作。它能够将系统的配置信息导出至配置管理数据库，推进基于信息技术基础设施库概念的流程。Cisco UCS Manager 的 XML API 还可促进与第三方配置工具之间的协调，以便在使用 Cisco UCS Manager 配置的服务器上部署虚拟机和安装操作系统与应用软件。

6.3.2 UCS 6100 初始化配置

通过 Console 口登录到 UCS 阵列互联器配置如下：

```
Enter the installation method (console/gui)? console
Enter the setup mode (restore from backup or initial setup) [restore/setup]? setup
You have chosen to setup a new switch. Continue? (y/n): y
Enter the password for "admin": adminpassword%958
Confirm the password for "admin": adminpassword%958
Do you want to create a new cluster on this switch (select 'no' for standalone setup or if
you want this switch to be added to an existing cluster)? (yes/no) [n]:no
```

如果是两台阵列交换机做集群，选择 yes，事先把两台阵列连到一起，连接的方法是在每个设备 Console 边上有两个 RJ45 网口，标记为 1 和 2，用两根直通网线以 1-1，2-2 的方式连好，输入 yes 后会提示：

```
Enter the admin password of the peer Fabric interconnect:
    Connecting to peer Fabric interconnect… done
```

```
    Retrieving config from peer Fabric interconnect… done
      Peer Fabric interconnect Mgmt0 IP Address：192.168.10.236
```
！另外一台阵列交换机地址
```
    Peer Fabric interconnect Mgmt0 IP Netmask：256.256.256.0
    Cluster IP address              ：192.168.10.234
Cluster IPv4 address ：192.168.10.254

Enter the system name：FABRIC
Mgmt0 IPv4 address：192.168.10.235
Mgmt0 IPv4 netmask：256.256.256.0
IPv4 address of the default gateway：192.168.10.254
Configure the DNS Server IPv4 address? (yes/no) [n]：yes
DNS IPv4 address：20.10.20.10
Configure the default domain name? (yes/no) [n]：yes
Default domain name：domainname.com
Following configurations will be applied：
Switch Fabric = A
System Name = FABRIC
Management IP Address = 192.168.10.235
Management IP Netmask = 255.255.255.0
Default Gateway = 192.168.10.254
DNS Server = 20.10.20.10
Domain Name = domainname.com
Apply and save the configuration (select 'no' if you want to re-enter)? (yes/no)：yes
```

这样就可以通过 Web 方式登录和管理设备了。

注意：如果需要重新初始化，设备加电后进入 admin 模式，然后 erase config 后重启。

6.3.3 Web 登录及基本配置

打开网页，输入 http://192.168.10.235，如图 6-18 所示，登录阵列互联器。

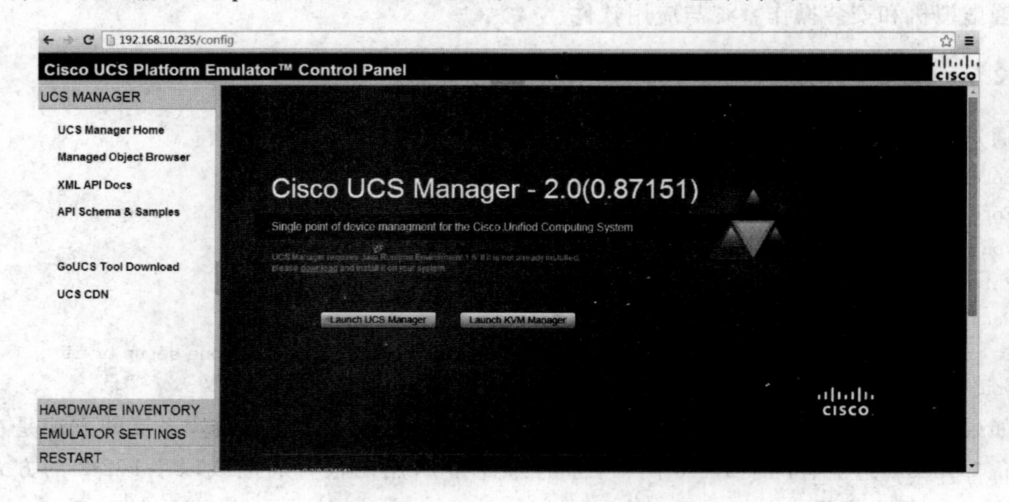

图 6-18 登录阵列互联器

单击“LUNCH UCS Manager”登录，右面的“Launch KVM Manager”是做服务器直接登

录的，当将所有的服务器配置成功后可以通过点选此按钮直接登录到服务器的 KVM 备选窗口，可方便地登录思科的各台刀片服务器。单击“Launch UCS Manager”登录后弹出如图 6-19 所示的界面。

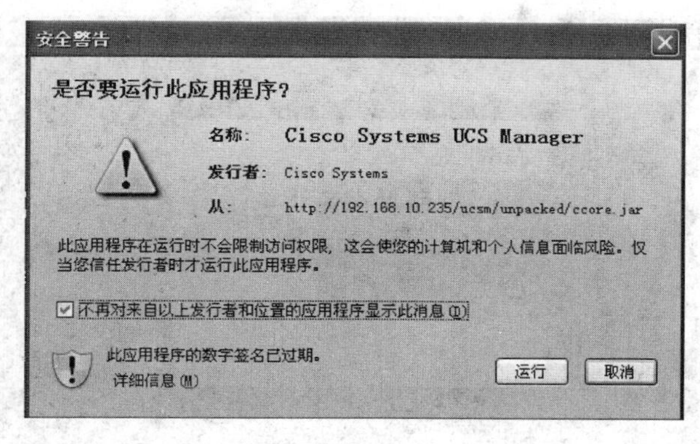

图 6-19　安装提示

选择“不再对来自以上发行者和位置的应用程序显示此消息(D)”，选择“运行”。然后配置界面会加载一个 Java 的小程序，如图 6-20 所示。如果管理主机上没有 Java 的运行环境将会提示错误，请登录 Java 的网站下载相关的 Java 环境并安装，登录地址：http://www.java.com/zh_CN/。

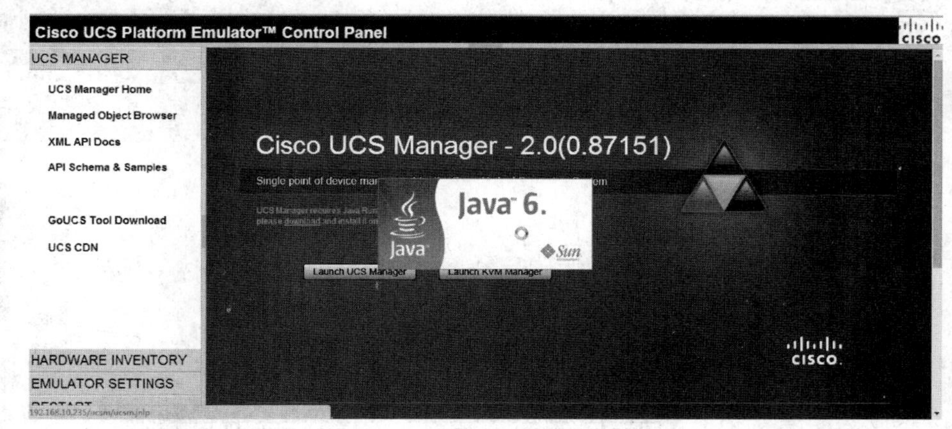

图 6-20　正在运行 Java

选择“免费 Java 下载”，然后按照默认的步骤完成相应的安装即可。

当 Java 环境加载成功后出现如下的界面，如图 6-21 所示。

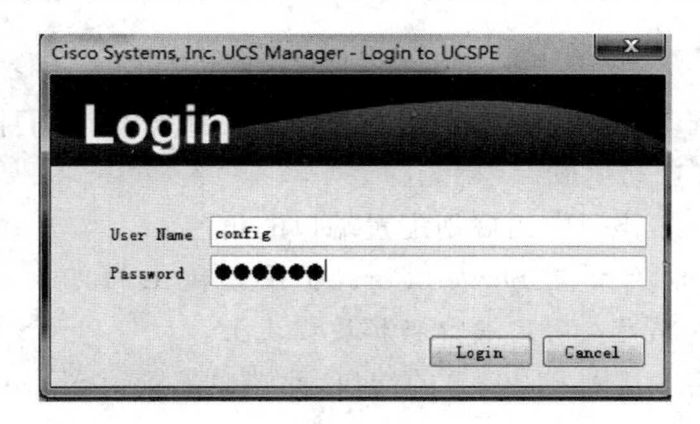

图 6-21　UCS Manager 登录认证

输入规划中的用户名和密码，这里应用缺省账号：config/config。可以在登录后再修改需要的账号，登录认证通过后的画面如图 6-22 所示。

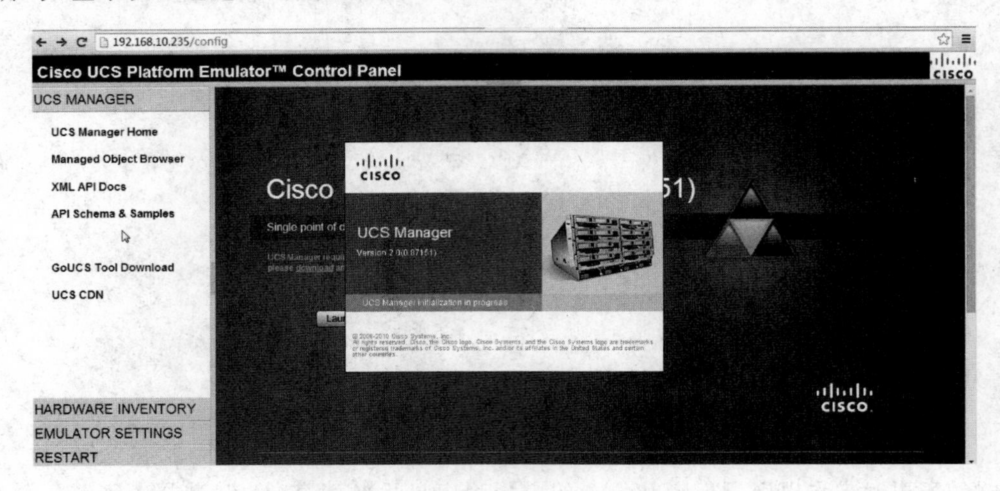

图 6-22　登录认证通过后的画面

加载成功后进入测试的 6120XP 的最终管理界面，如图 6-23 所示。

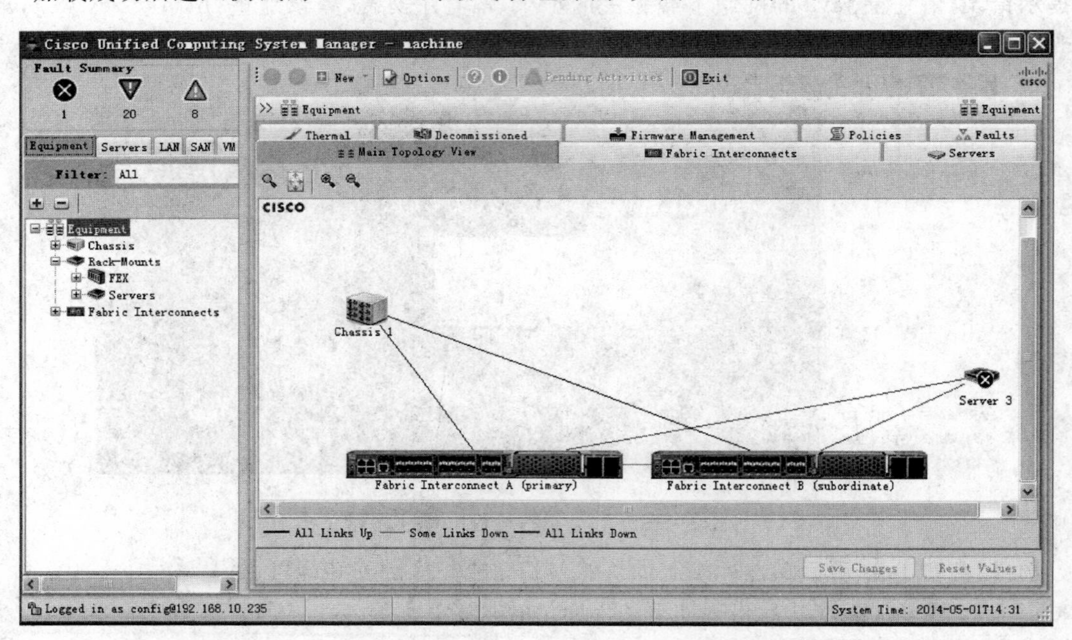

图 6-23　6120XP 的最终管理界面

第一次登录此界面的时候是不会出现图中的网络拓扑的，当我们做完基本的端口角色配置后，图 6-23 中的拓扑将自动生成。

接下来根据前面网络拓扑图中的规划配置端口的角色。

注意：只能在固定端口上配置 Server Port，扩展模块上的万兆端口不能配置为 Server Port。Up-Link Port 可以配置在固定端口和扩展模块上。

需要把所有刀片机箱连接到 6120XP 的端口，此端口配置为 Server Port，并选择一个空闲端口配置为 Up-Link Port。

在“导航窗口”,依次单击“Equipment”→“Fabric Interconnects”→“Fabric Interconnect_A”→“Fixed Module ”→“Unconfigured Ports”,选择连接到刀片机箱的端口(可多选),将选中的端口拖入“Server Ports” 组,再选择上联端口(可多选),将选择的端口拖入“Up Ethernet Ports”组,在“Fabric Interconnect_A”重复以上步骤,完成端口配置,如图 6-24 所示。

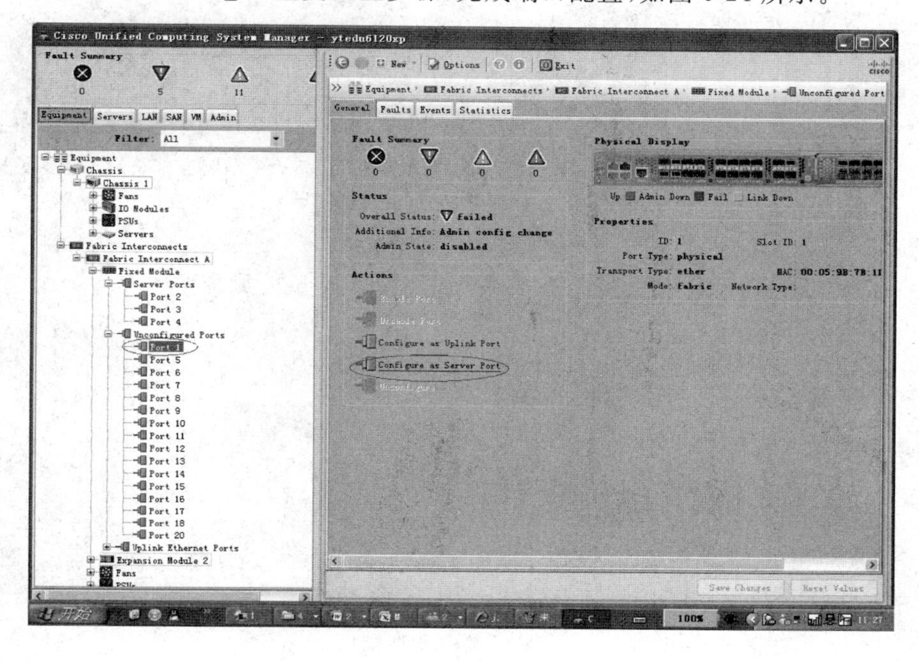

图 6-24　配置服务器端口

第一步,配置服务器端口,按照图 6-24 中的提示选择“Fix Module”中的“Port 1”,选中右边的“Configure as Server Port”,将此端口配置成服务器角色。

第二步,配置上联端口,单击左边的“Port 19”,将此端口的角色配置成上联端口,如图 6-25所示,配置完上联端口后的截图如图 6-26 所示。

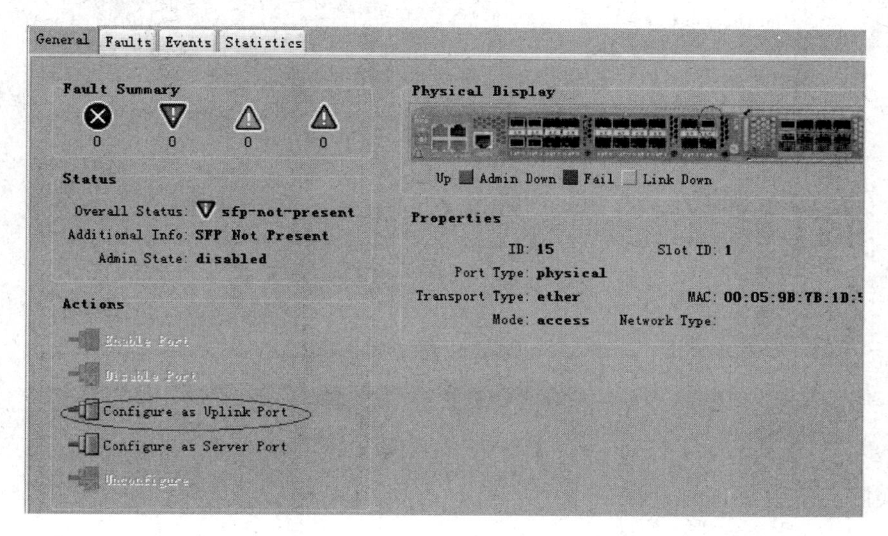

图 6-25　配置上联端口

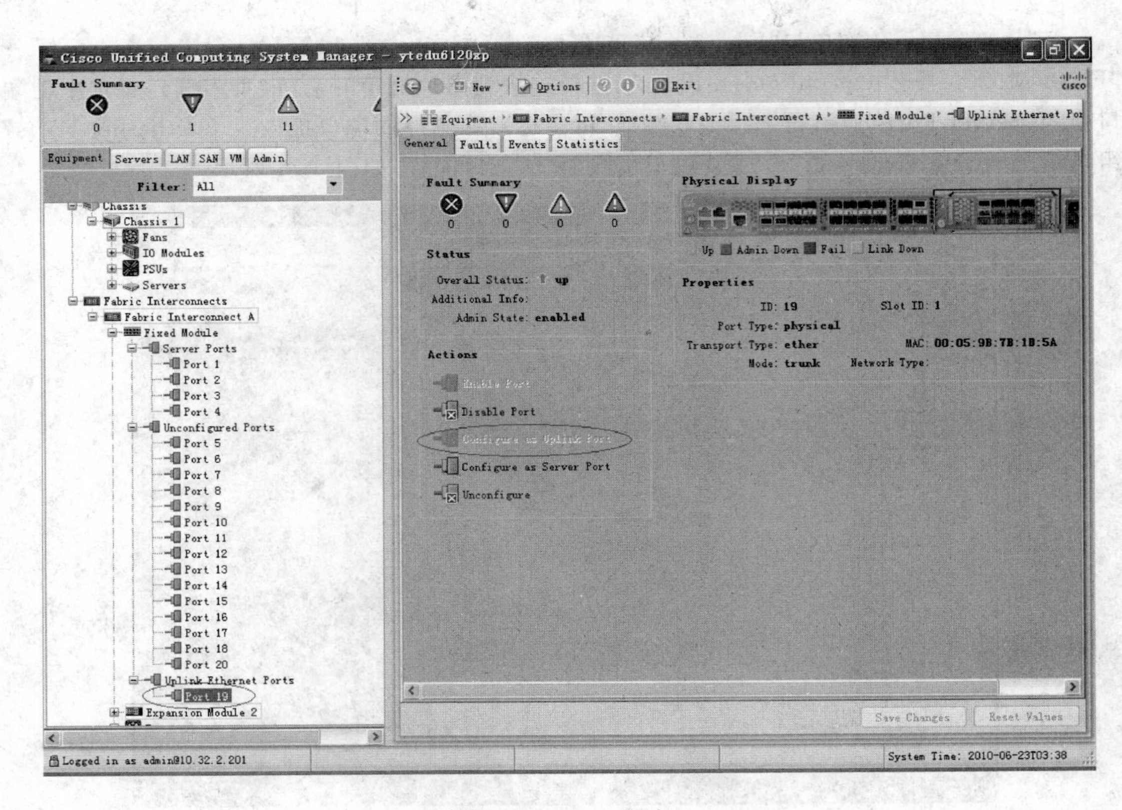

图 6-26　选择上联端口

第三步，配置连接思科 SAN 交换机的"Uplink FC Ports"，如图 6-27 所示，在"Expansion Module 2"中分别选择"FC Port1"和"FC Port2"，然后单击"Enable port"启用此接口即可，完成后的截图如图 6-28 所示。

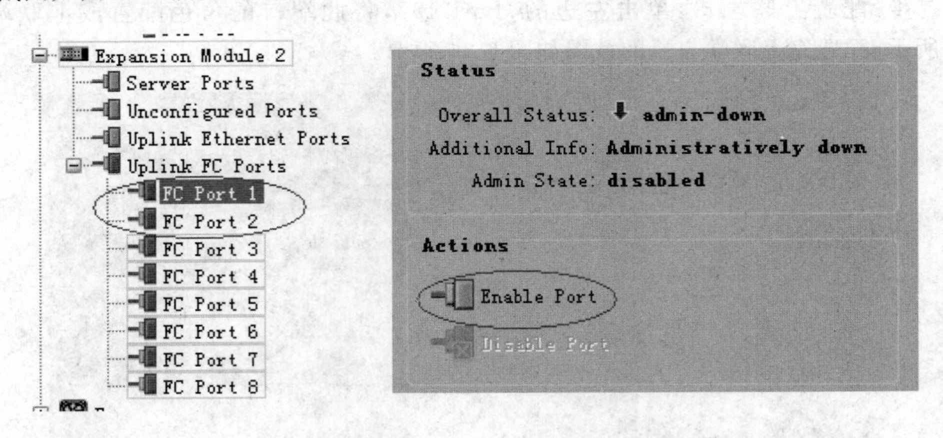

图 6-27　配置上联 FC 端口

第四步，定义 Chassis Discovery Policy。根据 UCS 6120 或 6248 UP 与 UCS 2104 或 2208XP 实际连接数量选择，如图 6-29 所示。

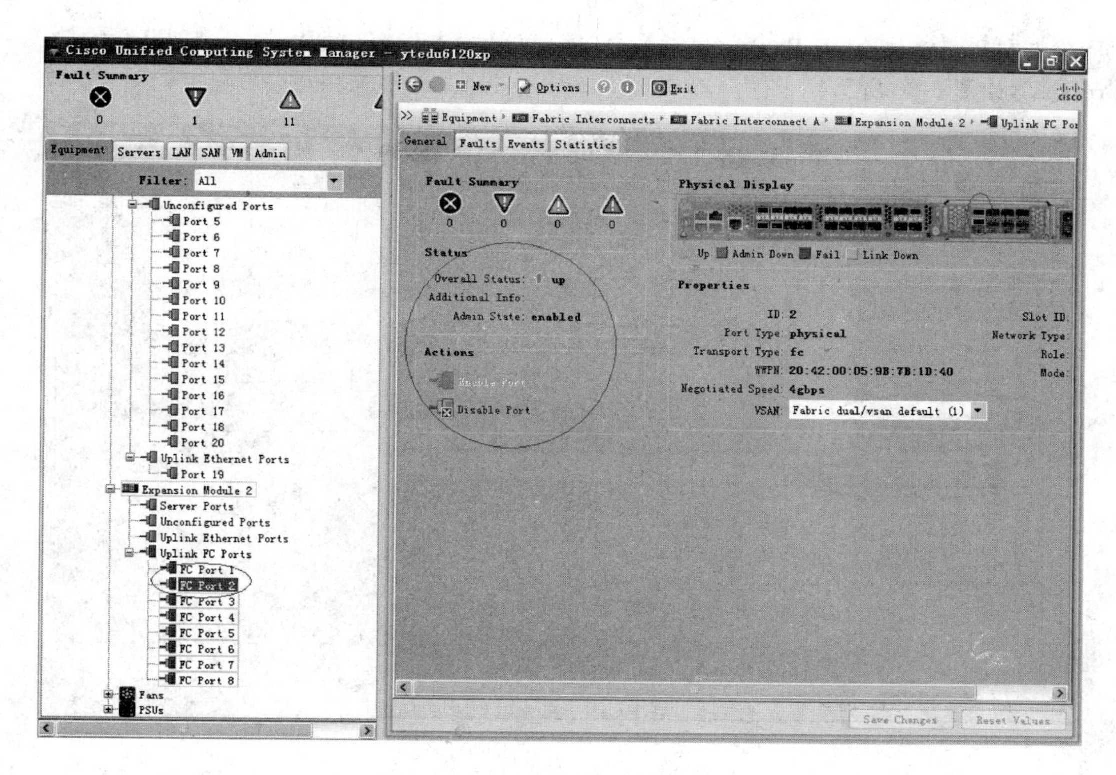

图 6-28　上联端口配置完成的情形

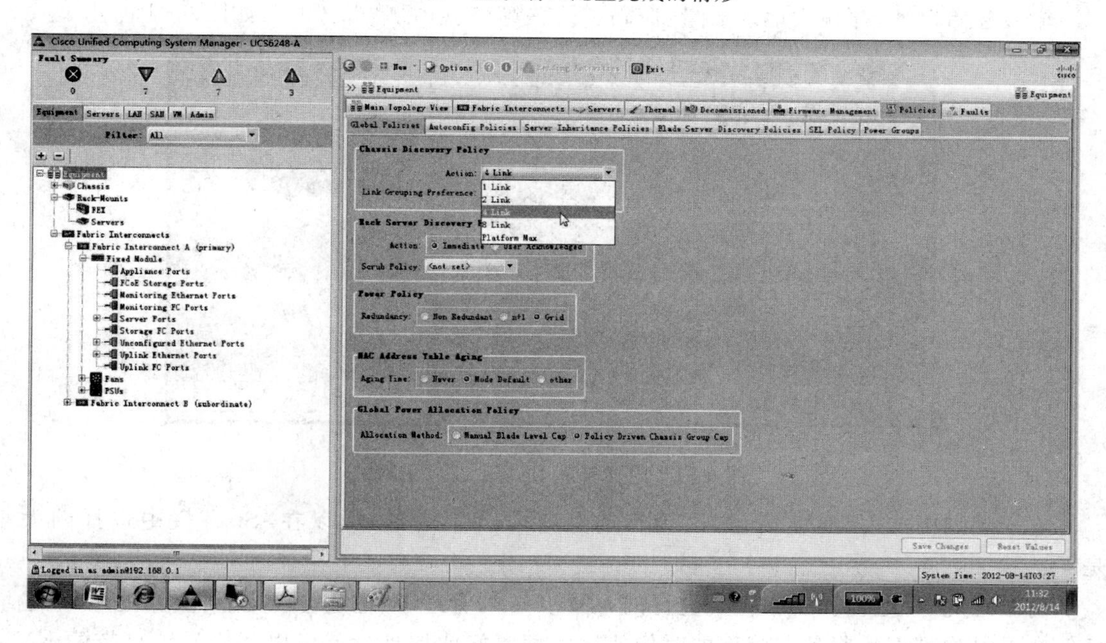

图 6-29　选择 Chassis Discovery Policy 连线数量

完成上述几个步骤后，实际的拓扑连接便可显示出来了。

6.3.4　配置 KVM 连接

Management IP 用于 UCSM 访问刀片服务器，登录进 UCSM，单击导航窗口的“Admin”

页,单击“Management IP Pool(ext-mgmt)”项,再单击工作窗口的“Create Block of IP Addresses”,如图 6-30 所示。

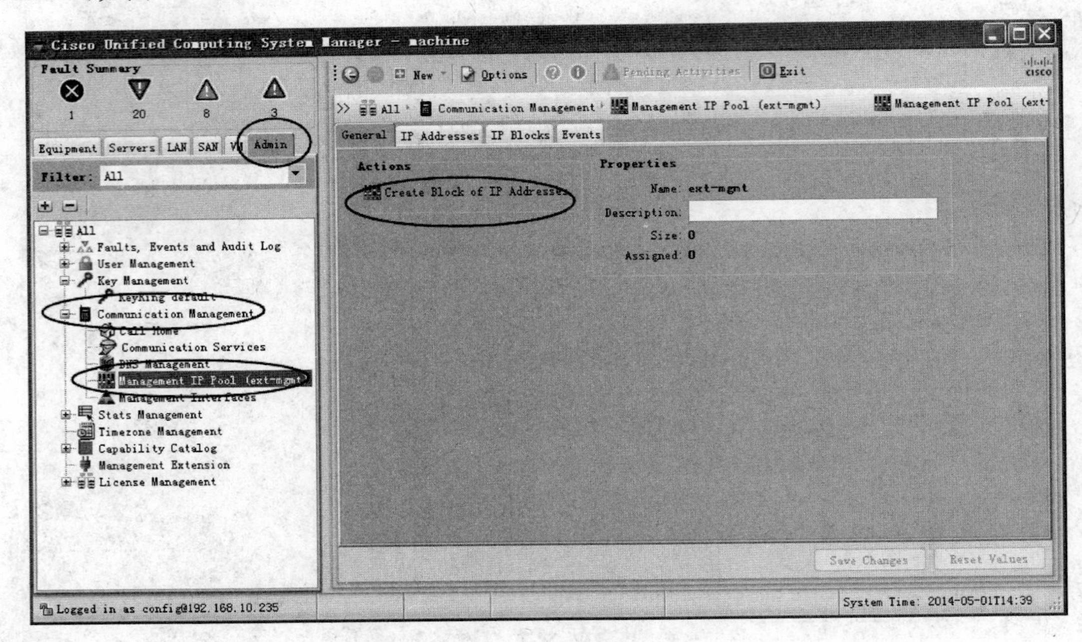

图 6-30　配置管理 IP

输入起始地址,并输入地址数量、掩码和网关等信息,并单击“OK”,如图 6-31 所示。

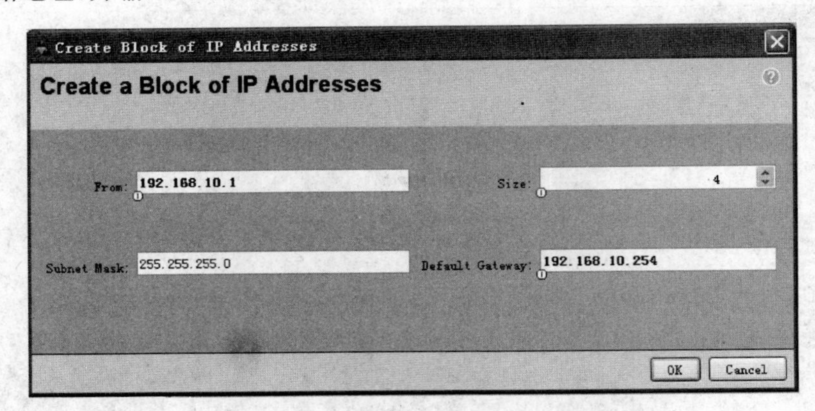

图 6-31　配置管理 IP 地址、掩码、网关

输入规划中的 IP 地址信息,因为目前共有 4 台服务器,所以在“Size”栏中选择的大小是 4。

请注意此 IP 地址和后来将要进行的 VMware 的管理 IP 地址可能是不同的,这个 IP 地址仅仅在思科物理刀片服务器管理的时候使用。配置好此地址后就可以利用系统自带的 KVM 登录刀片服务器,使用 vMedia 等功能。

单击导航窗口的“Equipment”页,单击“Server 1”,选择一个服务器,并单击工作窗口的“General”页,选择“KVM Console”,如图 6-32 所示。

出现 KVM 窗口,提供服务器的显示界面,同时系统自动将键盘和鼠标连接到 PC 上,如

图 6-33 所示。

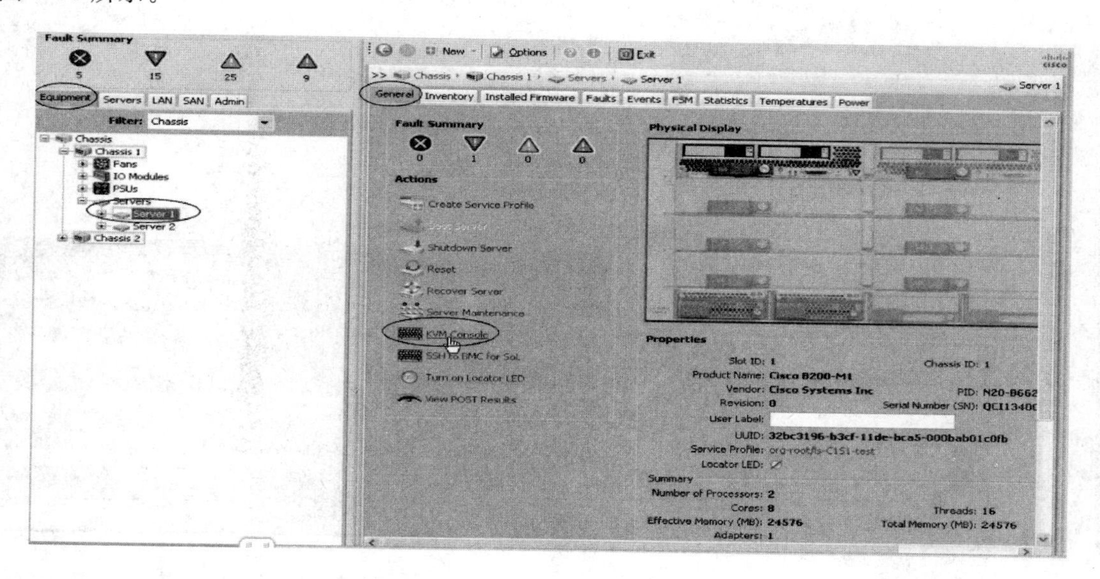

图 6-32　配置 KVM

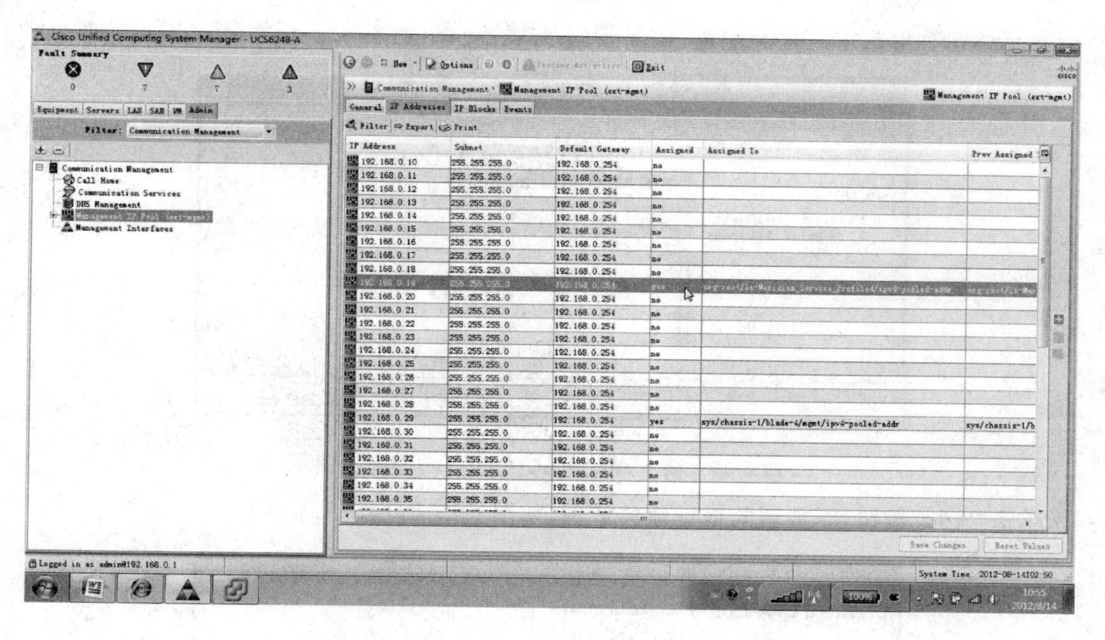

图 6-33　通过 KVM 访问服务器的界面

UCSM 为刀片服务器分配管理 IP，KVM Console 通过管理 IP 访问刀片服务器。

6.3.5　创建模板前的准备

1. 配置 MAC Pool

在导航窗口，选择“LAN”页，单击“Pools”项，单击并扩展“Root”项，右键单击“ MAC Pools”，然后选择“ Create MAC Pool”，如图 6-34 所示。输入我们建立 MAC Pool 的名字，如图 6-35 所示。

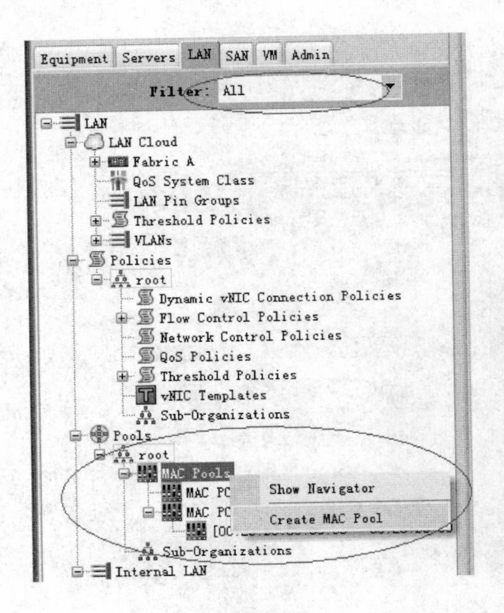

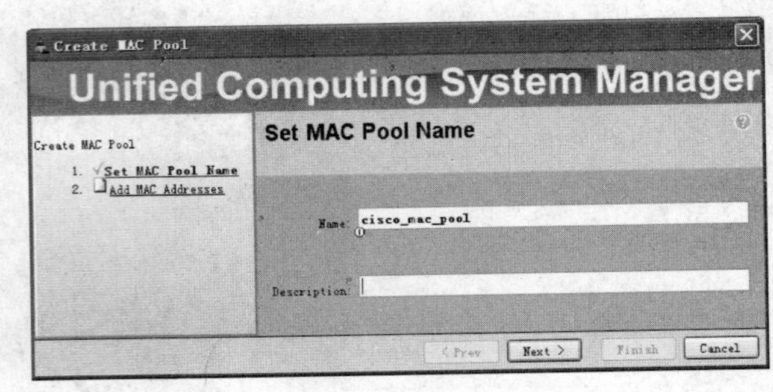

图 6-34　配置 MAC Pool　　图 6-35　输入 MAC Pool 名

单击“Next”进行下一步的配置，如图 6-36 所示。

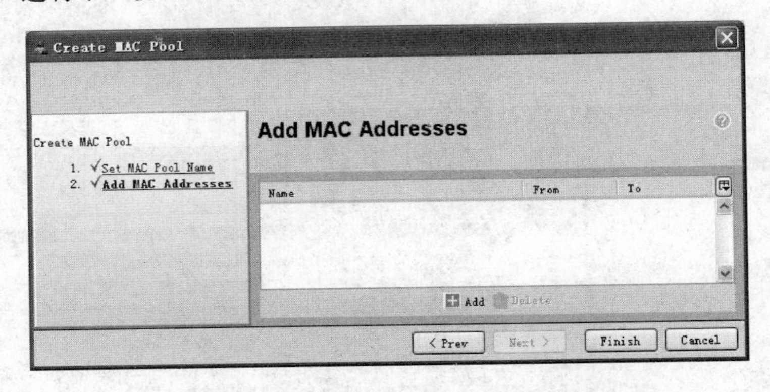

图 6-36　显示 Add MAC Addresses 的示意图

单击“Add”按钮，添加新的地址池的范围，如图 6-37 所示。

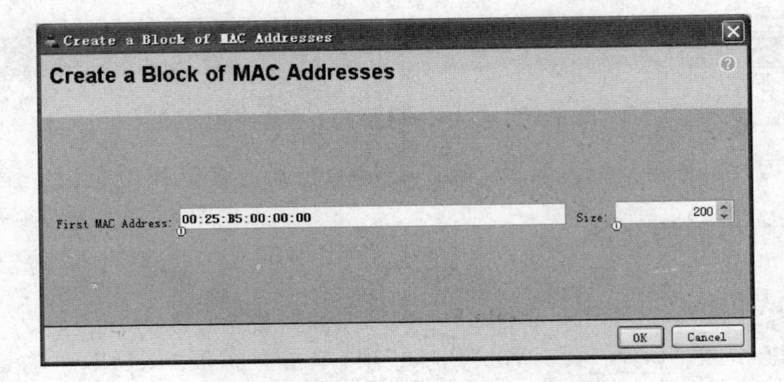

图 6-37　添加 MAC 地址

在此处给的"Size"的大小是200，这个值可以根据内部网络的实际情况来估算，完毕后单击"OK"，就可以从界面中看到创建的地址池，如图6-38所示。

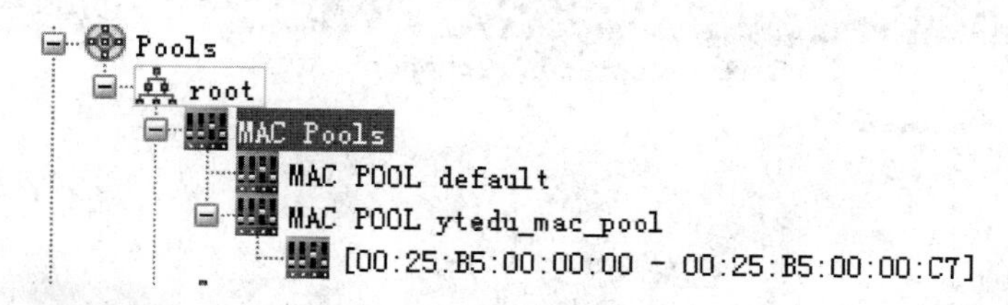

图6-38　创建好的地址池

注意：这个地址池是将来创建Profile模板的时候使用的，是用来分配给思科刀片服务器的网卡使用的，而不是给虚拟机的网卡用的。到这一步MAC地址池就已经创建好了。

2. 配置WWN Pool

WWN Pool是UCS里Fibre Channel vHBAs用到的WWN的集合，需要创建两个单独的Pool——WWPN Pool和WWNN Pool。WWPN(World Wide Port Number)是交换机给端口分配的地址。WWNN(World Wide Node Number)是64位的标志，厂商为每个产品分配的号码，类似MAC地址。

注意：一个WWN Pool里面的WWNN或WWPN的地址范围只能是从20:00:00:00:00:00:00:00到20:FF:FF:FF:FF:FF:FF:FF或从50:00:00:00:00:00:00:00到5F:FF:FF:FF:FF:FF:FF:FF，所有其他WWN范围都已被保留不能使用。

3. 配置WWN Pool

从SAN配置界面，在WWNN Pool上单击右键选择其中的"Creat WWNN Pool"，创建新的WWNN Pool池，如图6-39所示。单击后输入我们规划中的名字，如图6-40所示。

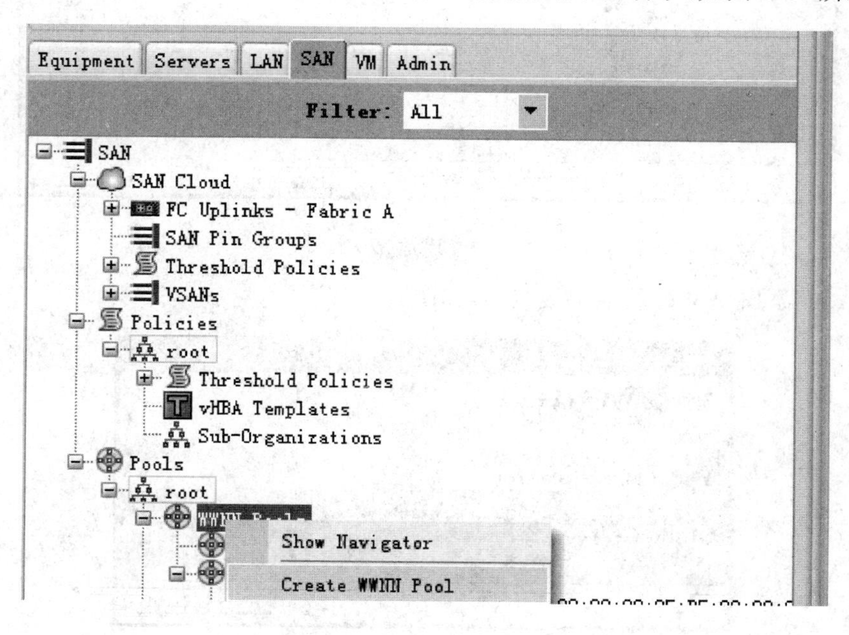

图6-39　添加WWNN Pool

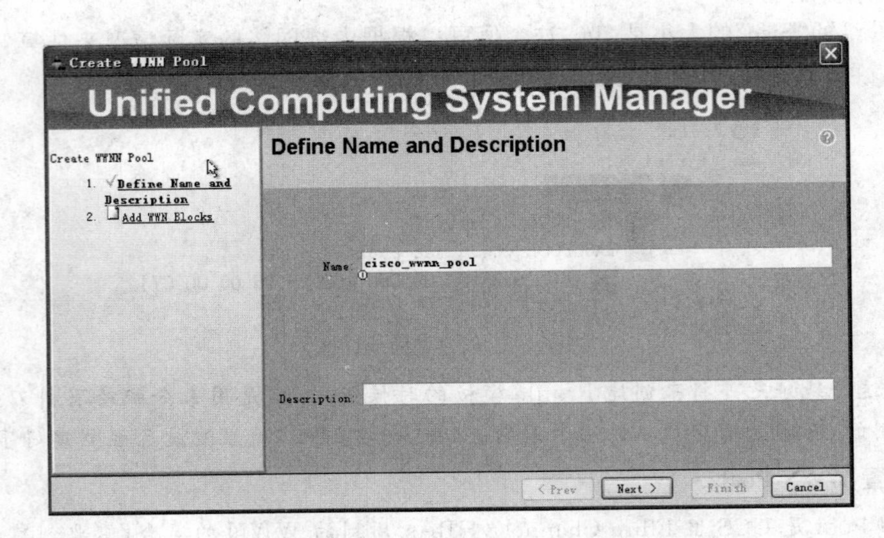

图 6-40　输入 WWNN 名字

单击“Next”继续下一步的配置，如图 6-41 所示。

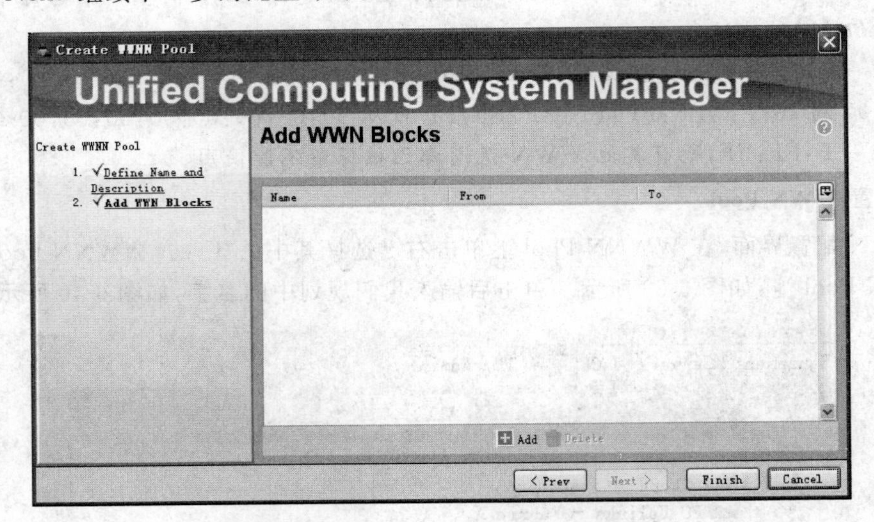

图 6-41　增加 WWNN

选择“Add”添加 WWN 的地址池范围，如图 6-42 所示。

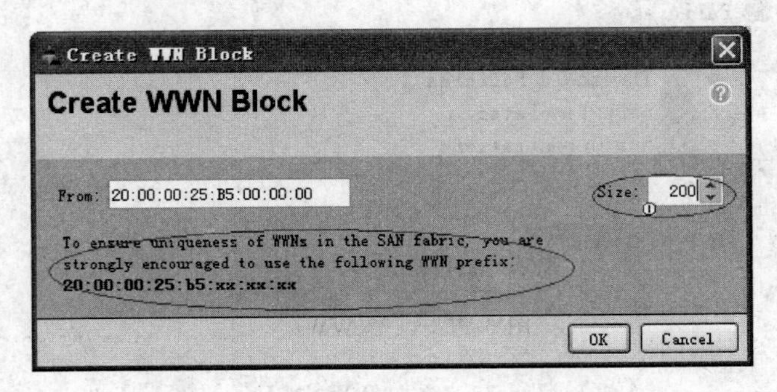

图 6-42　添加 MAC 地址

请注意图 6-42 圈中的提示，虽然 WWN 号码有个范围，但是在这个地方强烈建议别修改后面的推荐数值“25:B5:00”，前面的“20:00:00”可以酌情修改。因为在本次的应用中需要做 SAN BOOT，如果修改的话有可能在 EMC 的存储中不能看到刀片服务器 HBA 卡的 WWN 信息。

从“Size”中输入期望的 WWN 的大小，在部署中输入了 200，输入完毕后单击“OK”结束此配置。配置完毕后将看到已经建好的地址池，如图 6-43 所示。

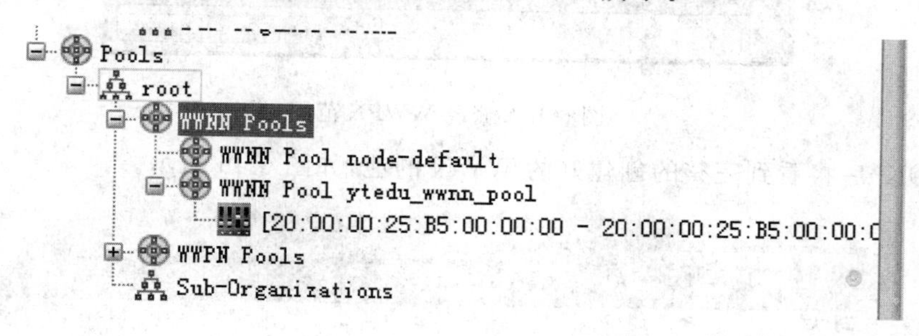

图 6-43　WWNN 添加完成的画面

按照相同的步骤创建 WWPN Pool。

4. 配置 WWPN Pool

如图 6-44 所示，操作步骤与配置 WWNN Pool 类似，在 SAN 配置界面，依次选择“Pools”→“WWPN Pools”→“Create WWPN Pool”。

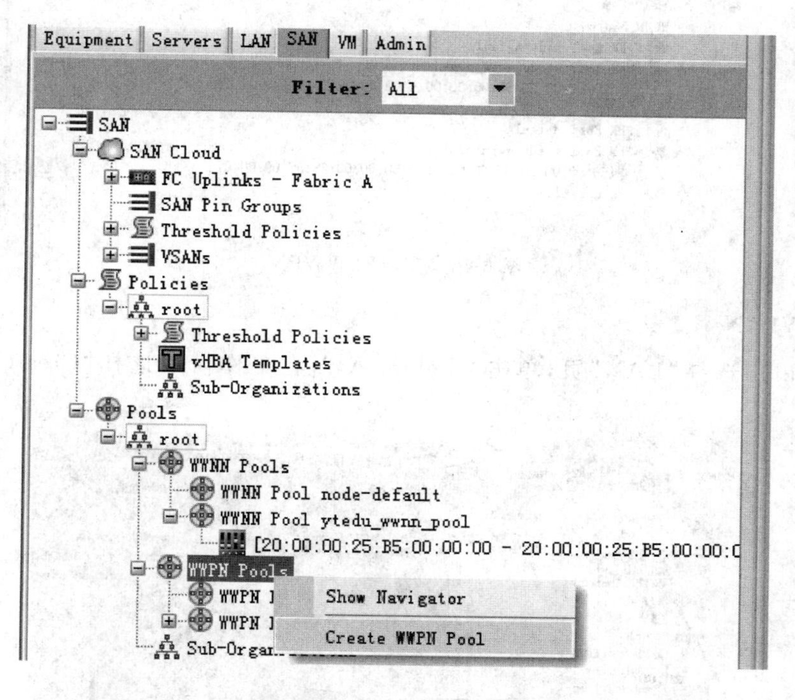

图 6-44　创建 WWPN

按照图 6-45 往下进行，需要注意的是在选择地址池的时候为了与上面的 WWNN Pool 有所区别，将地址池的范围修改为图 6-45 所示的值。

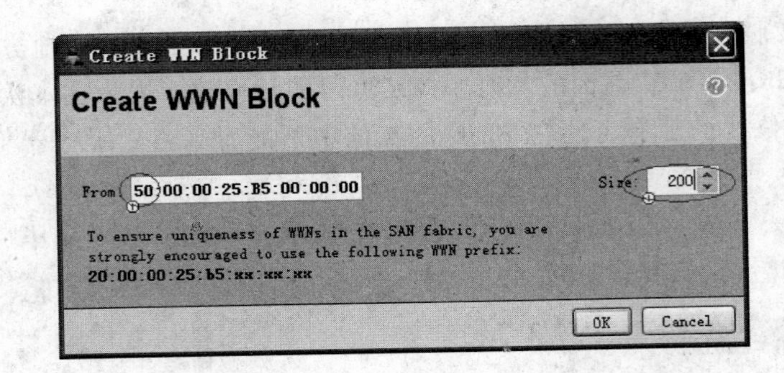

图 6-45　修改 WWPN 范围

单击“OK”后将看到完整的创建好的 WWN 的池，如图 6-46 所示。

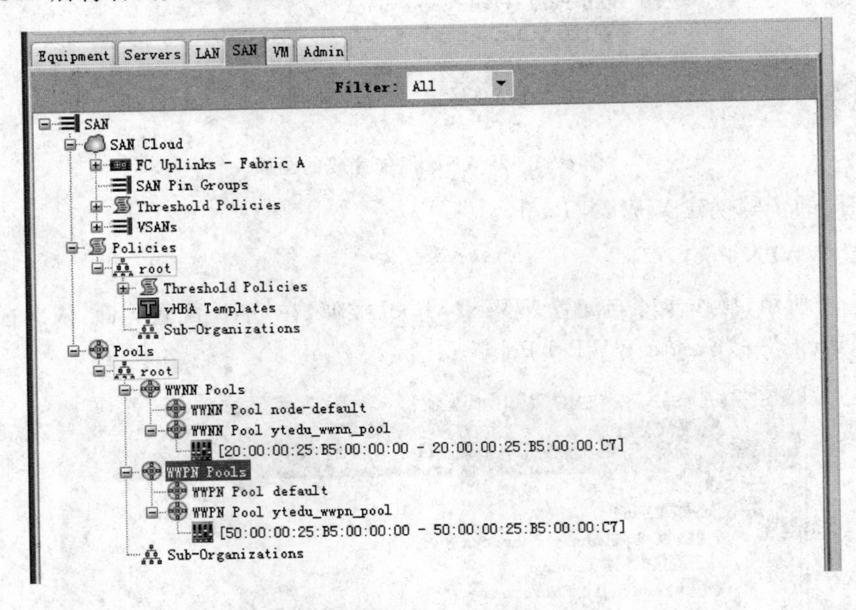

图 6-46　完成的 WWPN

5. 配置 VLAN

在导航窗口，选择“LAN”页，单击“ Fabric A ”，在工作窗口选择“Create VLAN”，如图 6-47 所示。

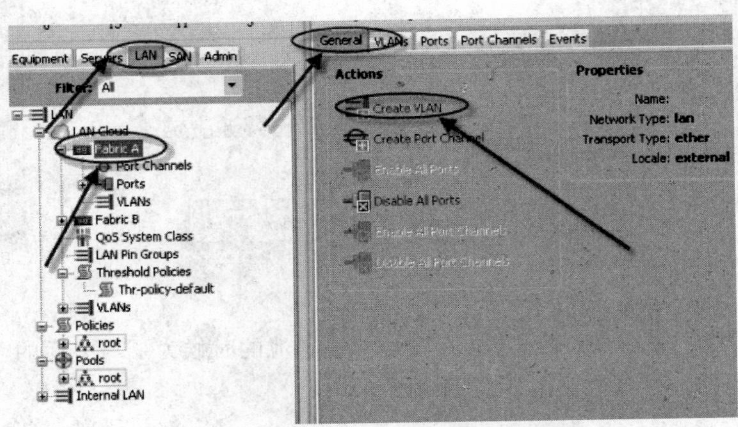

图 6-47　创建 VLAN

输入规划中的 VLAN 名称及 VLAN 号码，如图 6-48 所示，点击"Check Overlap"，确认无误后单击"OK"结束此 VLAN 的配置，重复上述步骤配置其他的 VLAN。

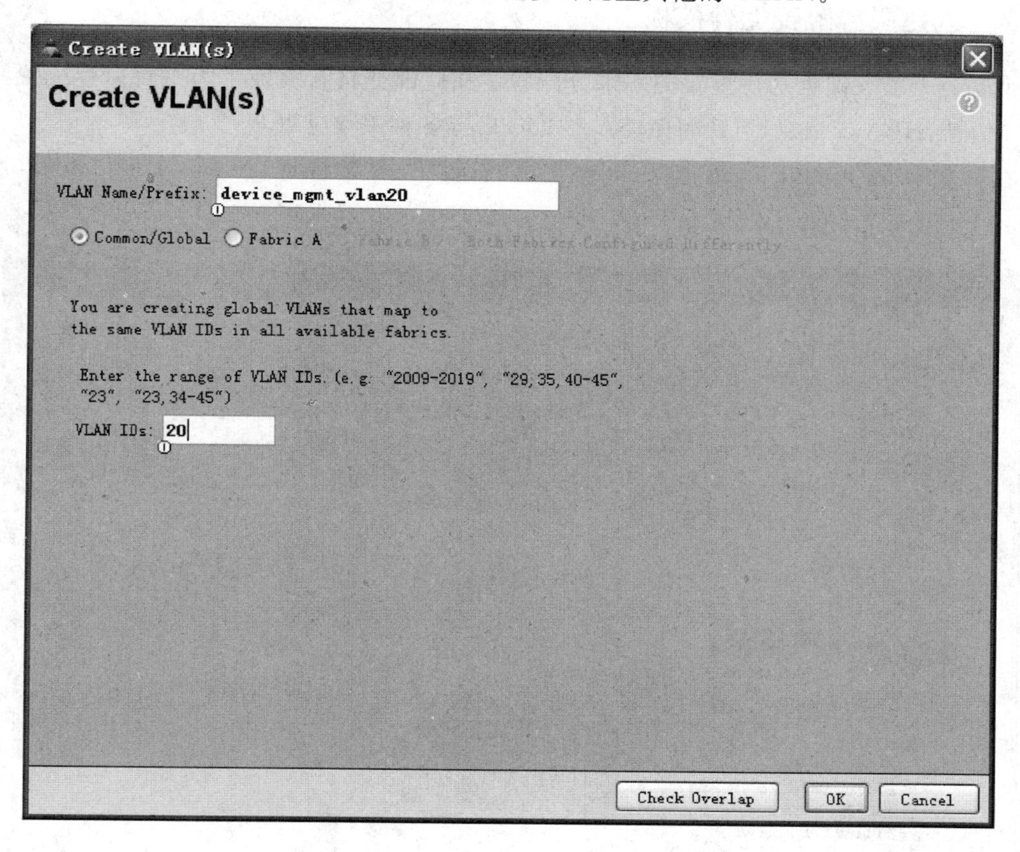

图 6-48　VLAN 的名字和 ID 号

按照目前的规划，将在虚拟机的操作系统中运行两个 VLAN，一个是 VLAN20 用作设备的管理，一个是 VLAN30 用于各种服务器。当日后需要其他 VLAN 的应用时，还可以在此处创建新的 VLAN。

全部 VLAN 配置结束后，将看到所有的 VLAN 信息，如图 6-49 所示。

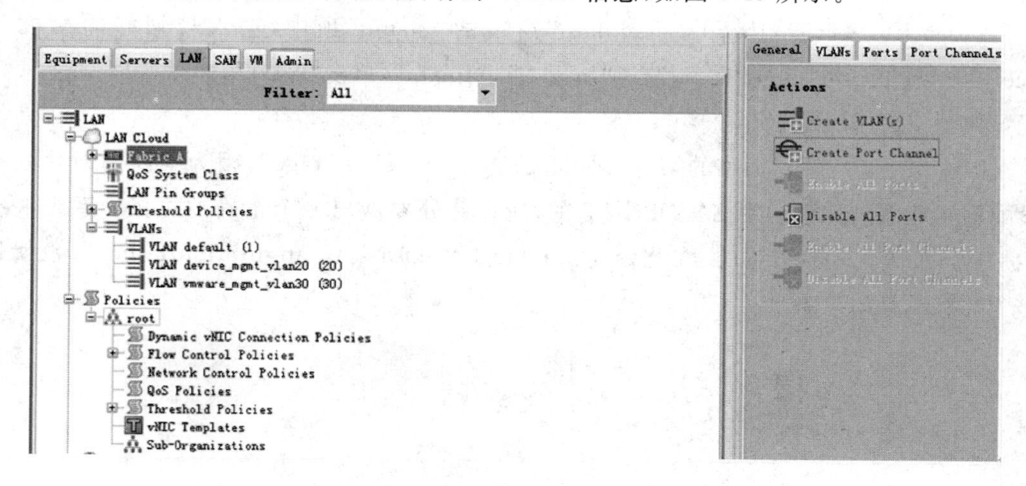

图 6-49　显示创建好的 VLAN 信息

至此，我们完成了 VLAN 的相关配置。

6.3.6 创建服务配置文件模板

此部分是最关键也是最重要的一部分配置。在此部分将在模板中为每台刀片配置网卡(NIC)信息、HBA 卡信息、启动顺序(SAN BOOT)等各种关键应用。

在图 6-50 的主菜单下，选择“Servers”页，右键单击“Service Profile Templates”，选择“Create Service Profile Template”。

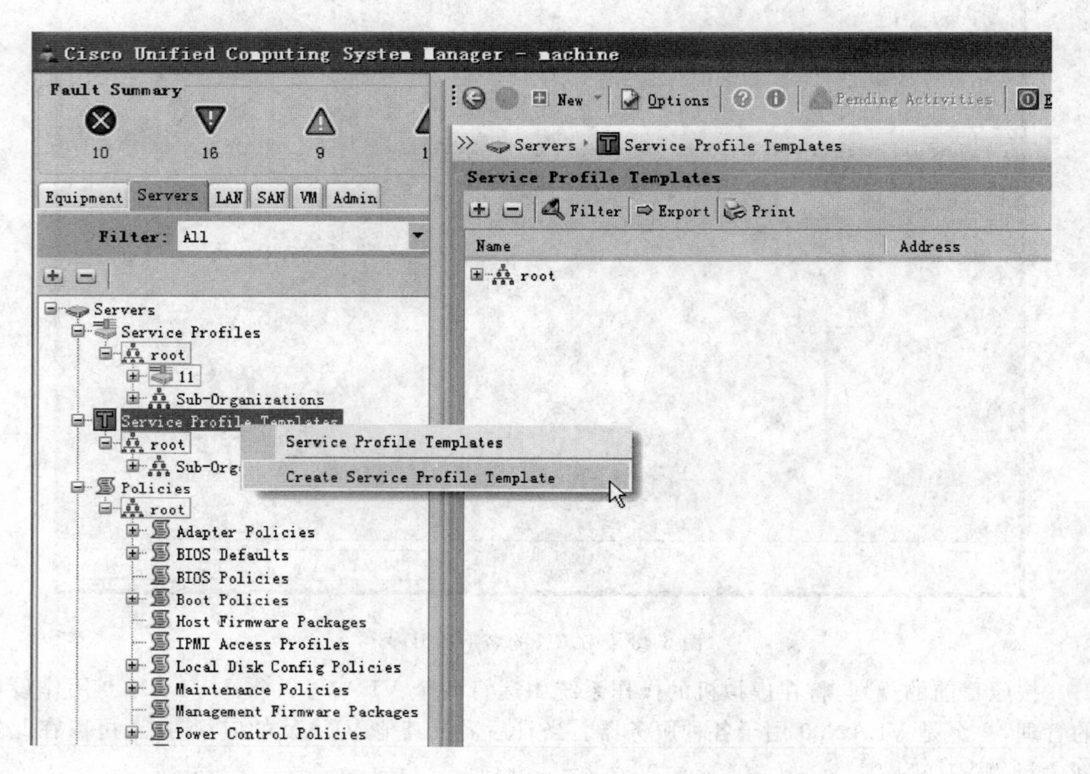

图 6-50 创建 Service Profile Template

在“Identify Service Profile Template”窗口的“Name”项中，输入规划好的模板名称“cisco_service_profile_template”，选择“Initial Template”，单击“Next”，如图 6-51 所示。模板的配置如图 6-52 所示。

在图 6-52 的界面中，面临的第一个问题是选择“Local Storage”，因为本期的刀片服务器是没配置本地硬盘打算利用 SAN BOOT 的，所以此处要做出一个正确的选择。系统缺省模式不是从 SAN BOOT 引导的，在此处要“Create Local Disk Configuration Policy”，也就是建立一个本次磁盘的策略。

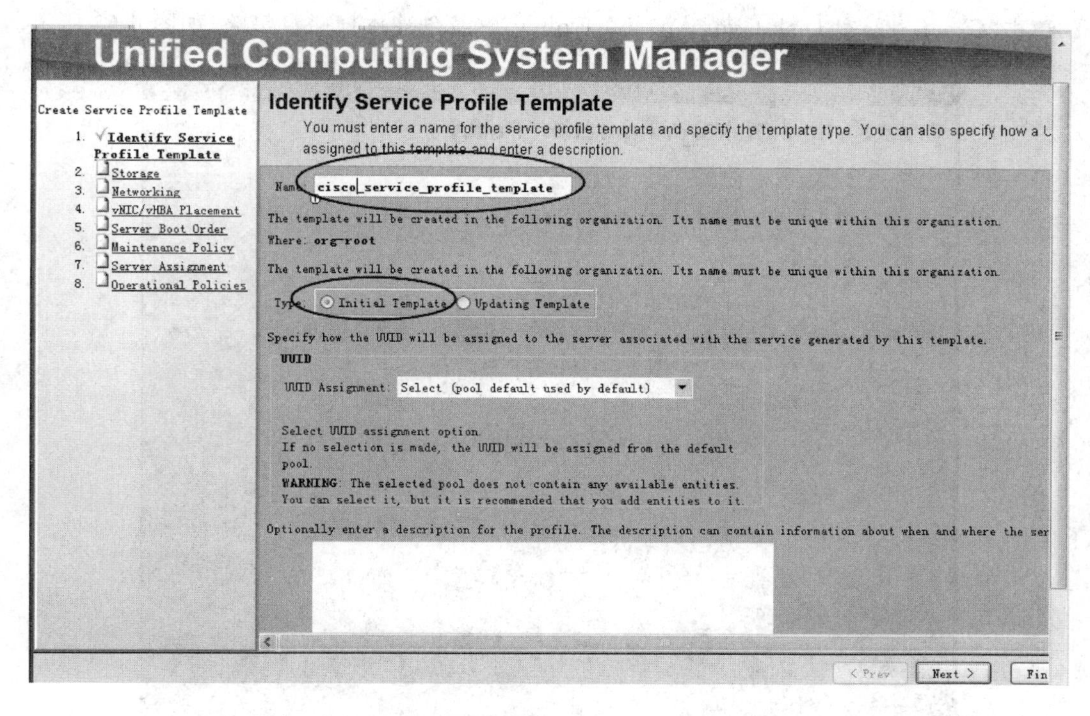

图 6-51　选择服务模板的名称、初始化模板

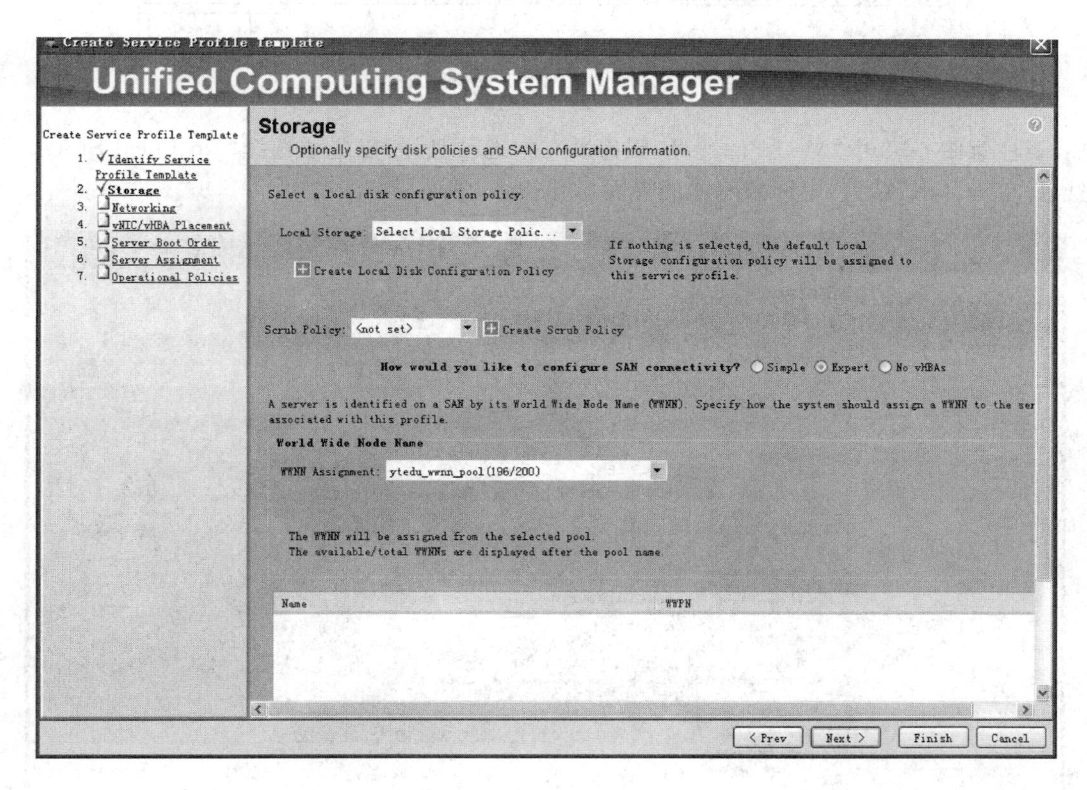

图 6-52　服务模板的配置

单击“Create Local Disk Configuration Policy”继续创建本地磁盘策略，如图 6-53 所示。

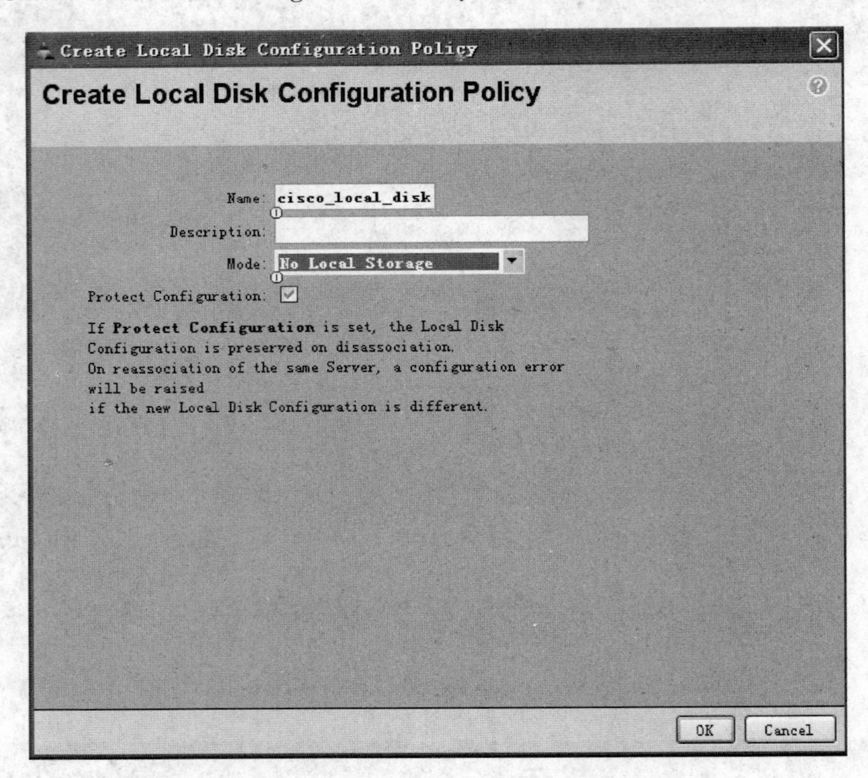

图 6-53　创建本地磁盘策略

要注意图 6-53 中“Mode”的选择，因为没有本地硬盘，所以应该选择“No Local Storage”，完毕后单击“OK”结束，返回到刚才的界面，如图 6-54 所示。

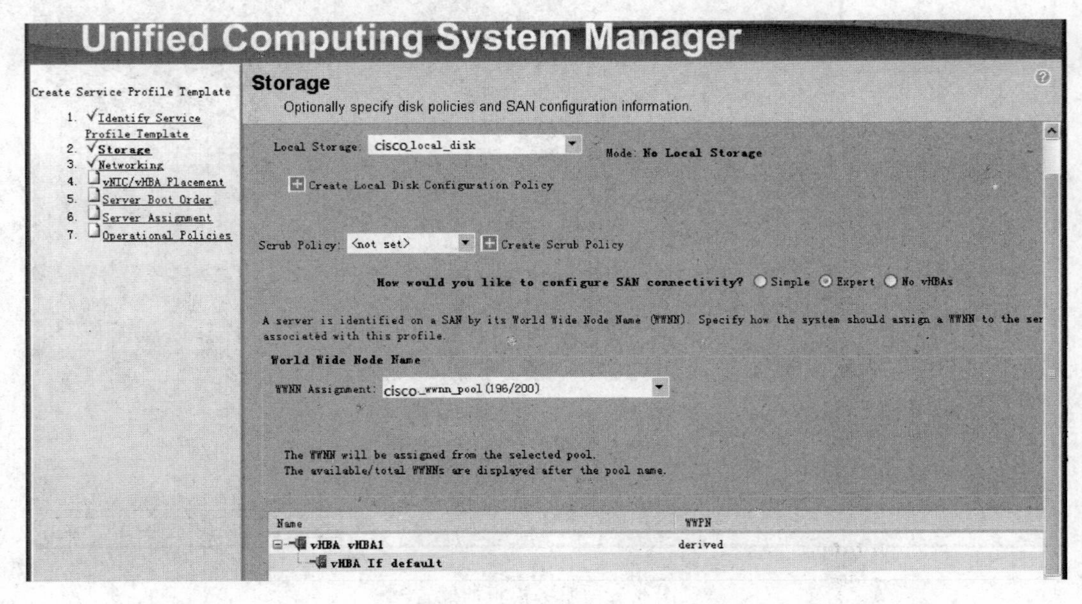

图 6-54　服务配置模板创建

在图 6-54 界面的“Local Storage”中选择刚才建立的策略“cisco_local_disk”，在下面的“WWNN Assignment”中选择在上节中创建好的 WWNN 池，名字是“cisco_wwnn_pool”。选中“How would you like to configure SAN connectivity?”中的“Expert”模式后，从图 6-55 添加 HBA 卡。

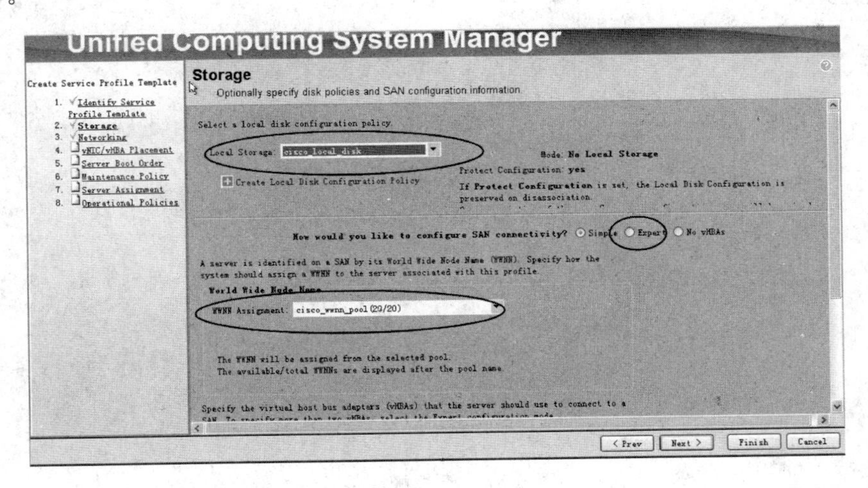

图 6-55 添加 HBA 卡

单击“add”，修改 HBA 相关参数，如图 6-56 所示。输入 HBA 卡的名字“vHBA1”，从“World Wide Port Name”中选择在前节创建好的 WWPN 地址池“cisco_wwpn_pool”，在下面的“Adapter Performance Profile”中的“Adapter Policy”中选择“VMWare”，以优化此 HBA 卡的性能，如图 6-56 所示。所有的选项配置完毕后单击“OK”，将返回到上一个界面，如图 6-57 所示。

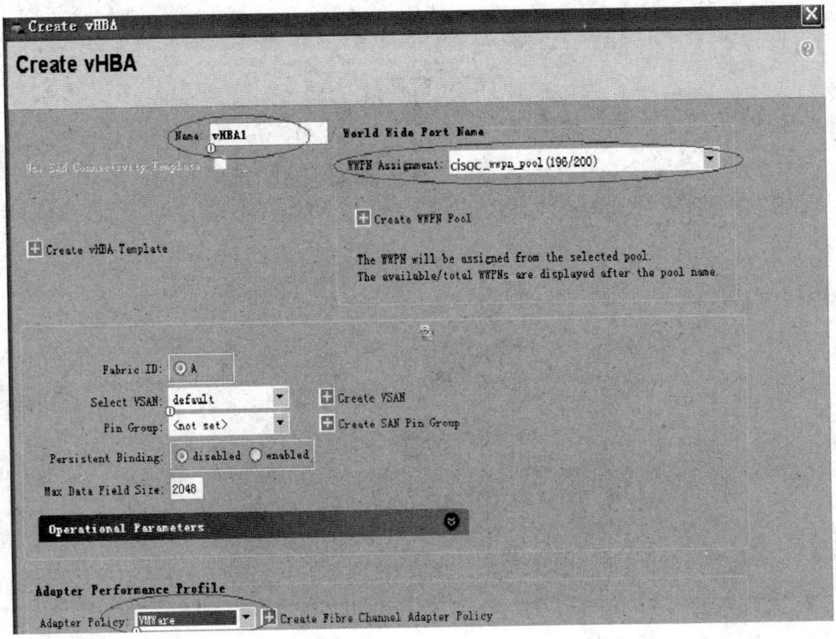

图 6-56 输入 HBA 卡参数

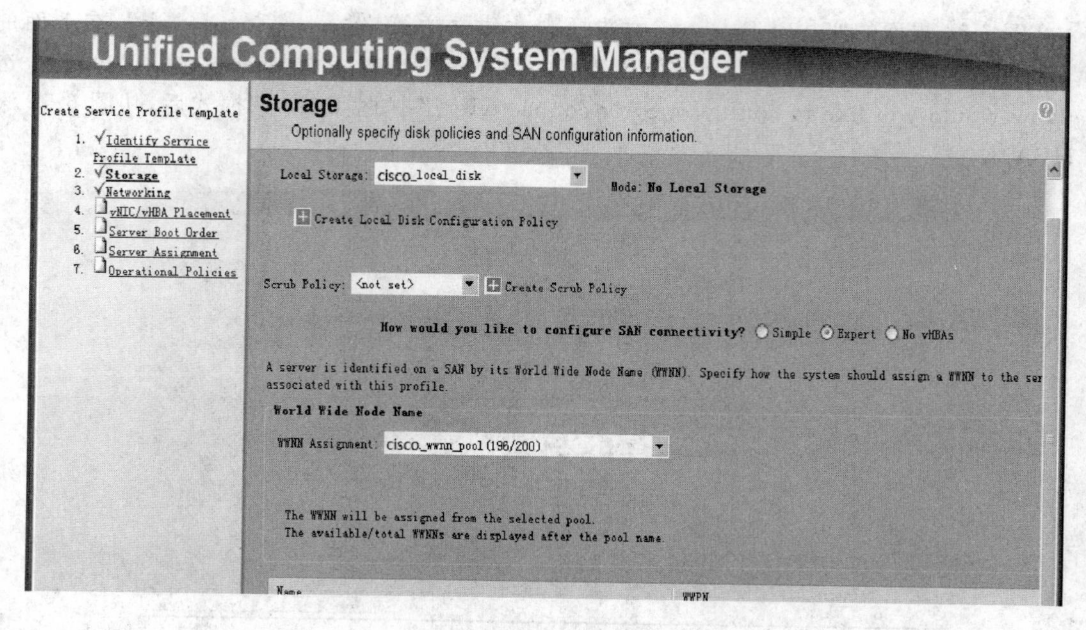

图 6-57 模板配置界面

在示例中,思科刀片配置一个 HBA 卡已经足够了,单击“Next”后将进行网卡的相关配置,网络的配置界面如图 6-58 所示。

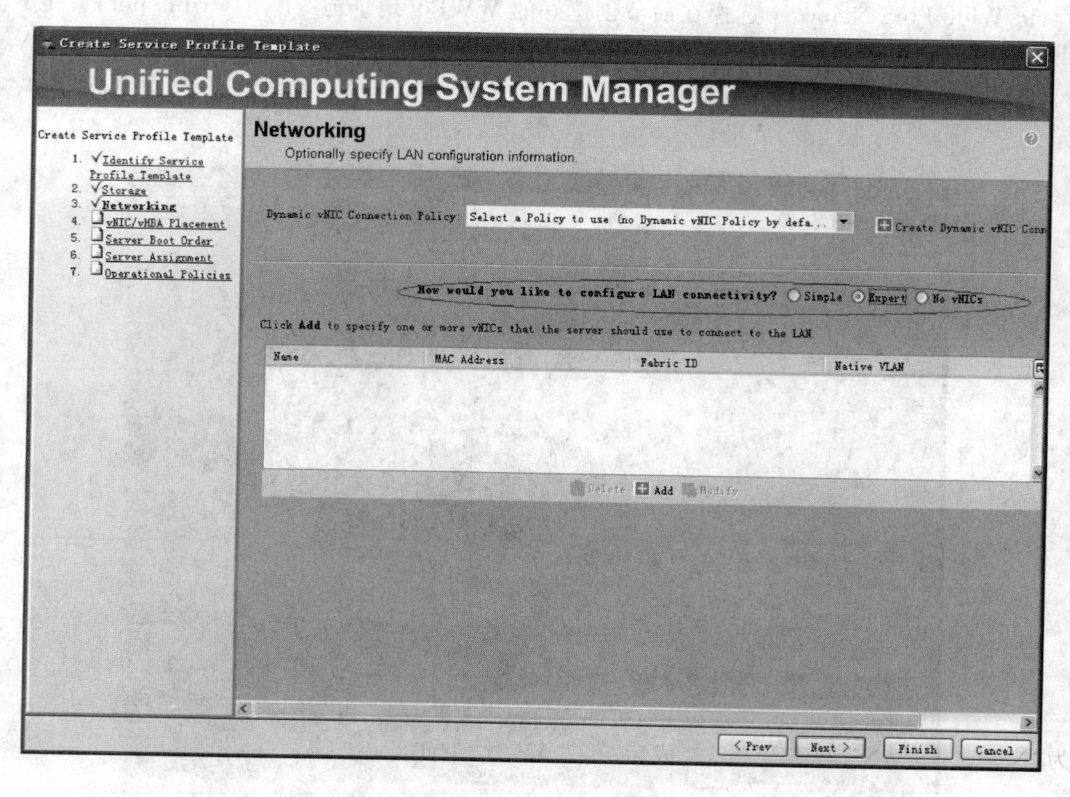

图 6-58 网络的配置界面

在网络的配置界面中，选择"How would you like to configure LAN connectivity?"中的"Expert"模式，单击"add"添加网卡，网卡的参数配置如图 6-59 所示。

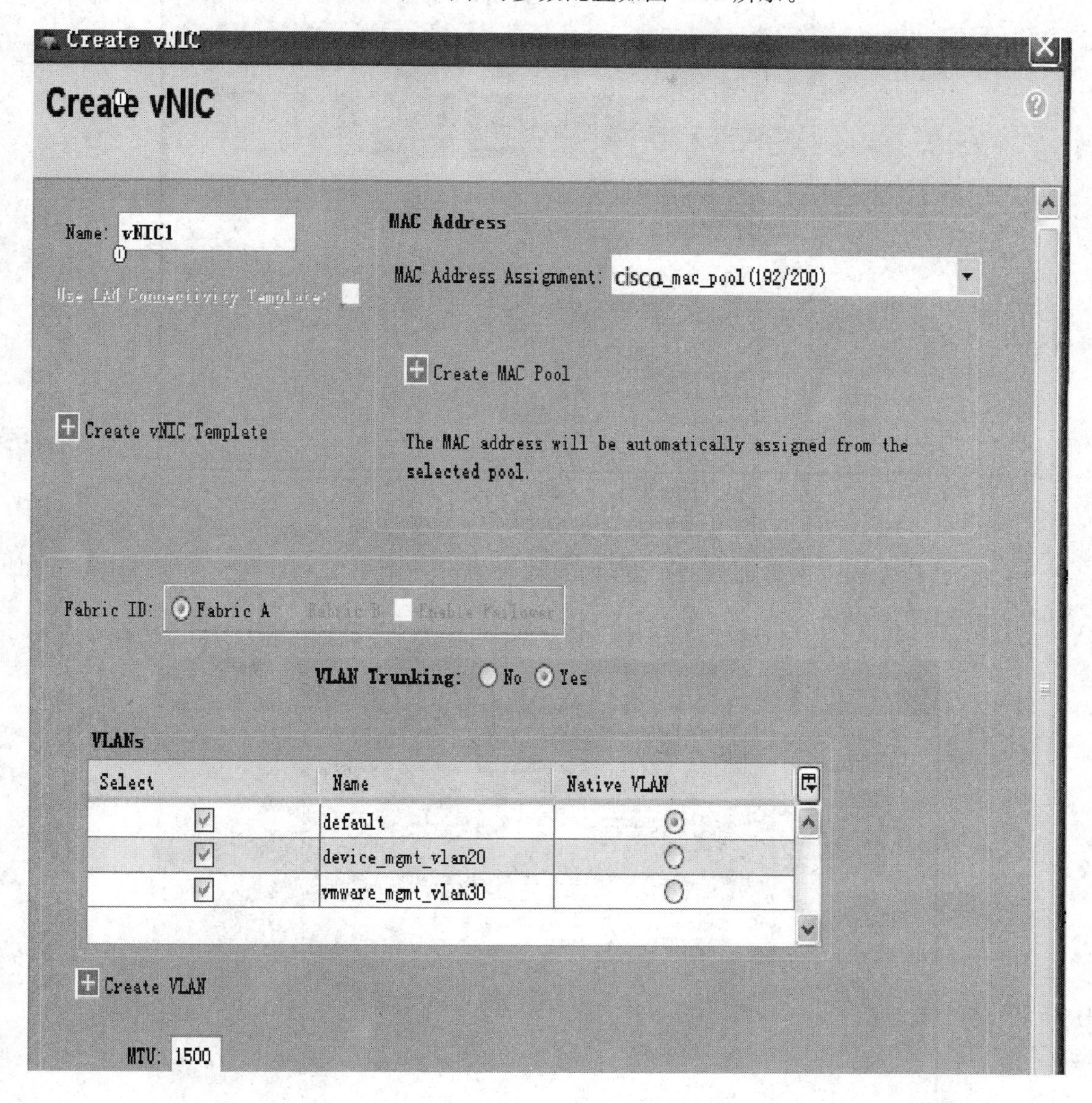

图 6-59　vNIC 配置参数

在图 6-59 的界面中，网卡的名字输入"vNIC1"，"MAC Address Assignment"选择在前节创建好的 MAC 地址池"cisco_mac_pool"，以便为创建的网卡自动分配 MAC 地址。在"VLAN Trunking"中选择"YES"，因为将来在思科刀片中部署的 VM 将可能属于好几个 VLAN，在下面的"VLANs"选项中选择在前节中已经创建好的 VLAN 号，并且选择"default"作为"Native VLAN"。全部选择完成后单击"OK"结束此部分配置。

利用相同的方式再创建第二块网卡，命名为"vNIC2"。全部完成后将看到已经配置好了的两块网卡，如图 6-60 所示。

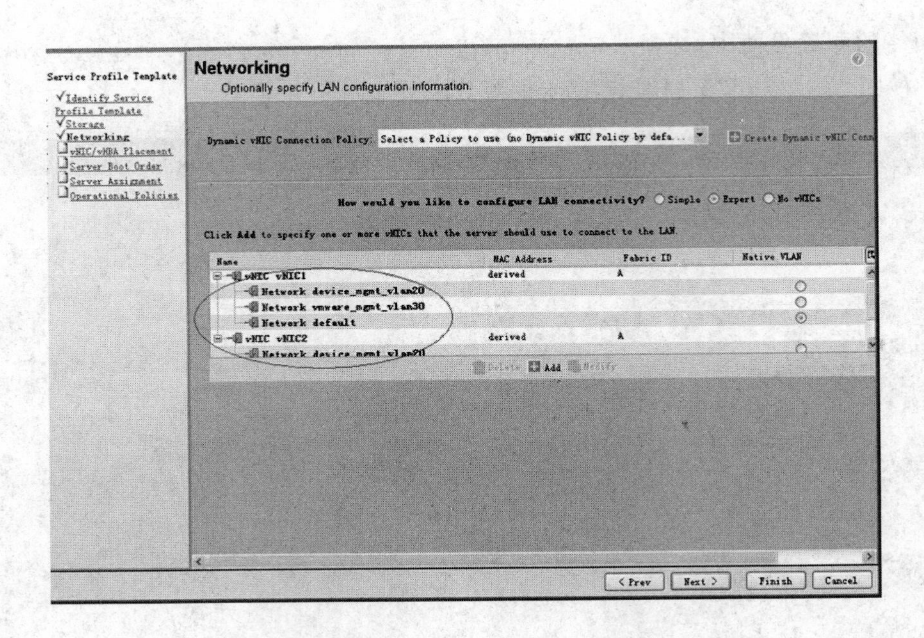

图 6-60　虚拟网卡配置完成

至此,已经完成了网络部分的配置,如图 6-61 所示。单击“Next”进行有关 HBA 卡和 NIC 位置放置的配置。

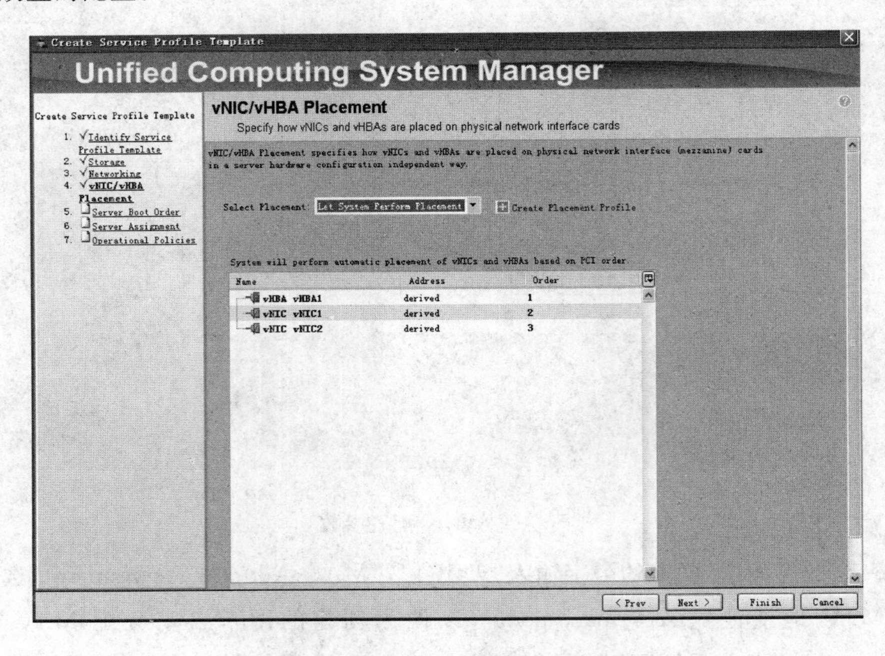

图 6-61　虚拟 HBA 卡和网卡都配置完成

在图 6-61 中已经默认将 HBA 卡放到了 Order 中的第一位,然后分别是两块网卡。此号码顺序是根据系统扫描 PCI 顺序来排列的,此处就保留这个配置不变即可。系统启动的时候将首先扫描到第一个设备也就是 HBA 卡,从而使之从 HBA 卡的启动成为可能,单击“Next”继续下面的配置。

下面的配置是有关启动顺序的配置，此部分的配置是本次配置的重点和关键。单击“Create boot policy”，建立一个启动顺序，如图 6-62 所示。

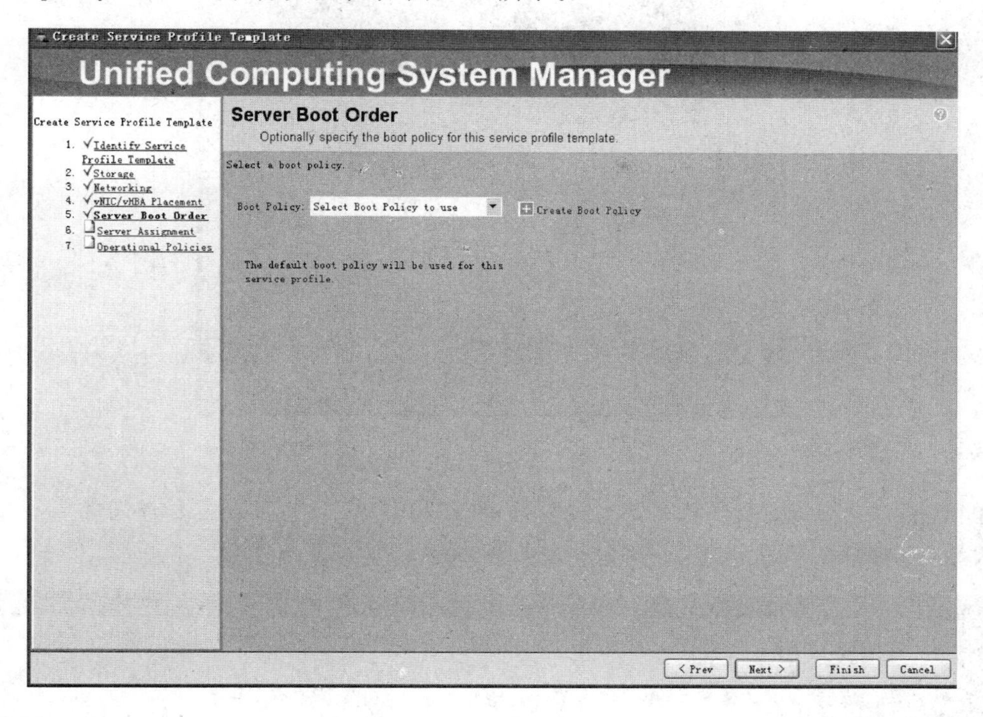

图 6-62　服务器引导顺序

输入规划中的名字“cisco_b_policy”，然后从“Local Devices”中选择“Add CD-ROM”，添加思科刀片启动的第一个设备“CD-ROM”，然后从 vHBAs 中选择 SAN BOOT 启动的相关选项(primary)，如图 6-63 所示。

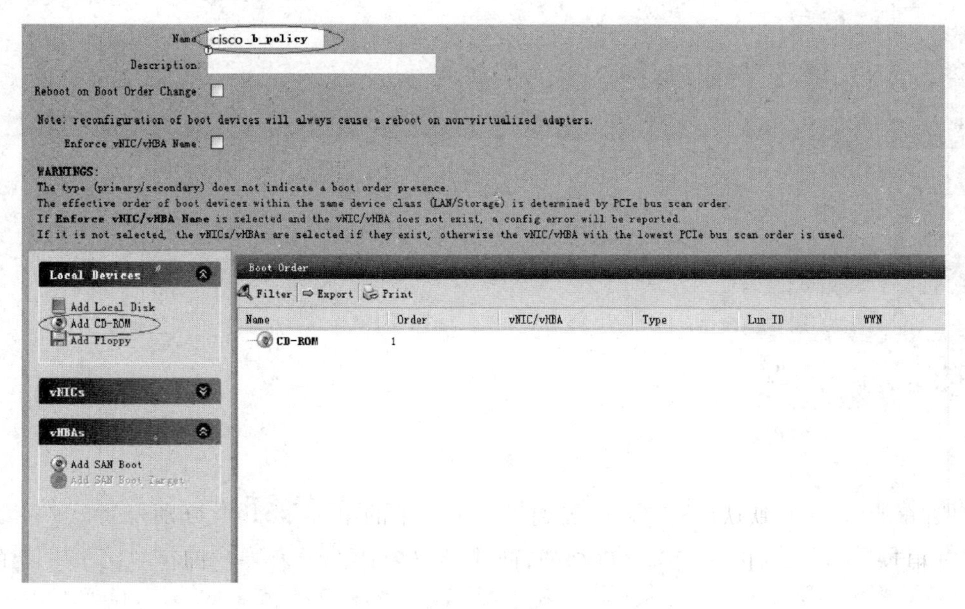

图 6-63　建立一个服务器启动顺序

单击“Add SAN Boot”后将出现如图 6-64 所示的界面，直接单击“OK”即可。

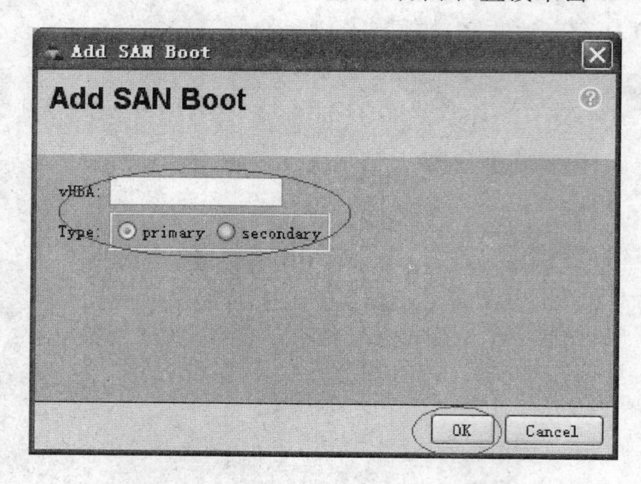

图 6-64 服务器引导顺序选项

单击左边的“Add SAN Boot Target”，选择 SAN BOOT 启动时需要查询的启动设备的 WWN 号后，如图 6-65 所示。

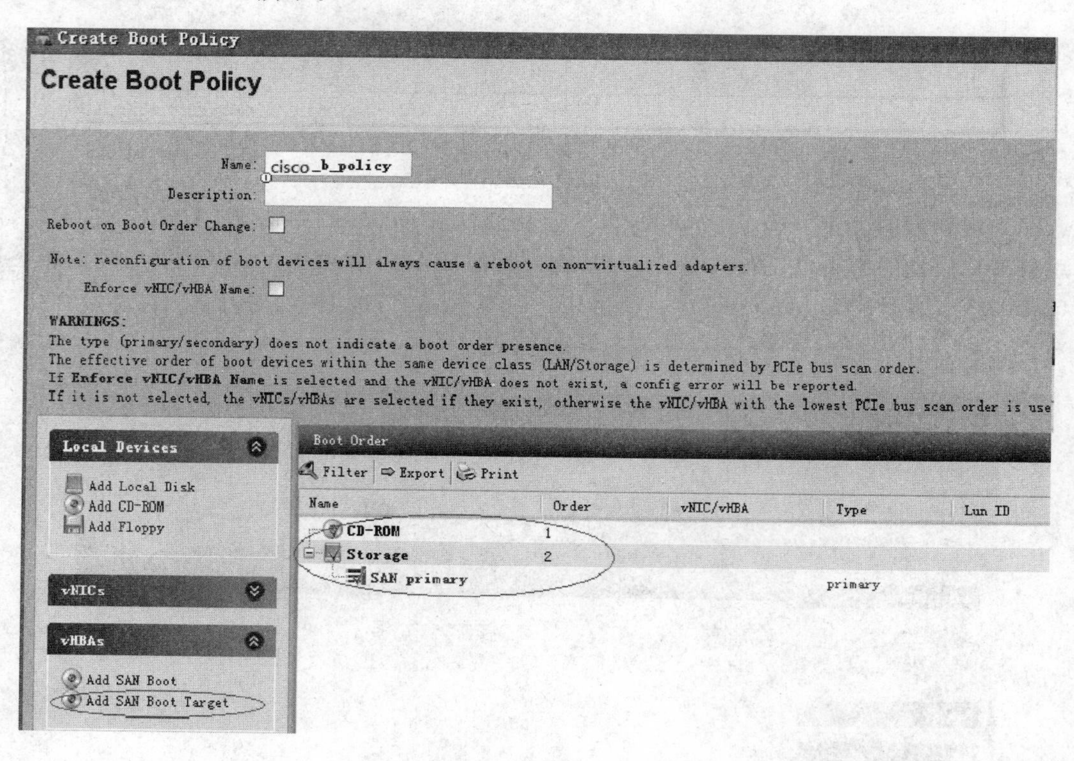

图 6-65 服务器引导顺序设置

说明：在此处将来 SAN BOOT 的时候其实是从 EMC 的 NS480 上的一个 LUN 启动的，那么选择目标的时候就应该选择 NS480 SP 上的 WWN 号码。NS480 上处于冗余的需求有两个 SP(SPA、SPB)，那么在此处需要添加两个 SP 的 WWN 号，已达到容错的目的。

在图 6-66 的配置界面中：

① Boot Target LUN 号码必须选择“0”;

② Boot Target WWPN 可以从 SAN 交换机上获取到,具体的获取方式将在 SAN 交换机配置章节做详细的说明,在此处将之先输入进去。

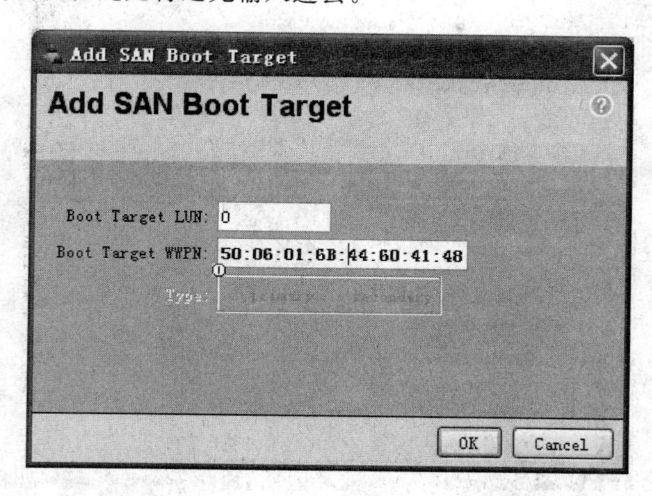

图 6-66　SAN 引导的界面

单击“OK”,配置完毕后的界面如图 6-67 所示。

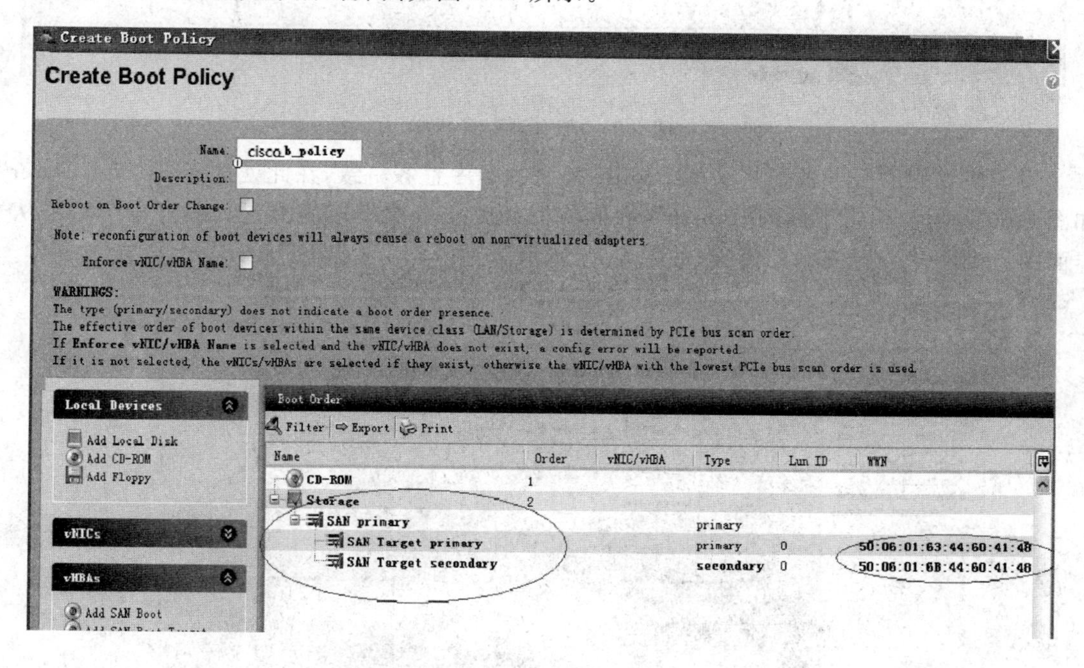

图 6-67　创建引导策略

从图 6-67 中看到已经配置好了启动信息及相关的目标 WWN 地址。单击“OK”后将返回到上一个界面。

从“Boot Policy”中选择刚才建立好的策略名字“cisco_b_policy”。至此完成了有关启动顺序及 SAN BOOT 的配置,单击“Next”进行下一步的配置,如图 6-68 所示。

在下一步的配置中,将配置此模板与物理服务器的关联情况。

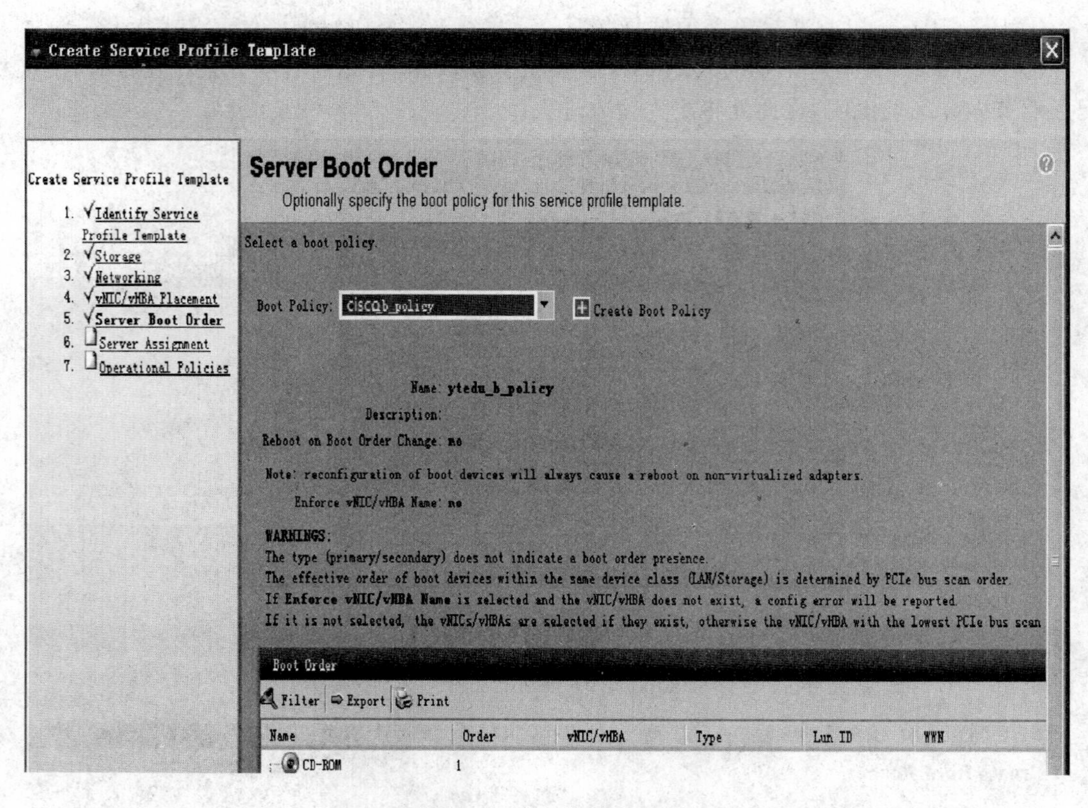

图 6-68 服务器引导顺序

为了使物理的服务器与将来的 Profile 关联的顺序整齐一致，在此处选择“Assign Later”，如图 6-69 所示。单击下面的“Finish”，我们将彻底完成本次模板的配置。配置完毕后可从主界面中看到刚才配置完毕的模板，如图 6-70 所示。

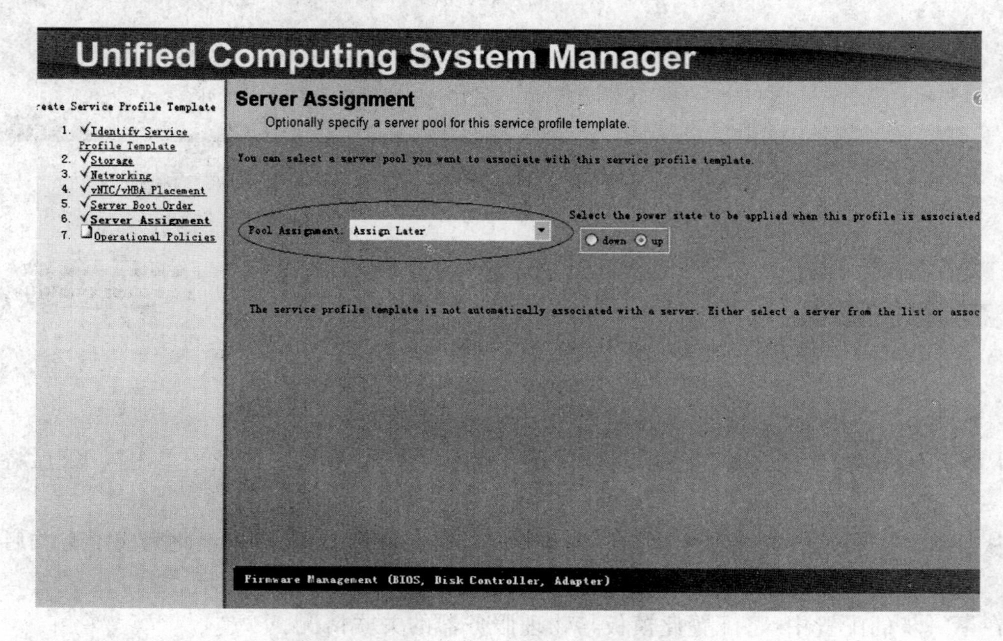

图 6-69 配置此模板与物理服务器的关联

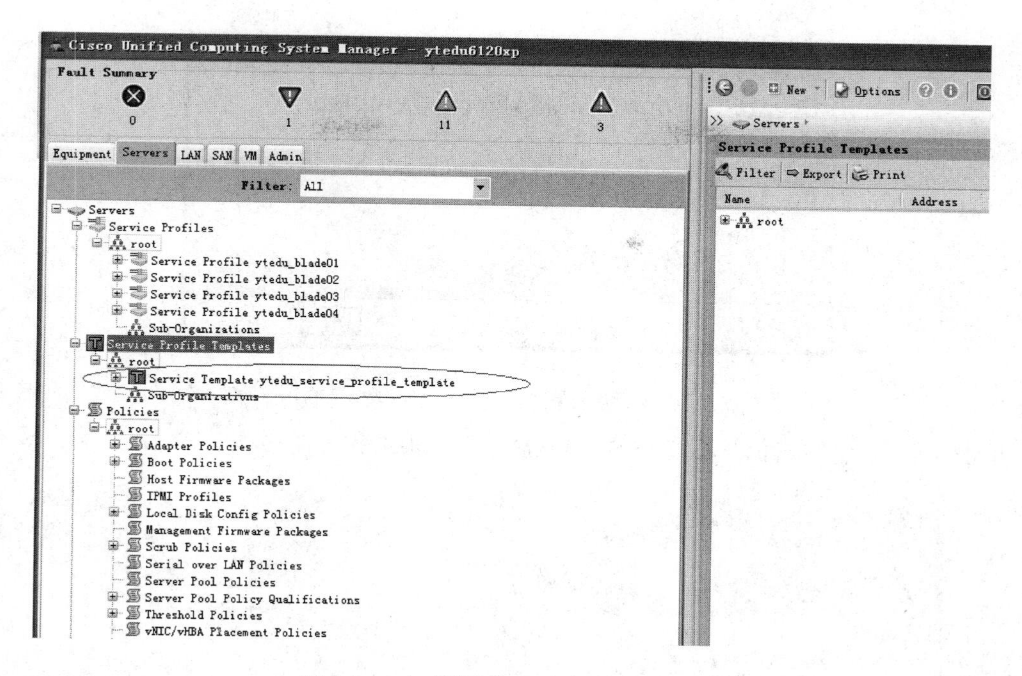

图 6-70　完成模板与物理服务器的关联

6.3.7　创建 Service Profile 并关联到刀片服务器

创建相应数量的服务配置文件，并关联到各刀片服务器。

在“Cisco Unified Computing System Manager”窗口，右键单击上一节创建的模板“cisco_service_profile_template”，选中“Create Service Profiles From Template”，如图 6-71 所示。

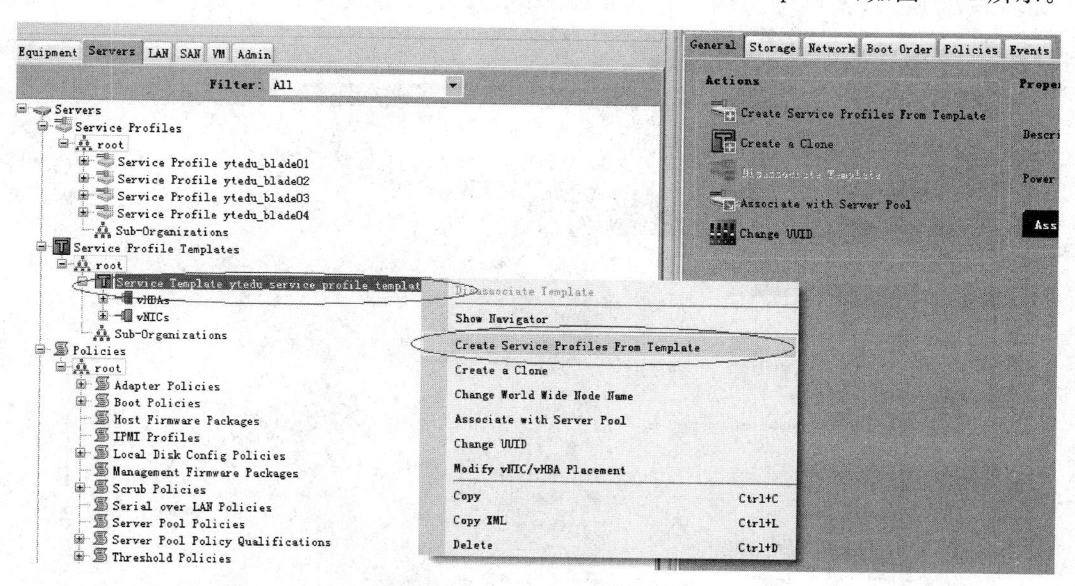

图 6-71　从模板创建服务文件

在“Create Service Profiles From Template”对话框中的“Naming Prefix ”项，输入“cisco-lbade0”，在“Number”项中，输入 4(因为有 4 台刀片服务器)，单击“OK”，如图 6-72 所示。

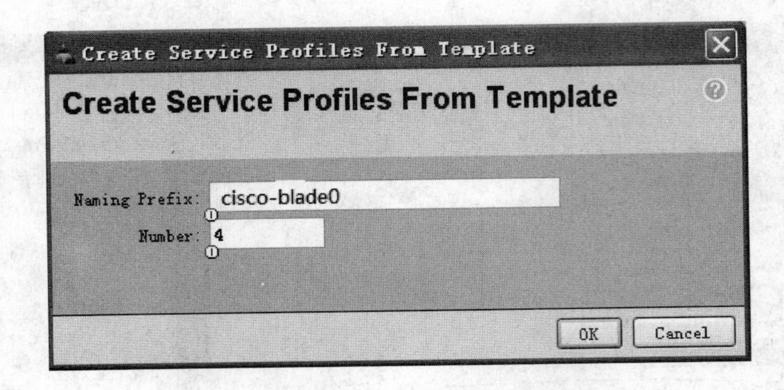

图 6-72　从模板创建服务文件(续上图)

单击“OK”后将自动的生成 4 个 Service Profile,如图 6-73 所示。

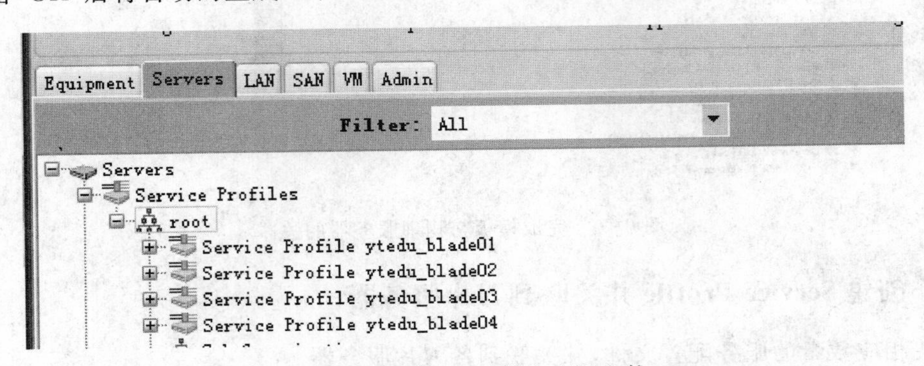

图 6-73　创建的 4 个服务配置文件

接下来将这些创建好的 Service Profile 关联到物理服务器,以第一个 Service Profile 为例,如图 6-74 所示,选中左边的“Service Profile ytedu_blade01”后,再选中右边的“Change Service Profile Association”。

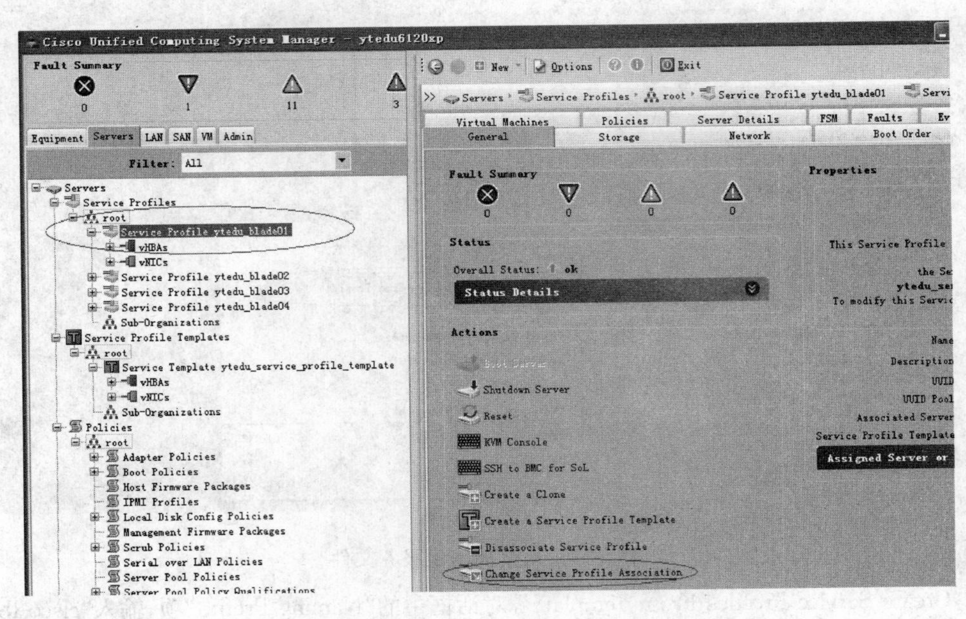

图 6-74　将服务文件关联到物理服务器

在“Associate Service Profile”界面选择“Select existing Server”，单击“OK”，如图 6-75 所示。

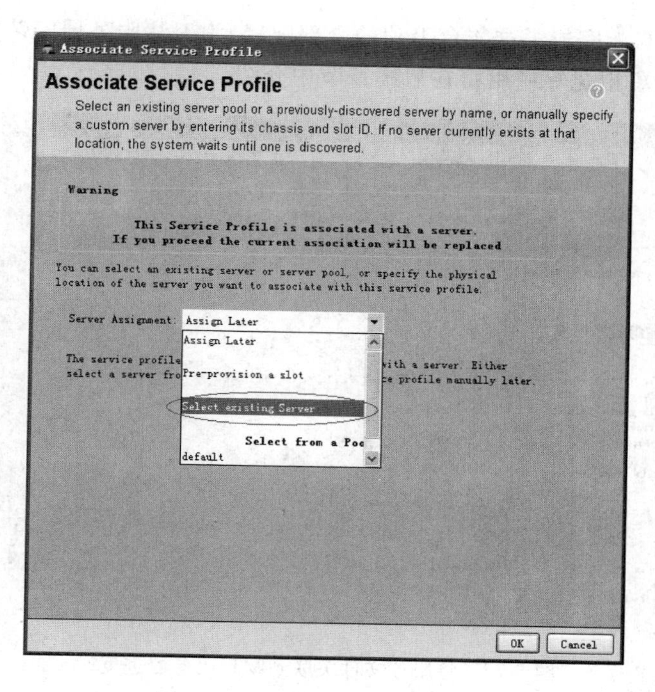

图 6-75　选择需要关联的服务器

从有效的服务器中选择第一台物理服务器，如图 6-76 所示，完毕后单击“OK”，此 Service Profile 将开始与此思科物理刀片服务器相关联，此关联过程需要一段时间。可利用 FSM 来查看进度，如图 6-77 所示。

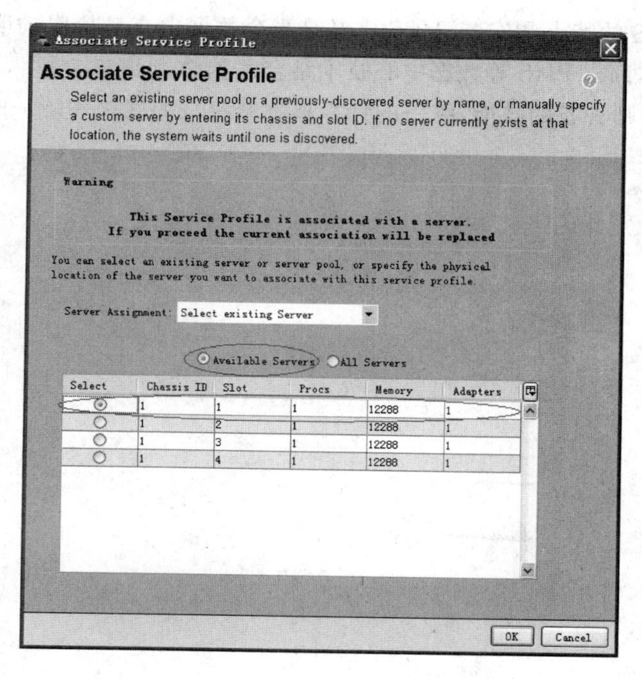

图 6-76　将服务文件关联到物理服务器

在其进度过程中，可按照相同的步骤完成其他服务器的关联工作，此进度需要一点时间耐心等候直到状态到达了 100%。

至此已经在 6120 上完成了这 4 台刀片的全部配置，为刀片服务器安装操作系统已经做了介绍，配置 VMware 虚拟化等应用将在后续章节中介绍。

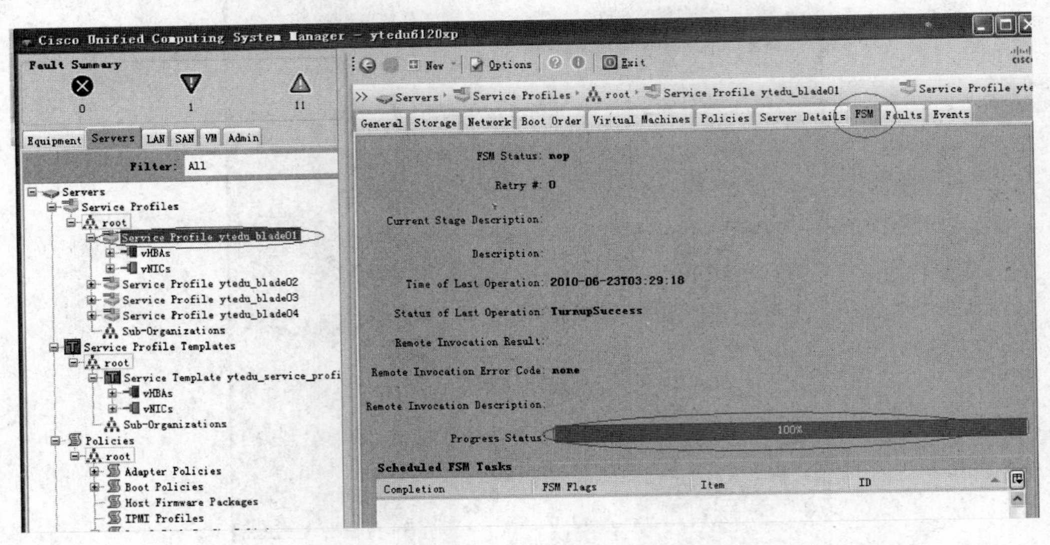

图 6-77 将服务文件关联到物理服务器进度

6.4 小 结

本章介绍了 UCS 的特点和安装，UCS 已经是当今数据中心最抢眼的明星，当然其卓越的性能要与 VN-LINK，FraicPath 等数据中心技术结合才能体现出来，对 UCS 的安装使用已经是一个数据中心工程师必备的技能，统一的计算离不开统一的网络和统一的存储。

第 7 章　服务器虚拟化安装与配置

7.1　实验环境

数据中心很难离开 VMware 的虚拟化产品，下面以笔者的笔记本电脑安装的环境介绍如何搭建一个基本的测试环境，电脑软硬件配置如下：ThinkPad X230i，16G 内存，500G 硬盘，Intel 酷睿 i3，操作系统 Windows 7。电脑里面安装 VMware WorkStation 8，在 VMware WorkStation 8 里，用一个虚拟机安装 Openfiler 版本 2.9 模拟共享存储，另外两个虚拟机模拟 ESXi 主机，测试版本是 VSphere 5.1，第四个虚拟机安装 Windows Server 2008，在 2008 Server 里面安装 vCenter，就可以在电脑上模拟 vMotion 了，完全达到学习的目的，如图 7-1 所示。

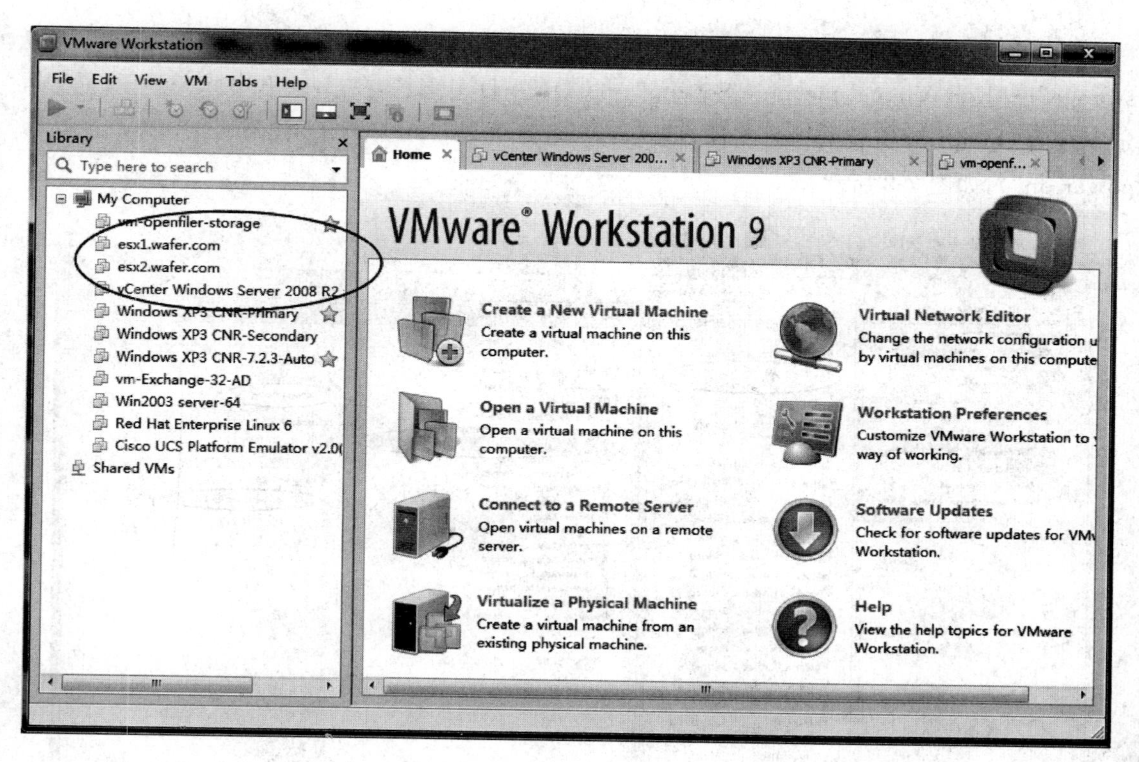

图 7-1　虚拟化测试需要安装的项目

7.2 虚拟存储安装

7.2.1 Openfiler 介绍

Openfiler 能把标准 x86/64 架构的系统变成一个强大的 NAS 存储、SAN 存储和 IP 存储网关，为管理员提供一个强大的管理平台，并能应付未来的存储需求。如 VMware，VirtualIron 和 Xen 服务器虚拟化技术，Openfiler 也可部署为一个虚拟机实例。Openfiler 这种灵活高效的部署方式，确保存储管理员能够在一个或多个网络存储环境下使系统的性能和存储资源得到最佳的利用和分配。Openfiler NAS/SAN Appliance，version 2.99 (Final Release)下载地址：https://downloads.sourceforge.net/project/openfiler/openfiler-distribution-iso-2.99-x64/openfileresa-2.99.1-x86_64-disc1.iso。

有关 Openfiler 的功能介绍和详细说明请参考其官方网站。

7.2.2 Openfiler 安装与配置

在 VMware Workstation 中，首先新建一个虚拟机，依次单击该虚拟机的“vm-openfiler-storage”→“Edit Virtual machine Setting”→“CD/DVD(IDE)”，在“Use IOS image files”中指定下载的 Openfiler 安装文件的路径，重启 VM 时就可以自动安装了(注意：选中“Connect at power on”)，如图 7-2 所示。

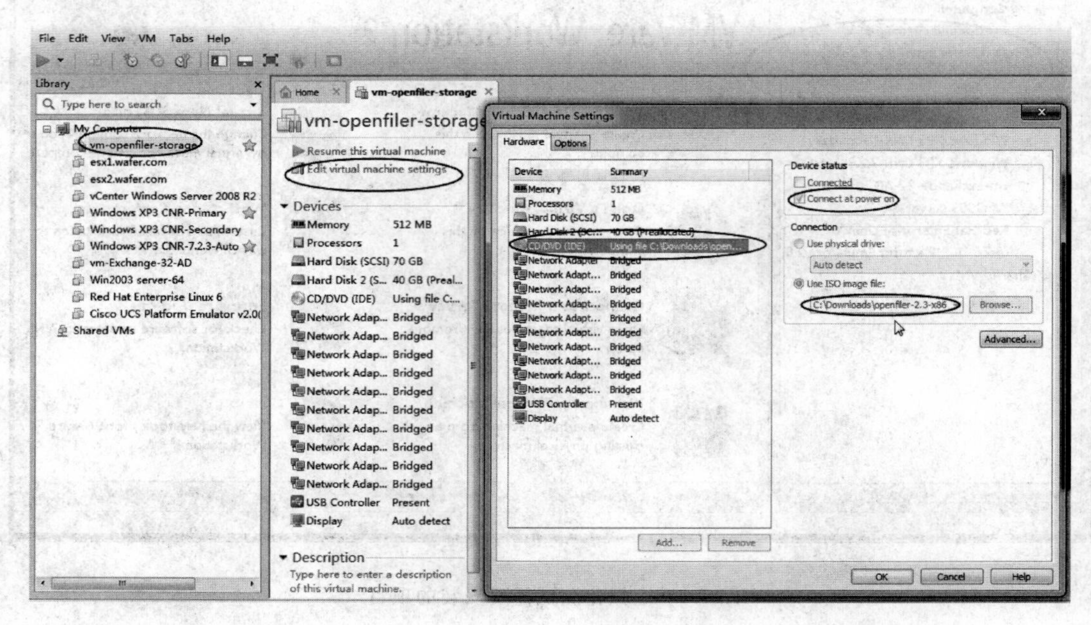

图 7-2 Openfiler 安装完成的截图

Openfiler 安装程序有图形和文本两种安装模式，这里选择图形安装模式。

安装完成后就是配置任务了，打开 IE 浏览器，访问 Openfiler 网页管理界面地址：https://x. x. x. x：446/（x. x. x. x 为安装 Openfiler 时设置的地址，缺省账号：openfiler/password）。

在 Web 页面下完成后续的配置任务，Openfiler 登录界面如图 7-3 所示。

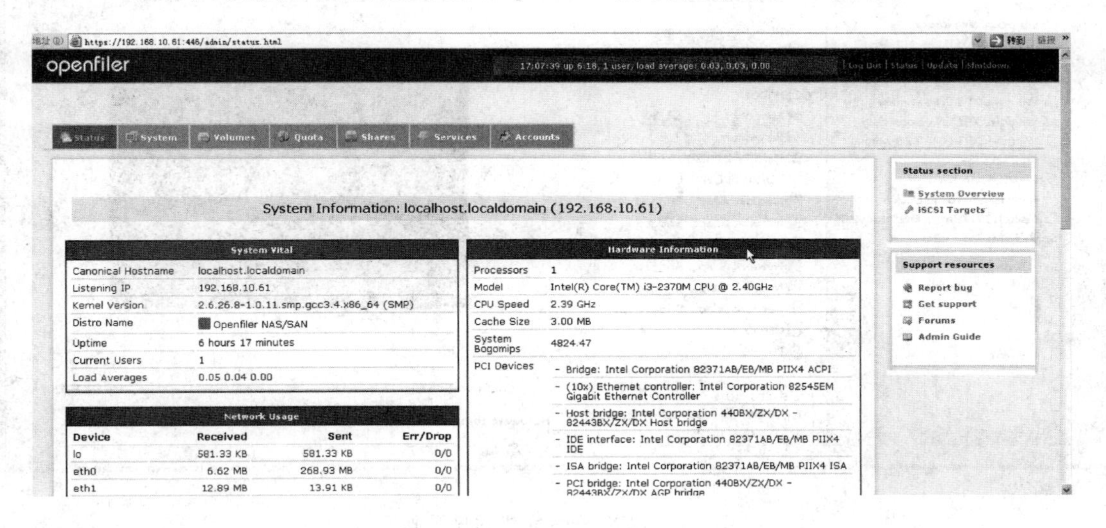

图 7-3　Openfiler 登录界面

由于这一部分不是本书重点，请读者自行参照 Openfiler 官方网站完成后续安装与配置步骤。

7.3　ESXi 安装和配置

1. 系统要求

确保主机符合 ESXi 5.0 以上支持的最低硬件配置和系统资源，要安装和使用 ESXi 5.0，硬件和系统资源必须满足下列要求：

① ESXi 5.0 将仅在安装有 64 位 x86 CPU 的服务器上安装和运行；

② ESXi 5.0 要求主机至少具有两个内核；

③ ESXi 5.0 仅支持 LAHF 和 SAHF CPU 指令；

④ ESXi 至少需要 2 GB 的物理 RAM；

⑤ 要支持 64 位虚拟机，x64 CPU 必须能够支持硬件虚拟化；

⑥ 一个或多个千兆或 10 GB 以太网控制器；

⑦ 一个或多个基本 SCSI 控制器或 RAID 控制器的任意组合。

交互方式安装就是通过 ESXi 安装光盘进行安装。

在“VMware Workstation”窗口中，在虚拟机“esx1. wafer. com”页面中，单击“Power on this virtual machine”，如图 7-4 所示。虚拟机将从 ESXi 安装光盘启动，如图 7-5 所示。

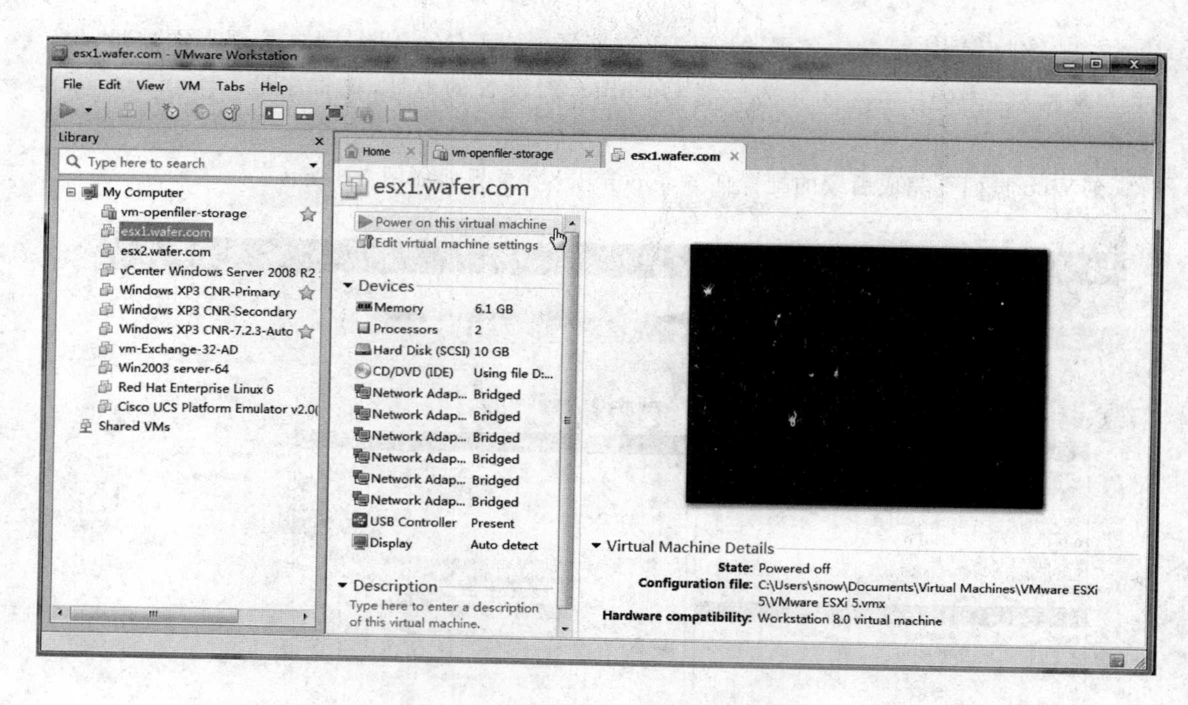

图 7-4　ESXi 安装完毕的显示画面

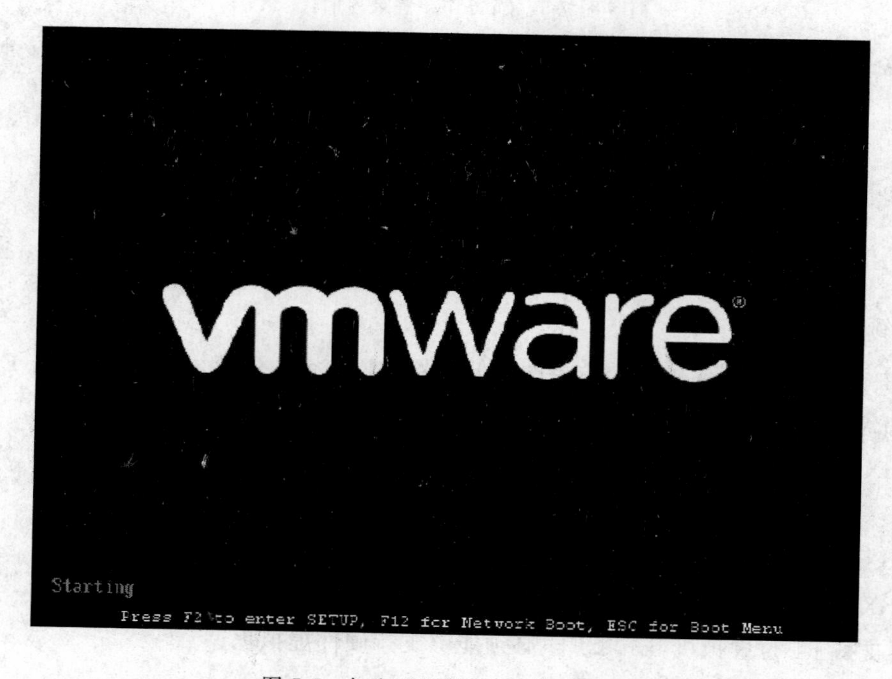

图 7-5　加电要安装 ESXi 的虚拟机

启动 ESXi 安装光盘引导菜单，默认选择了“ESXi-5. 1. 0-799733-standard Installer”，按回车键或倒计时 8 s 进入，如图 7-6 所示。

打开“Loading ESXi installer”程序，按回车键或倒计时 5 s 引导，如图 7-7 所示。

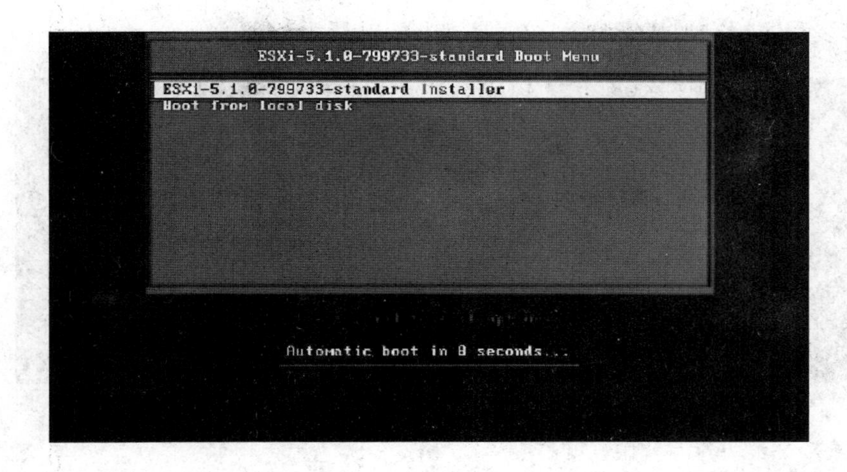

图 7-6　选择 ESXi 软件

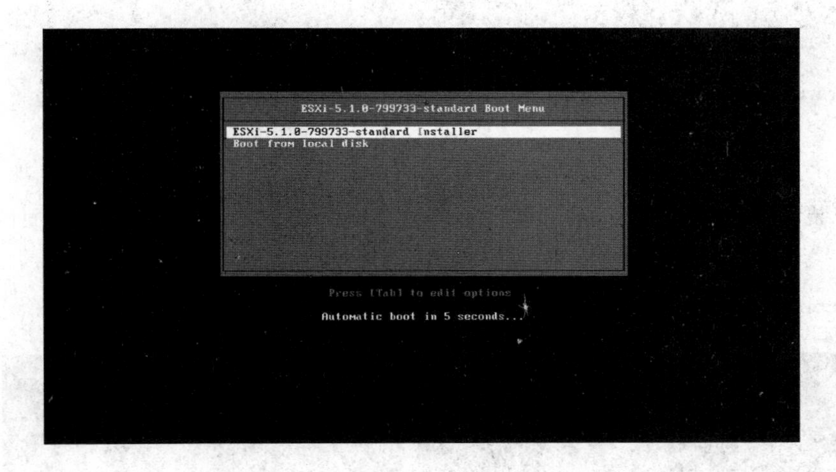

图 7-7　开始安装 ESXi

ESXi 安装程序正在调入安装程序文件，如图 7-8 所示。

图 7-8　ESXi 安装过程中

ESXi 安装程序正在加载安装程序模块，如图 7-9 所示。

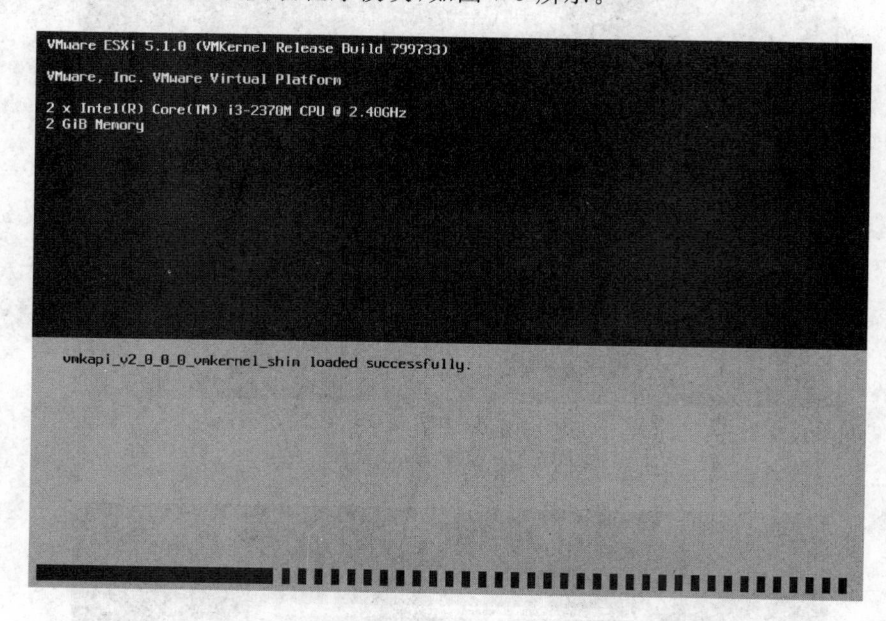

图 7-9 ESXi 安装过程中

ESXi 安装程序已成功启动，主机硬件不符合系统要求，是无法进入这一界面的，如主机上没有网卡。

在“Welcome to the VMware ESXi 5.1.0 Installation”页，按回车键继续，如图 7-10 所示。

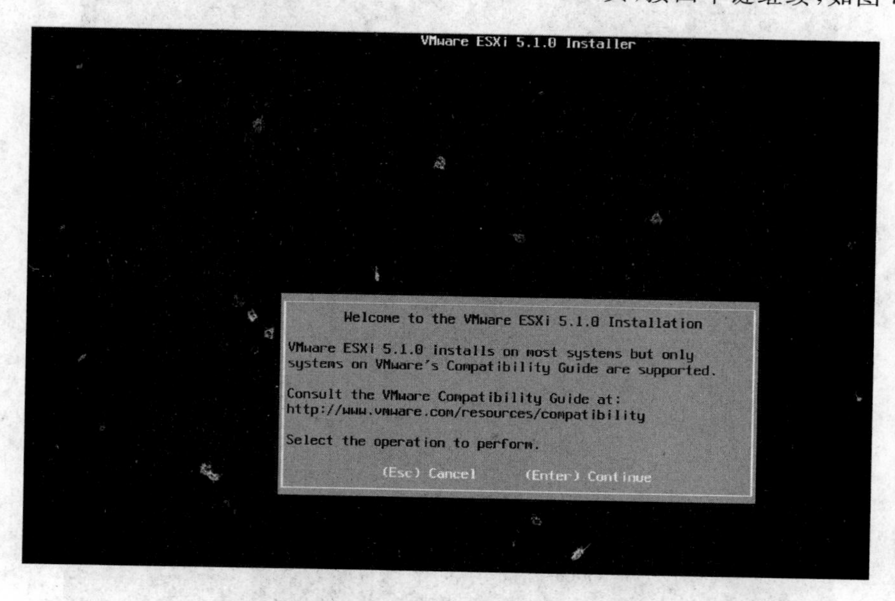

图 7-10 ESXi 安装过程中

在“End User License Agreement (EULA)”页，按 F11 键，同意用户许可协议并继续，如图 7-11 所示。

安装程序正在搜索可用的磁盘驱动器，如图 7-12 所示。

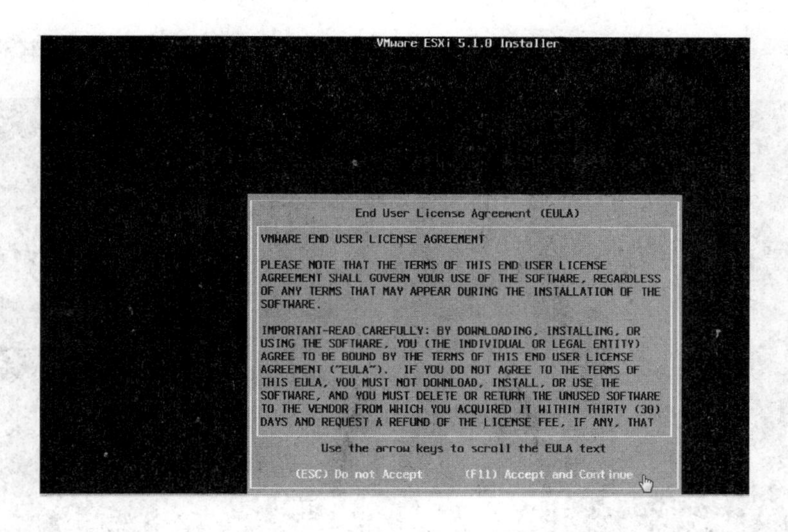

图 7-11　ESXi 安装过程中接受许可

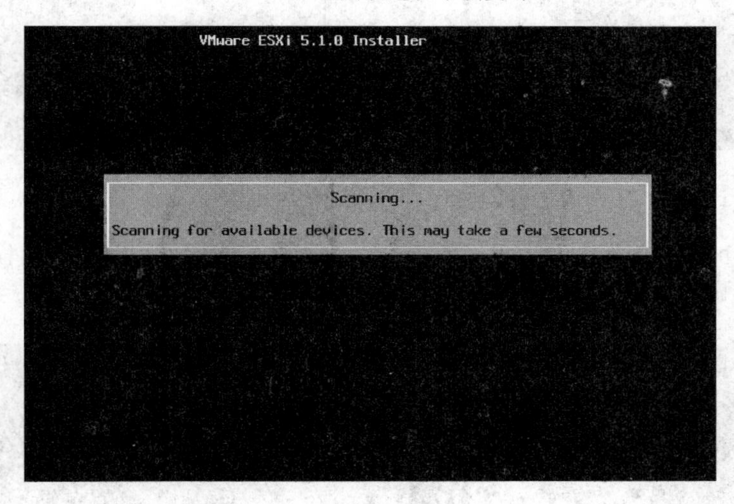

图 7-12　ESXi 安装过程中，搜索磁盘驱动器

安装程序找到了一个 40 GB 的 Local 本地存储驱动器，按回车键继续，如图 7-13 所示。

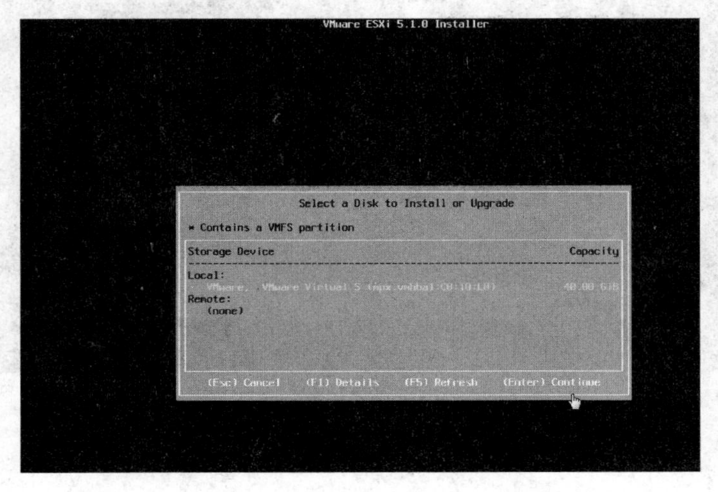

图 7-13　ESXi 安装过程中，发现本地磁盘驱动器

在“Please select a keyboard layout”页，已默认选择“US Default”，按回车键继续。

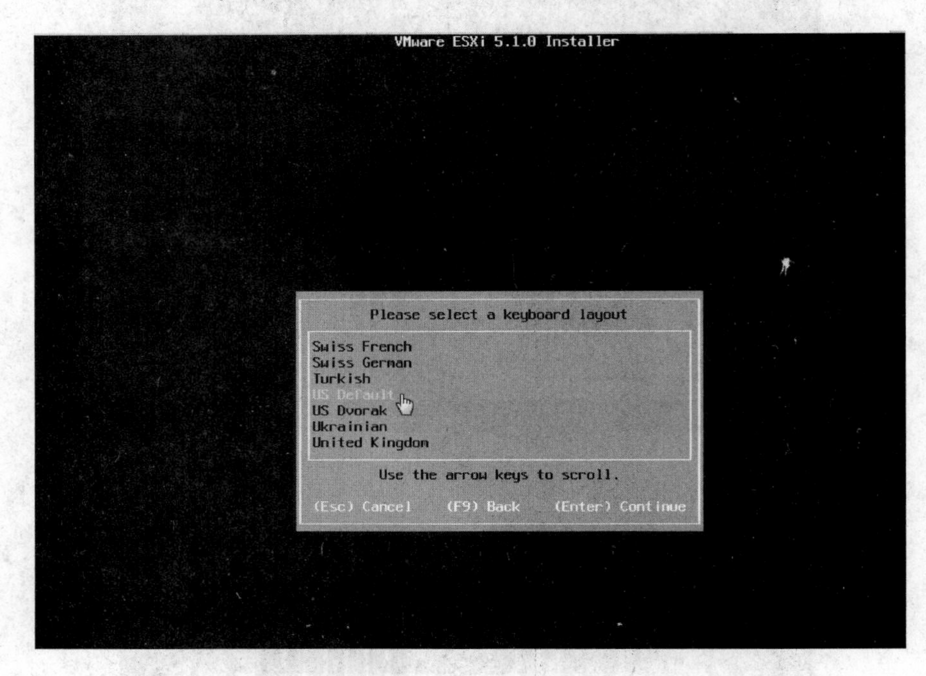

图 7-14　ESXi 安装过程中，选择键盘标准

在“Please enter a root password”页，输入 root 用户的密码，密码要求不少于 7 个字符，并满足复杂性规则，按回车键继续，如图 7-15 所示。

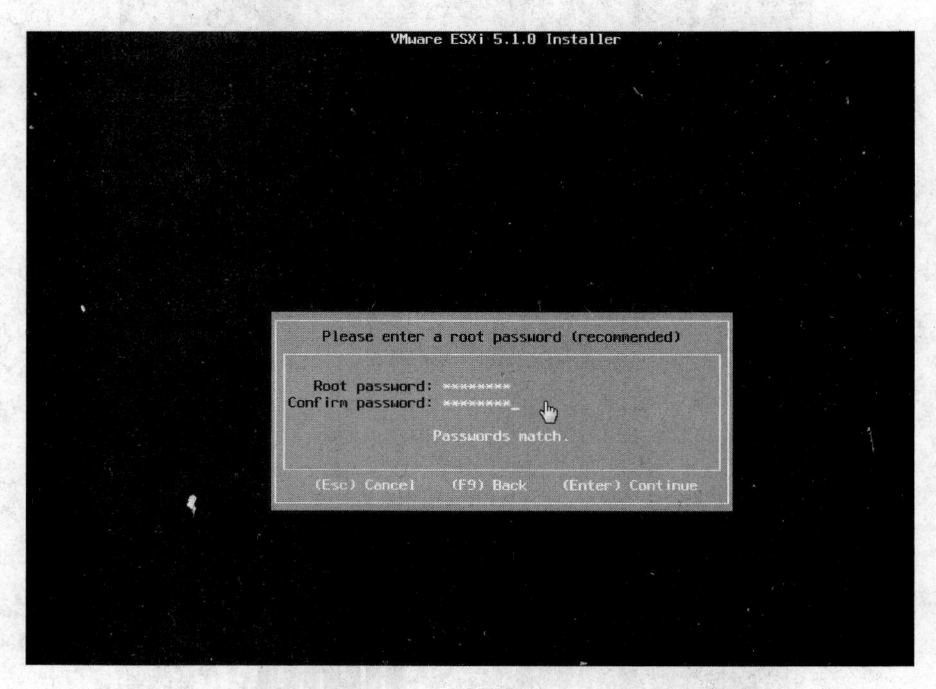

图 7-15　ESXi 安装过程中，输入管理员账号

安装程序正在收集系统信息，如图 7-16 所示。

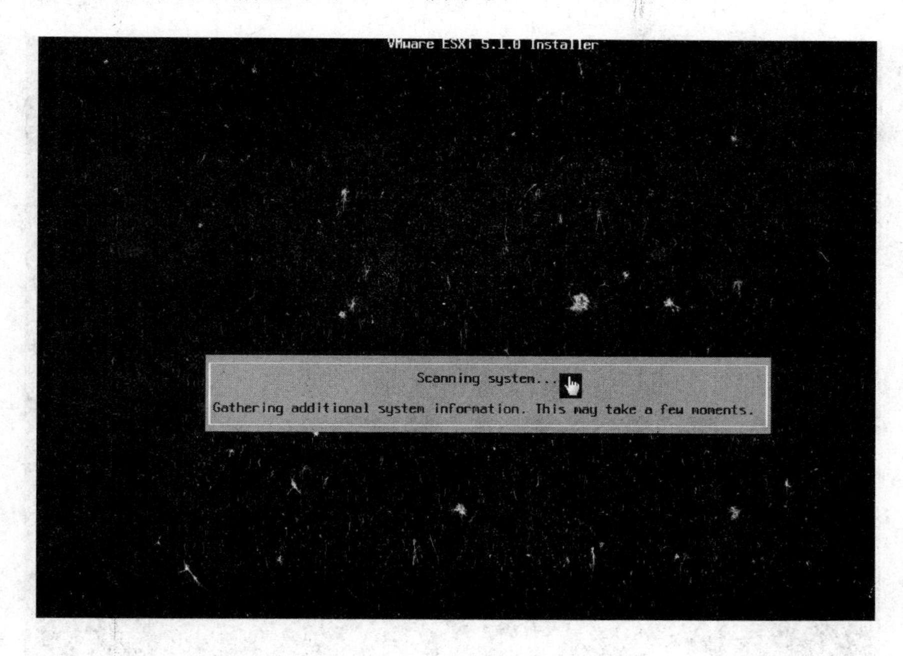

图 7-16 ESXi 安装过程中，收集信息

在"Confirm Install"页，显示 ESXi 将要安装的目标驱动，按回车键开始安装，如图 7-17 所示。

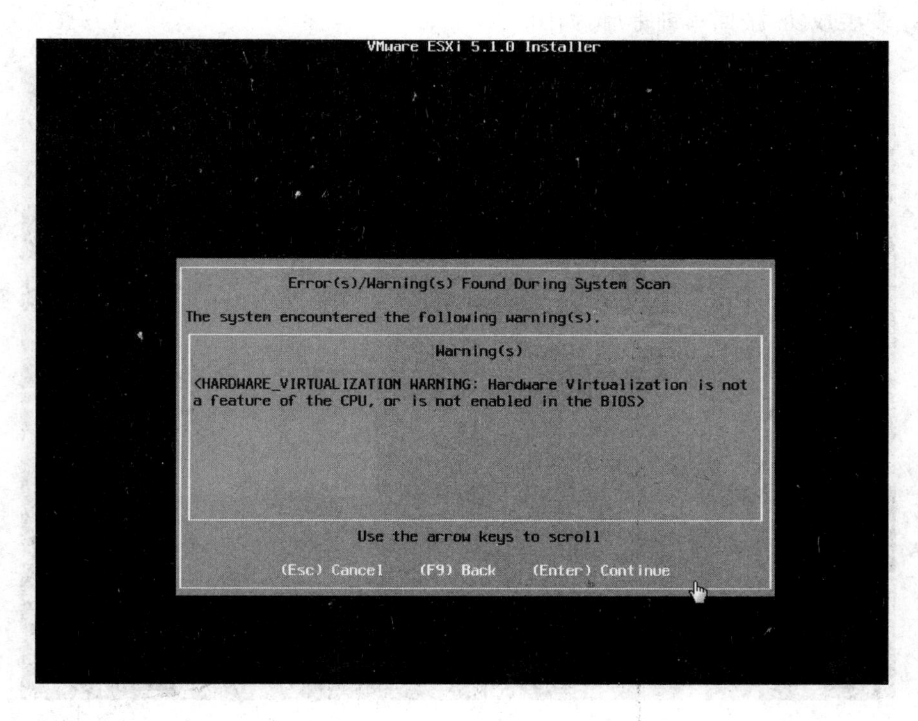

图 7-17 ESXi 安装过程中，选择目标驱动

如图 7-18 所示，ESXi 正在安装，可查看安装进度。

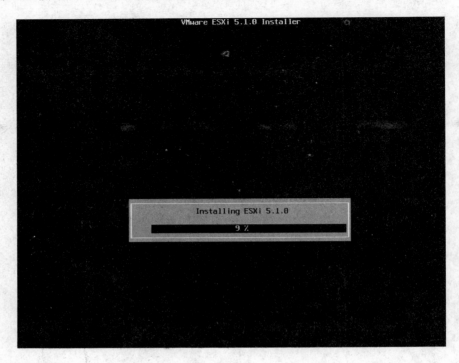

图 7-18　ESXi 安装过程中

ESXi 安装成功，按回车键重启，如图 7-19 所示。

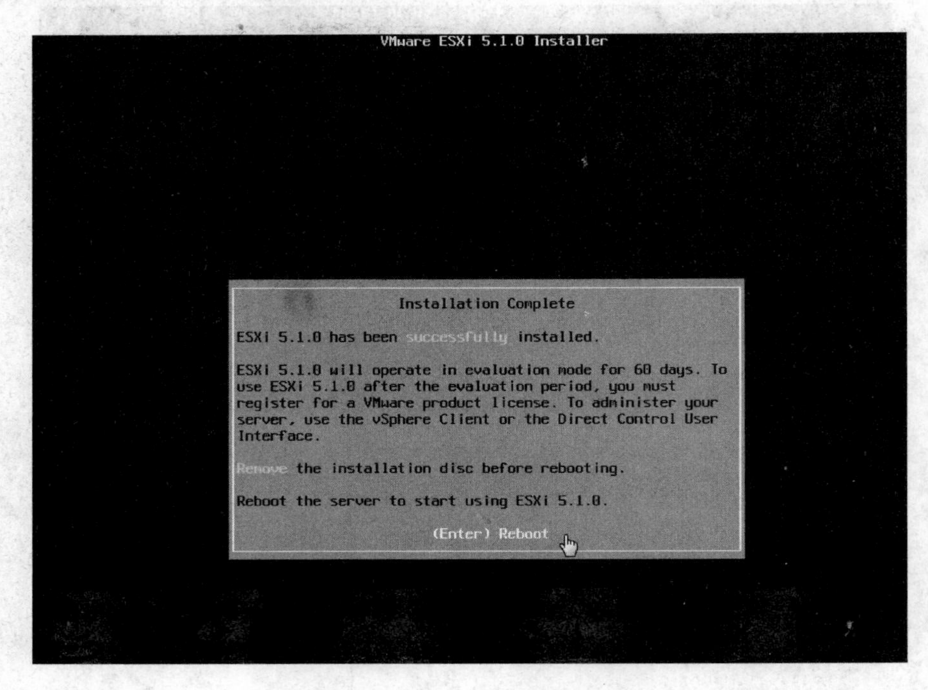

图 7-19　ESXi 安装过程完成，回车重启。

系统正在重启，如图 7-20 所示。

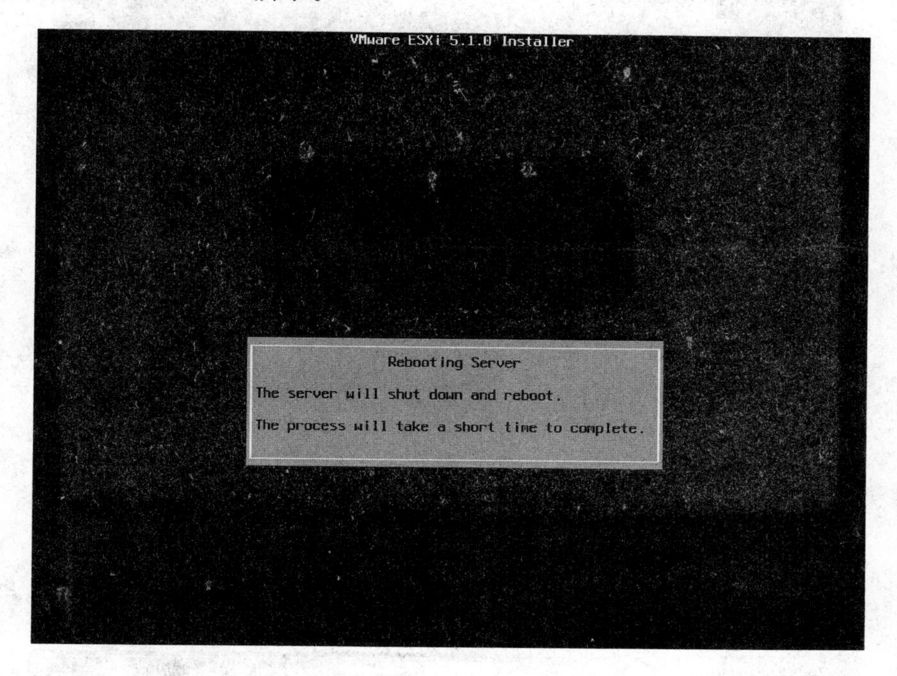

图 7-20　ESXi 安装过程后重启

2. 配置 ESXi

在 ESXi 主界面按 F2 键进入系统配置，如图 7-21 所示。

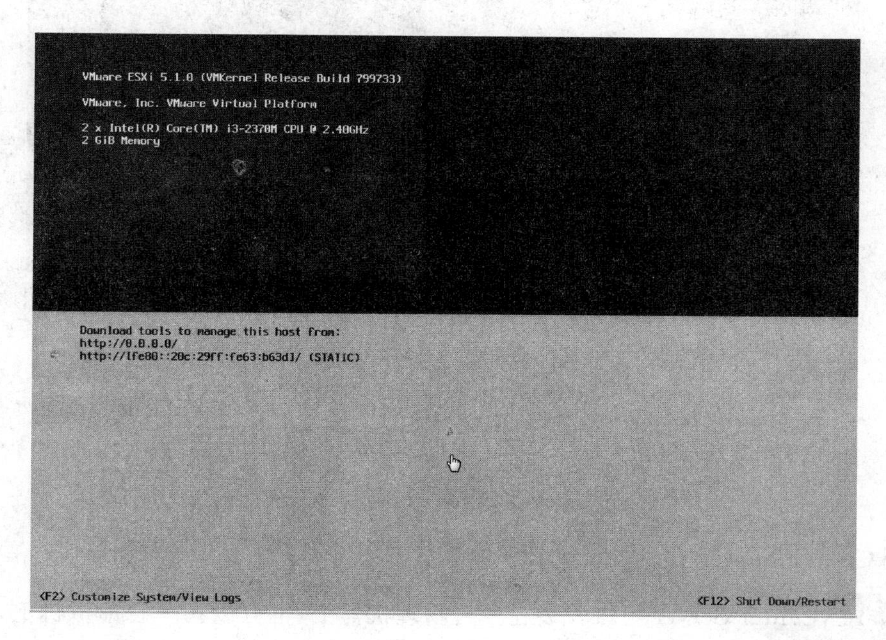

图 7-21　按 F2 进入 ESXi 配置

在“Authentication Required”页，输入登录的用户名和密码，按回车键确定，如图 7-22 所示。

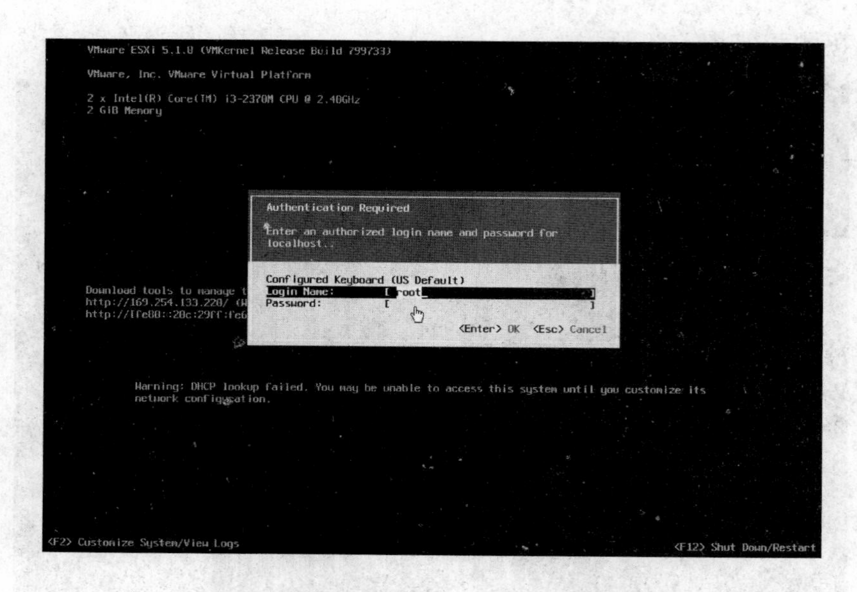

图 7-22　ESXi 登录账号

打开“System Customization”配置界面，在这里有众多的配置选项栏，如图 7-23 所示。

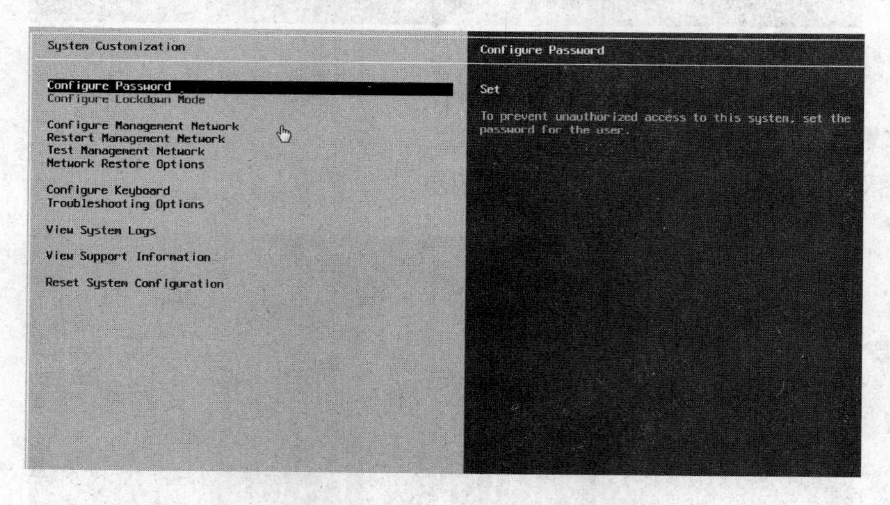

图 7-23　ESXi 系统配置界面

7.4　vCenter 安装和使用

7.4.1　vCenter Server 的基本要求

VMware vCenter Server 可以安装在一台物理机器上，也可以在一台虚拟机上，无论安装在哪里，都要满足 vCenter Server 以下两方面的要求。

(1) vCenter Server 硬件要求：两个 64 位 CPU 或一个 64 位双核 CPU，CPU 主频为 2.0 GHz 或 2.0 GHz 以上；4 GB 内存，如果数据库运行在同一台计算机上，则对内存的要求更高；4 GB 的存储空间；千兆的网络。安装 vCenter Server 主要看内存的大小，内存的大小影响可以控制虚拟机的台数。小清单(主机 1～100 台或虚拟机 1～1 000 个)内存至少为 1 GB;中等清单(主机 100～400 台或虚拟机 1 000～4 000 个)内存至少为 2 GB;大清单(主机超过 400 台或虚拟机超过 4 000 个)内存至少为 3 GB。

(2) vCenter Server 软件要求:vCenter Server 要求使用 64 位操作系统,最好是 Windows 2008 R2,vCenter Server 需要使用 64 位系统 DSN 以连接到其数据库。虽然 vCenter Server 在安装过程中会捆绑 Microsoft SQL Server 2008 R2 Express 数据库软件包,但是考虑到安全性和数据可恢复性,最好不要将数据库和 VC 同时安装在一个系统之上。建议不选择安装其数据库,而是使用 64 位系统 DSN 以连接到其他数据库。VC 支持的数据库有 IBM DB2 9.5, IBM DB2 9.7,Microsoft SQL Server 2008 R2 Express,Microsoft SQL Server 2005,Microsoft SQL Server 2008,Microsoft SQL Server 2008 R2,Oracle 10g R2 和 Oracle 11g,vCenter Server 还需要 Microsoft .NET 3.5 SP1 Framework。如果要使用和 vCenter Server 捆绑在一起的 Microsoft SQL Server 2008 R2 Express 数据库,则系统上需要安装 Microsoft Windows Installer 5.5 版。

7.4.2 安装 vCenter Server

通过使用 VMware vCenter Simple Install 安装 vCenter Single Sign On,vCenter Inventory Service 和 vCenter Server。

可以在单个主机上通过使用"vCenter Server Simple Install"选项一起安装 vCenter Single Sign On、vCenter Inventory Service 和 vCenter Server。下面以小型部署选项为例讲解 vCenter Server 安装过程。

vCenter Server 5.1。必须按此顺序安装这些组件:vCenter Single Sign On,Inventory Service 和 vCenter Server。

vc.cisco.com 在安装 VMware vCenter 前,已经加入到 cisco.com 域中。在 VMware vCenter 安装程序窗口中,提供了众多安装选项,在这里选择 VMware。在"vCenter Simple Install"栏,单击"安装",如图 7-24 所示。

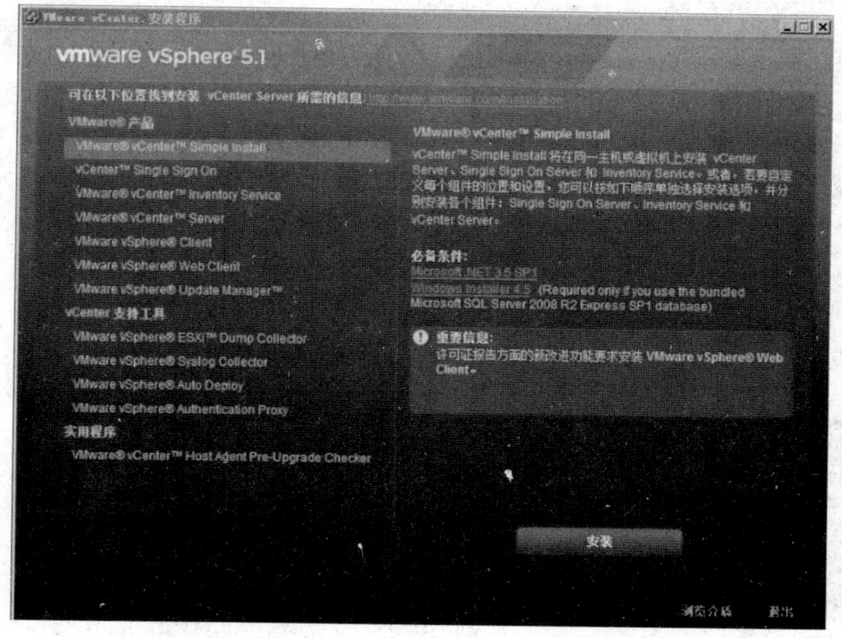

图 7-24　单击安装 vCenter Simple Install

1. vCenter Single Sign On 的安装

安装程序首先启动 vCenter Single Sign On 安装部分，如图 7-25 所示。

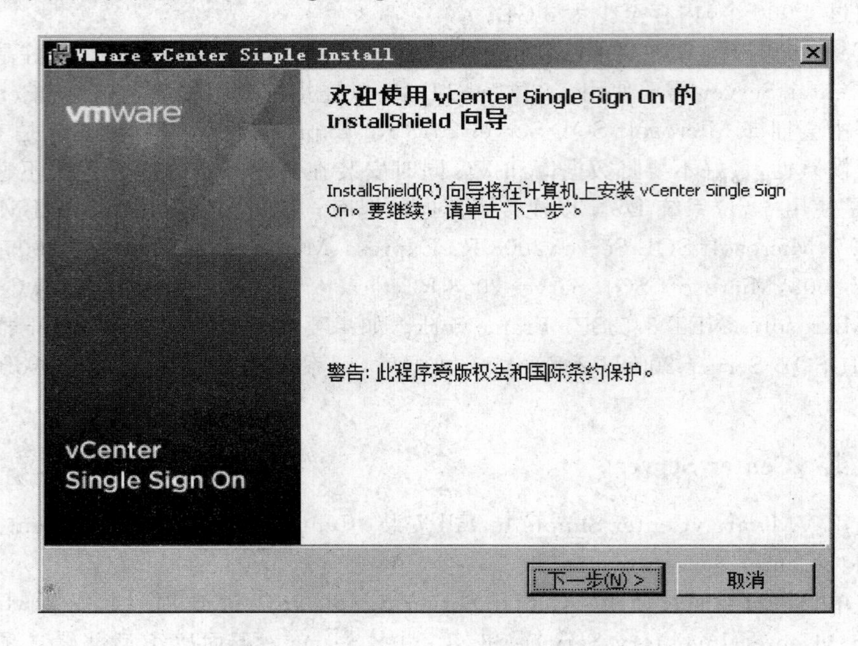

图 7-25　单击"下一步"开始安装

在"最终用户专利协议"窗口中，单击"下一步"，如图 7-26 所示。

图 7-26　安装 vCenter Simple Install

在"最终用户许可协议"窗口中，选择"我接受许可协议中的条款(A)"，单击"下一步"，如图 7-27 所示。

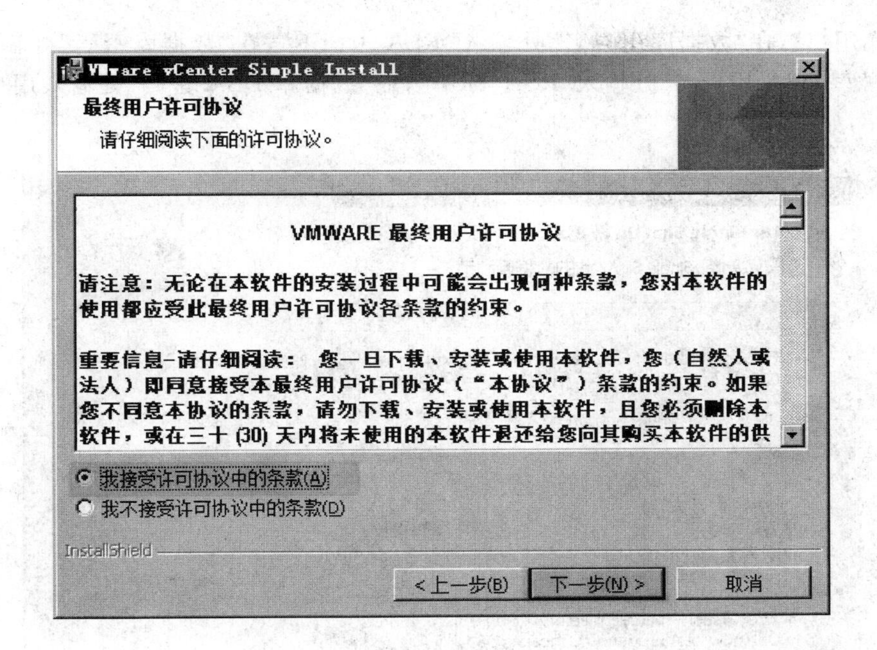

图 7-27　安装 vCenter Simple Install，接受许可

在“vCenter Single Sign On”信息窗口中，根据要求输入管理员密码，单击“下一步”，如图 7-28 所示。

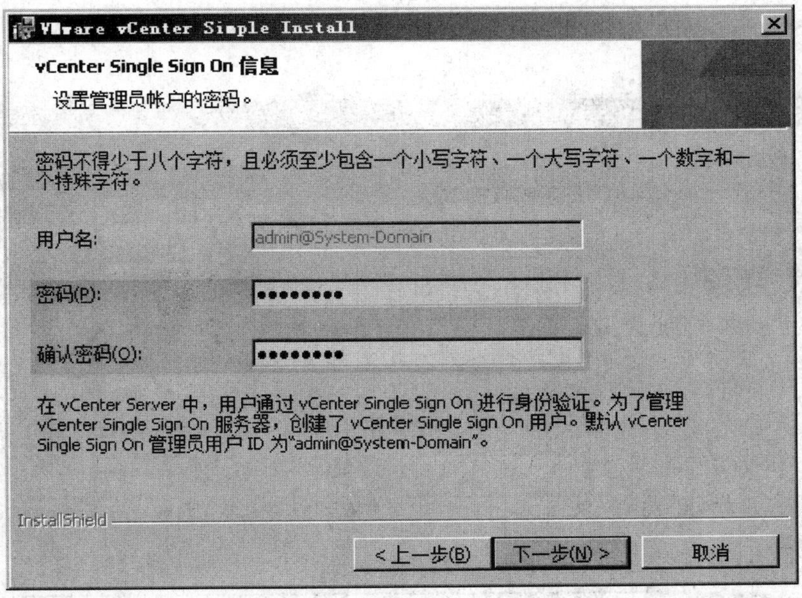

图 7-28　安装 vCenter Simple Install，输入账号

在“vCenter Single Sign On”数据库窗口中，选择“现有支持的数据库”，单击“下一步”，如图 7-29 所示。

在“数据库信息”窗口中，在“数据库类型”处选择“Mssql”，在“数据库名称”处输入“RSA”，在“主机名或 IP 地址”处输入 SQL 服务器名称，端口默认是 1433。勾选“使用手动创建的数据库用户”,在“数据库用户名”处输入“RSA_USER”,在“数据库密码”处输入用户密码,在“数据库 DBA 用户名”处输入“RSA_DBA”,在“数据库 DBA 密码”处输入用户密码,单击“下一步”,如图 7-30 所示。

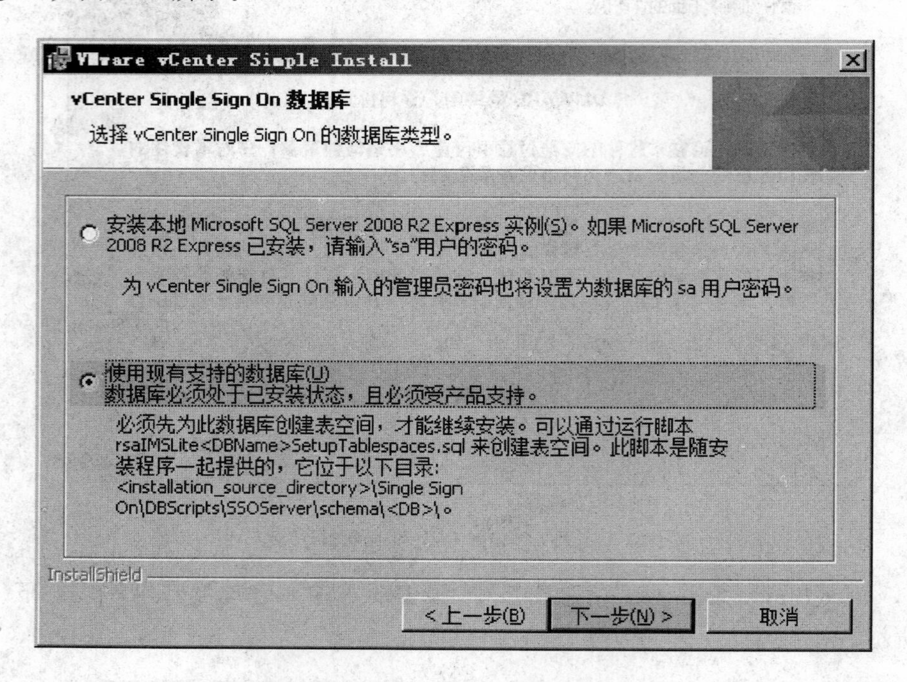

图 7-29　安装 vCenter Simple Install,选择数据库类型

图 7-30　配置数据库参数

在“本地系统信息”窗口中，在“完全限定域名或 IP 地址”处输入名称，单击“下一步”，如图 7-31 所示。

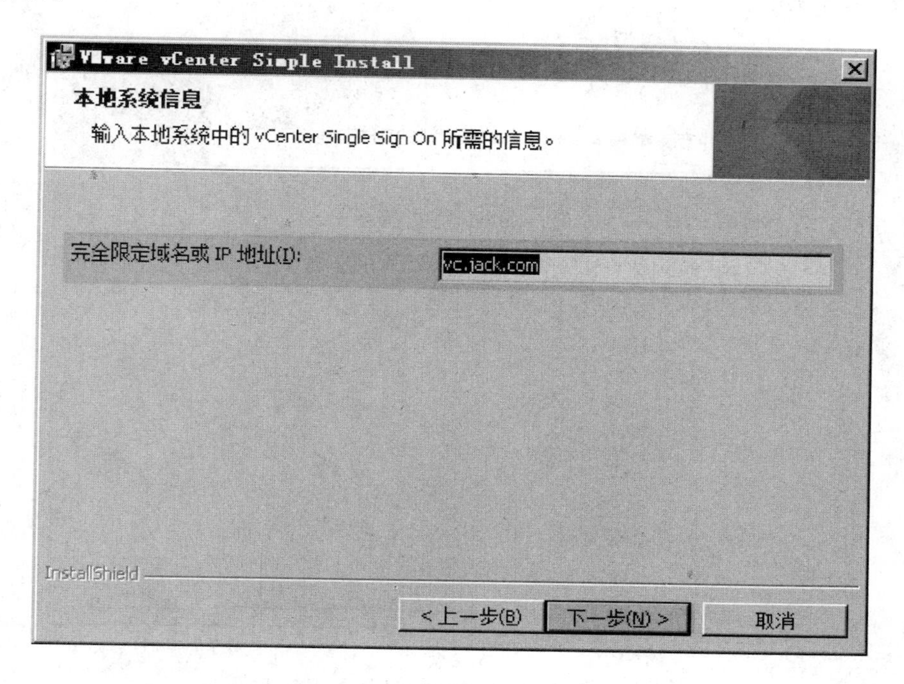

图 7-31　设置域名信息

在“安全支持提供者接口服务信息”窗口中，勾选“使用网络服务账户”，单击“下一步”，如图 7-32 所示。

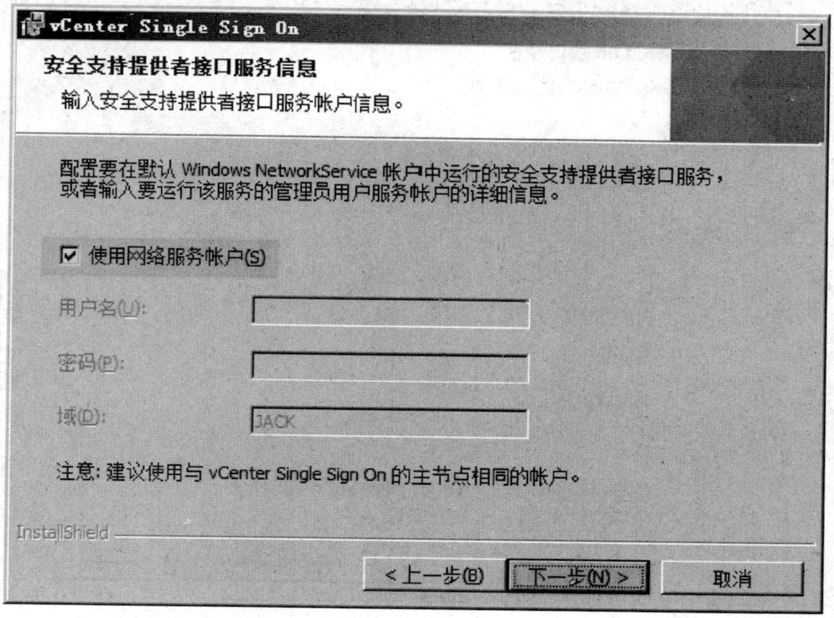

vCenter Single Sign On
安全支持提供者接口服务信息
输入安全支持提供者接口服务帐户信息。
配置要在默认 Windows NetworkService 帐户中运行的安全支持提供者接口服务，或者输入要运行该服务的管理员用户服务帐户的详细信息。
☑ 使用网络服务帐户(S)
用户名(U):
密码(P):
域(D): JACK
注意：建议使用与 vCenter Single Sign On 的主节点相同的帐户。
InstallShield
< 上一步(B)　下一步(N) >　取消

图 7-32　安装 vCenter Simple Install，使用网络服务账户

在“目标文件夹”窗口中，可使用 VMware vCenter Simple Install 默认安装位置，单击“下一步”，如图 7-33 所示。

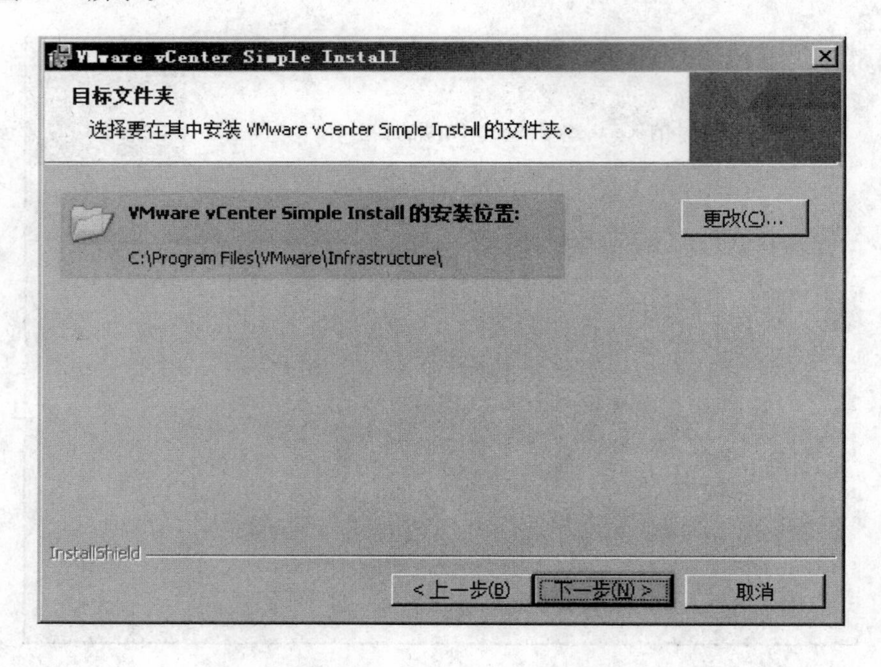

图 7-33 安装 vCenter Simple Install 默认位置

在“vCenter Single Sign On 端口设置”窗口中，使用 HTTPS 默认端口，单击“下一步”，如图 7-34 所示。

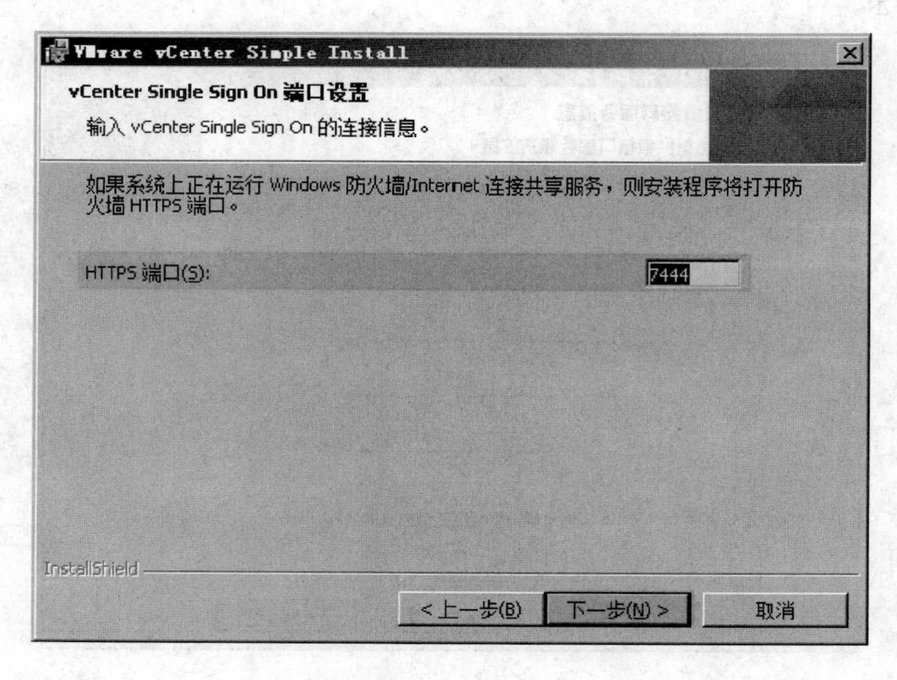

图 7-34 安全 Web 访问端口设置

在“准备安装”窗口中，单击“安装”，如图 7-35 所示。

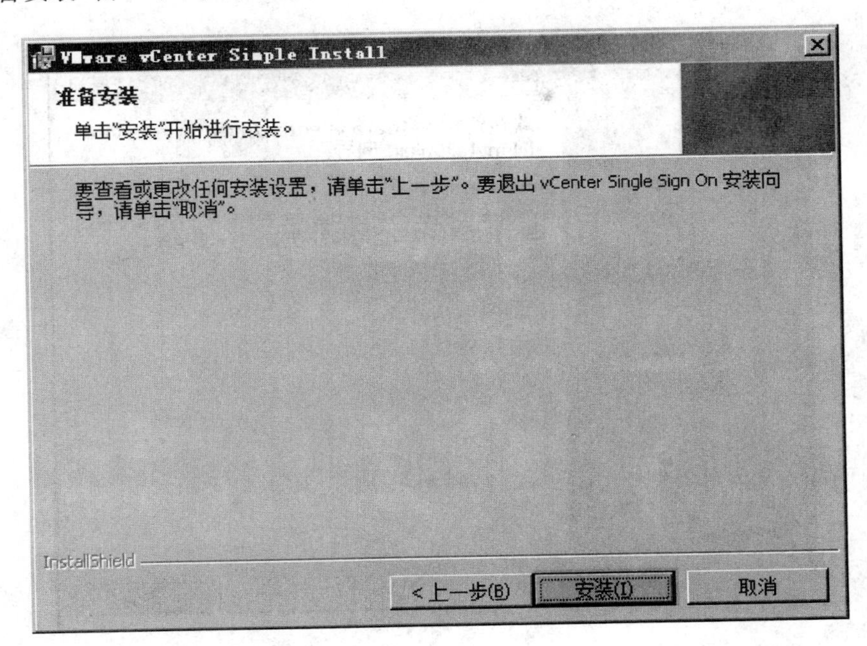

图 7-35 开始安装 vCenter Simple Install

在“安装 vCenter Single Sign On”窗口中，显示安装正在进行，如图 7-36 所示。

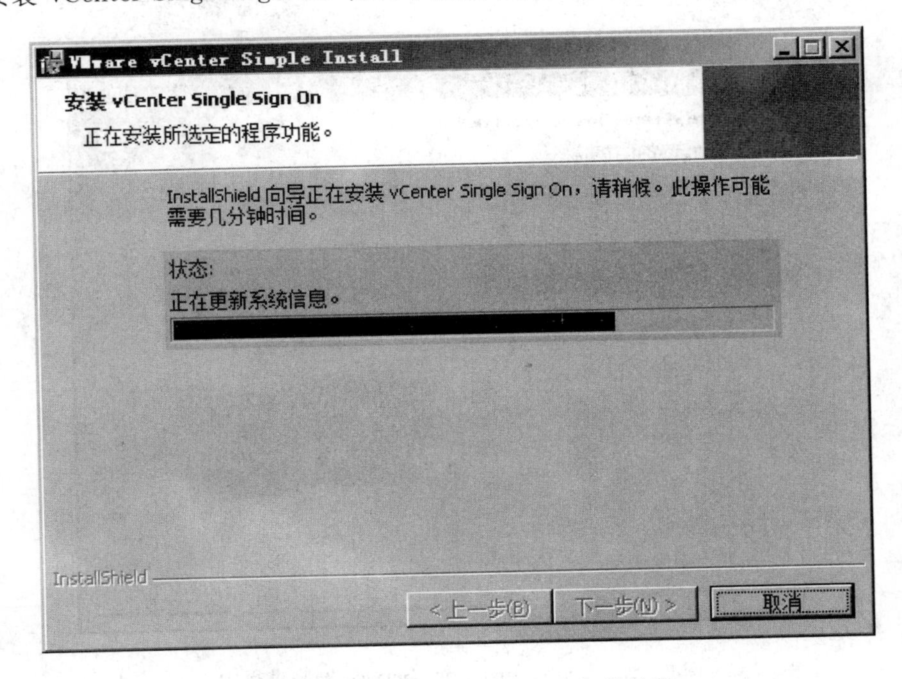

图 7-36 安装 vCenter Simple Install 过程

至此，vCenter Single Sign On 安装已完成。

2. VMware vCenter Inventory Service 的安装

安装程序继续 vCenter Inventory Service 安装部分。

vCenter Inventory Service 安装程序会自动完成安装过程，而不需要进行选择操作，如图 7-37 所示。

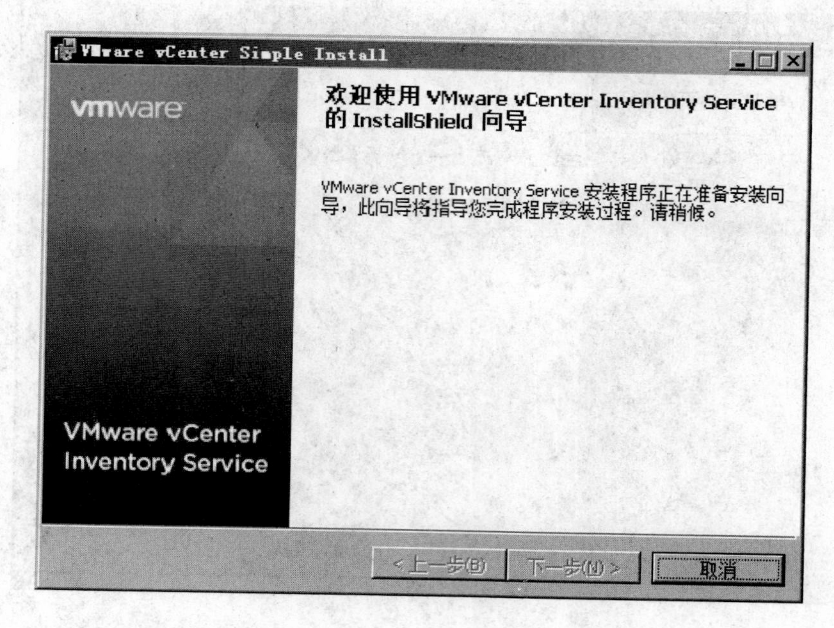

图 7-37 vCenter Inventory Service 安装

在"安装 VMware vCenter Inventory Service"窗口中，显示安装正在进行，如图 7-38 所示。

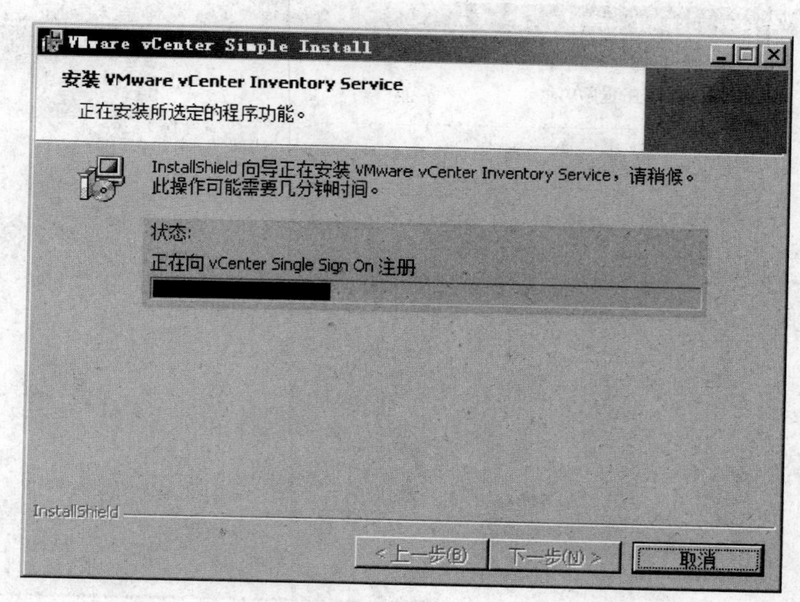

图 7-38 vCenter Inventory Service 安装进行中

至此，vCenter Inventory Service 安装已完成。

3. VMware vCenter Server 的安装

安装程序继续 VMware vCenter Server 安装部分，如图 7-39 所示。

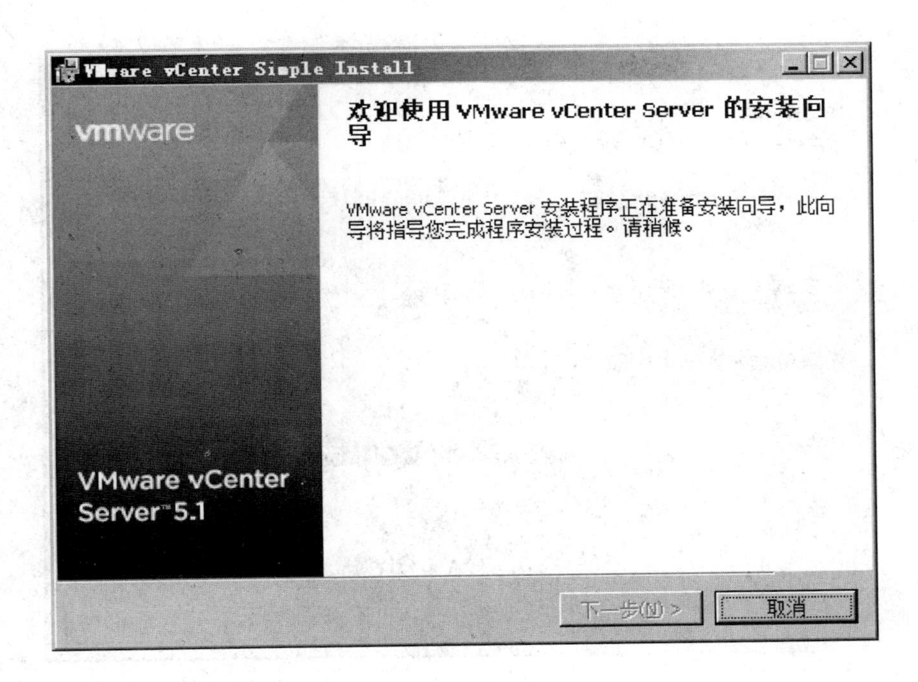

图 7-39 VMware vCenter Server 安装

在“许可证密钥”窗口中，在“许可证密钥”处输入密钥，也可不输入许可证密钥使用评估模式，单击“下一步”，如图 7-40 所示。

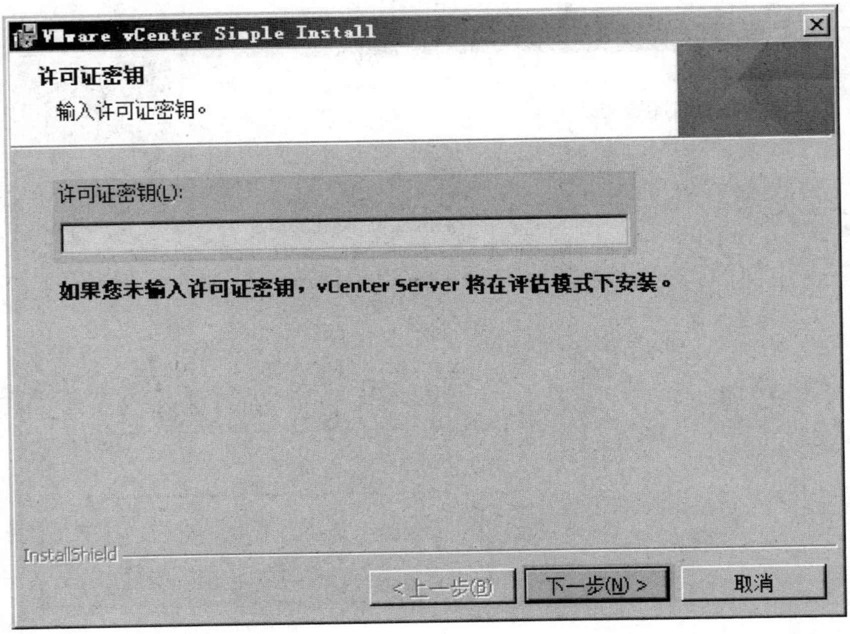

VMware vCenter Simple Install
许可证密钥
输入许可证密钥。
许可证密钥(L):
如果您未输入许可证密钥，vCenter Server 将在评估模式下安装。
InstallShield
< 上一步(B)
下一步(N) >
取消

图 7-40 VMware vCenter Server 安装输入许可证

在“数据库选项”窗口中，选择使用现有的受支持数据库，在“数据源名称(DSN)”处选择创建的 64 位 DSN，单击“下一步”，如图 7-41 所示。

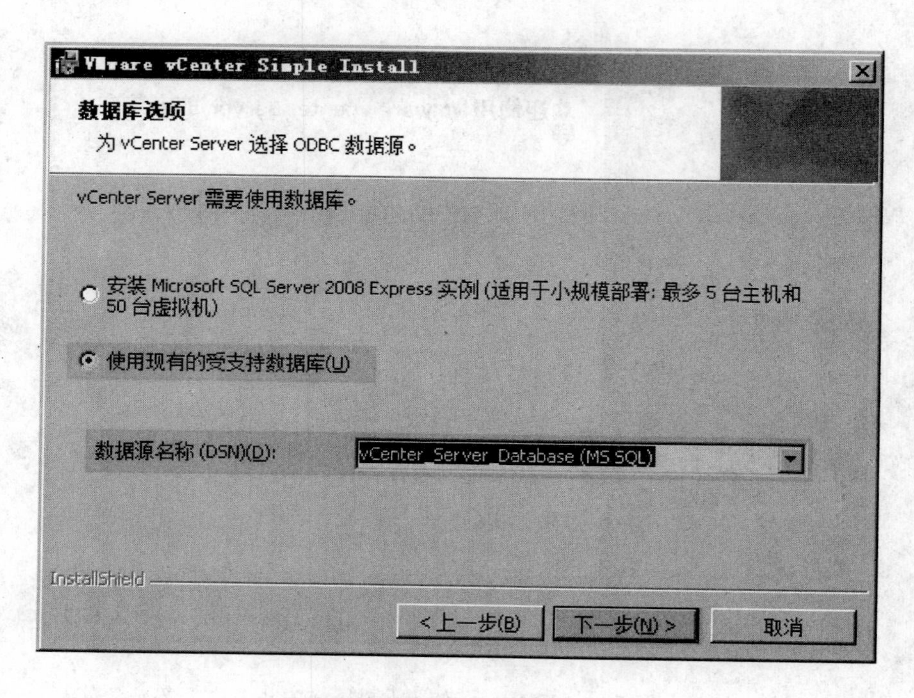

图 7-41　选择数据源名称

在“数据库选项”窗口中，输入数据库用户名和数据库密码，单击“下一步”，如图 7-42 所示。

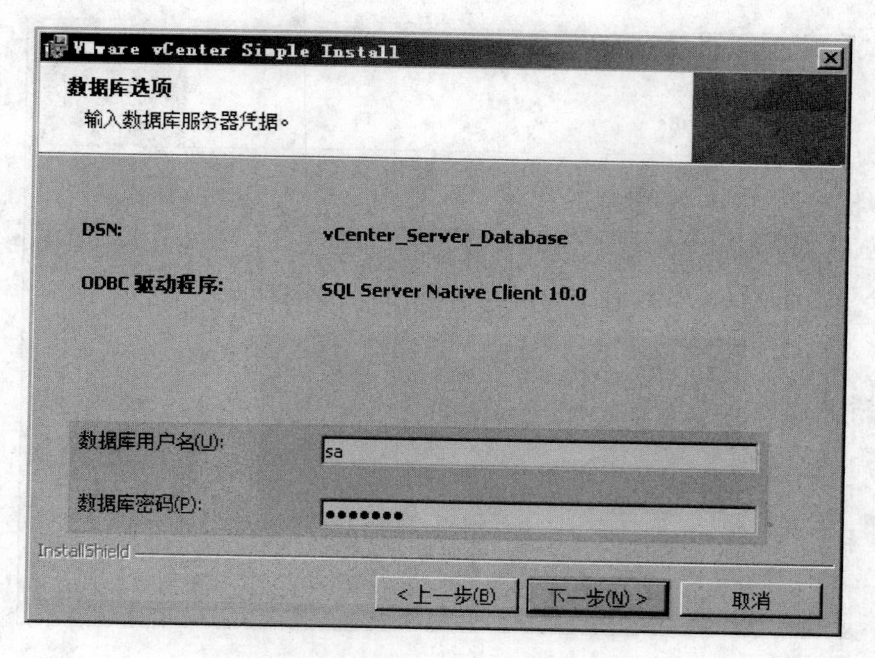

图 7-42　输入数据库用户名和数据库密码

在“vCenter Server 服务”窗口中，勾选“使用 SYSTEM 账户”，单击“下一步”，如图 7-43 所示。

VMware vCenter Server
vCenter Server 服务
输入 vCenter Server 服务帐户信息。
将 vCenter Server 服务配置为以默认的 Windows LocalSystem 帐户运行，或者输入要运行 vCenter Server 服务的管理员用户服务帐户的详细信息。
使用 SYSTEM 帐户(U)
帐户名称(A):
帐户密码(P):
完全限定域名(F):
vc.jack.com
安全建议: vCenter Server 安装程序会向用户指定的帐户授予"作为服务登录"权限。
InstallShield
< 上一步(B)
下一步(N) >
取消

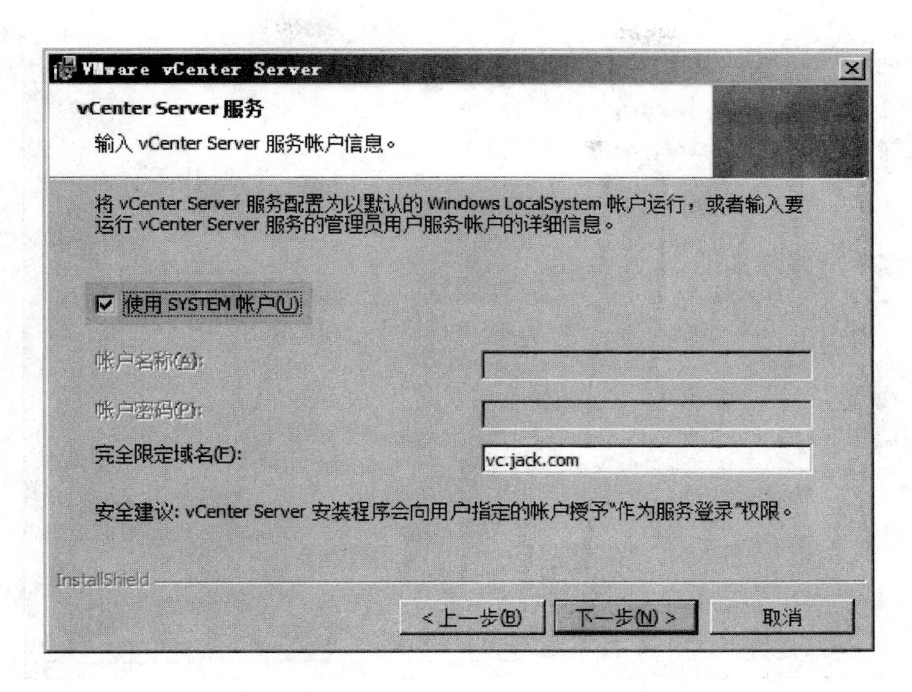

图 7-43　使用 SYSTEM 账户

在“配置端口”窗口中，保持默认端口设置，单击“下一步”，如图 7-44 所示。

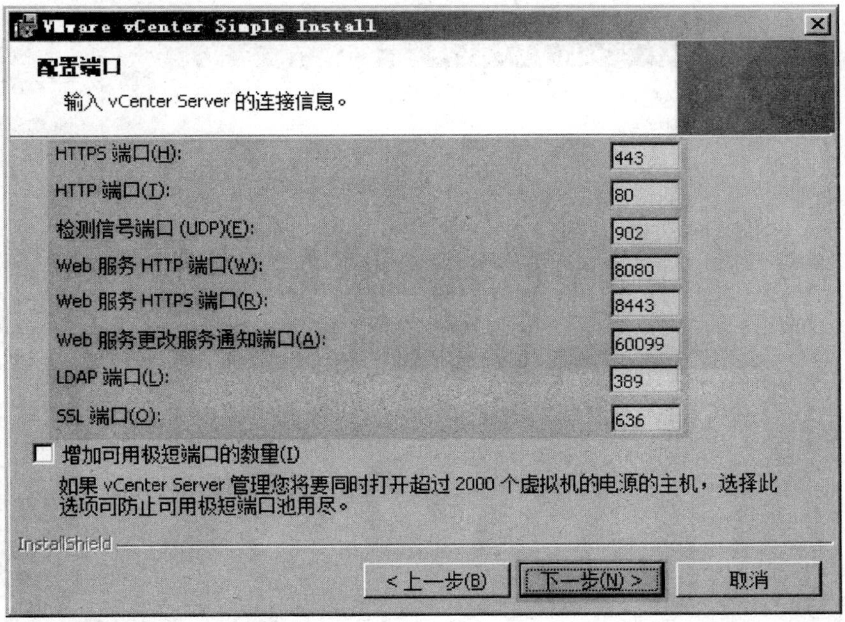

VMware vCenter Simple Install
配置端口
输入 vCenter Server 的连接信息。
HTTPS 端口(H):
443
HTTP 端口(T):
80
检测信号端口 (UDP)(E):
902
Web 服务 HTTP 端口(W):
8080
Web 服务 HTTPS 端口(R):
8443
Web 服务更改服务通知端口(A):
60099
LDAP 端口(L):
389
SSL 端口(O):
636
增加可用极短端口的数量(I)
如果 vCenter Server 管理您将要同时打开超过 2000 个虚拟机的电源的主机，选择此选项可防止可用极短端口池用尽。
InstallShield
< 上一步(B)
下一步(N) >
取消

图 7-44　配置端口

在“vCenter Server JVM 内存”窗口中，在“清单大小”处选择“小型”，单击“下一步”，如图 7-45 所示。

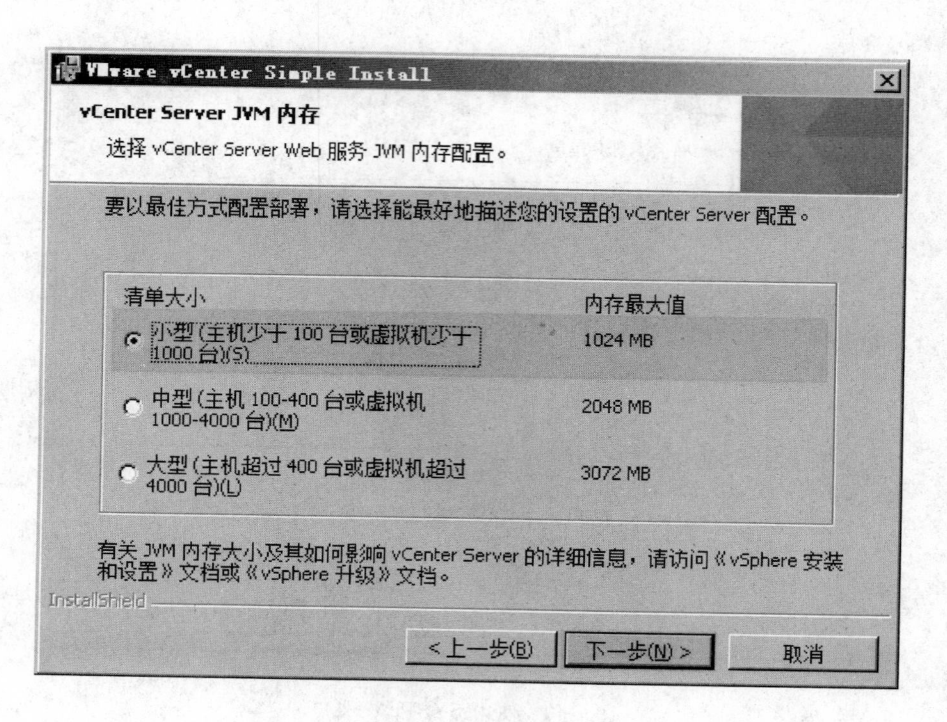

图 7-45 选择主机类型

在“准备安装程序”窗口中，单击“安装”，如图 7-46 所示。

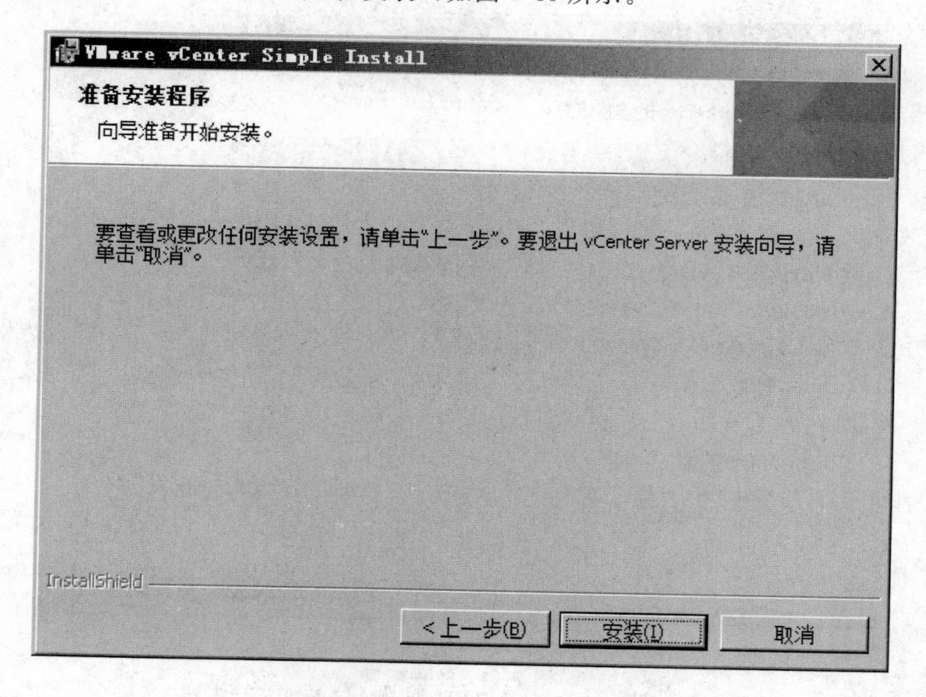

图 7-46 开始安装 vCenter Server

在“安装 VMware vCenter Server”窗口中，显示安装正在进行，如图 7-47 所示。

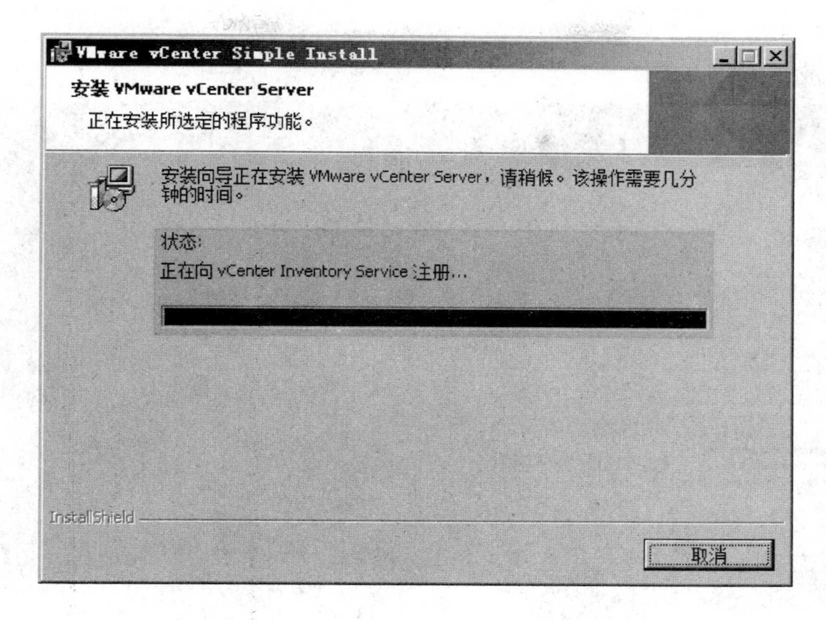

图 7-47　安装 VMware vCenter Server

VMware vCenter Server 安装成功，单击“完成”，如图 7-48 所示。

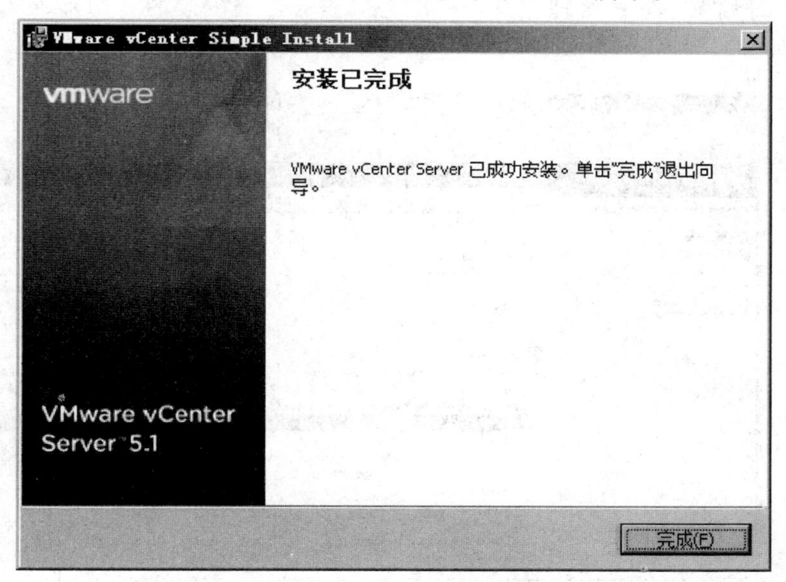

图 7-48　安装 VMware vCenter Server 完成

至此，VMware vCenter Simple Install 的安装已全部完成，如图 7-49 所示。

图 7-49　安装 VMware vCenter Server 后的画面

依次打开“开始菜单”→“所有程序”→“VMware”，查看已经安装的 VMware 程序菜单，如图 7-50 所示。

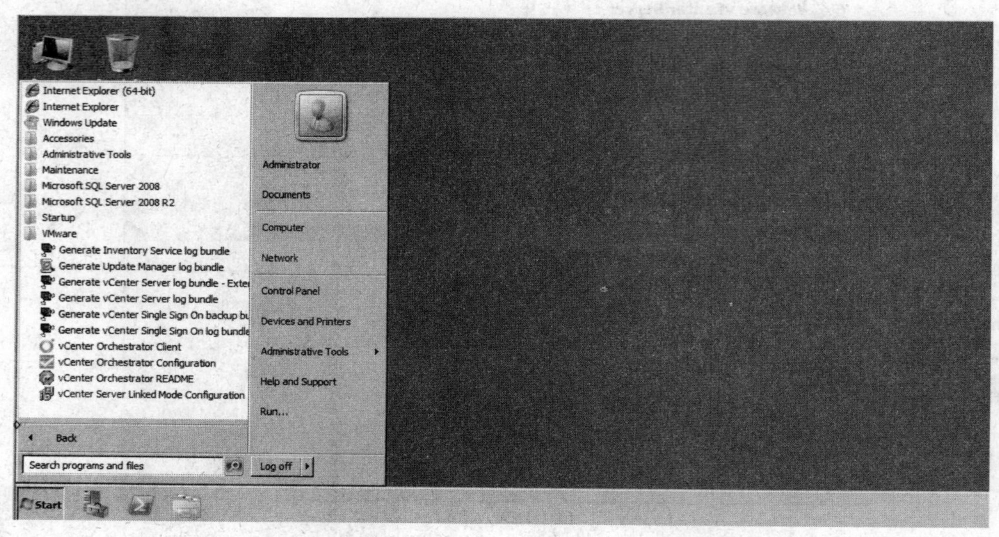

图 7-50　查看安装的 VMware vCenter Server

依次打开“开始菜单”→“管理工具”→“服务”，查看已经安装的 VMware 程序相关服务，如图 7-51 所示。

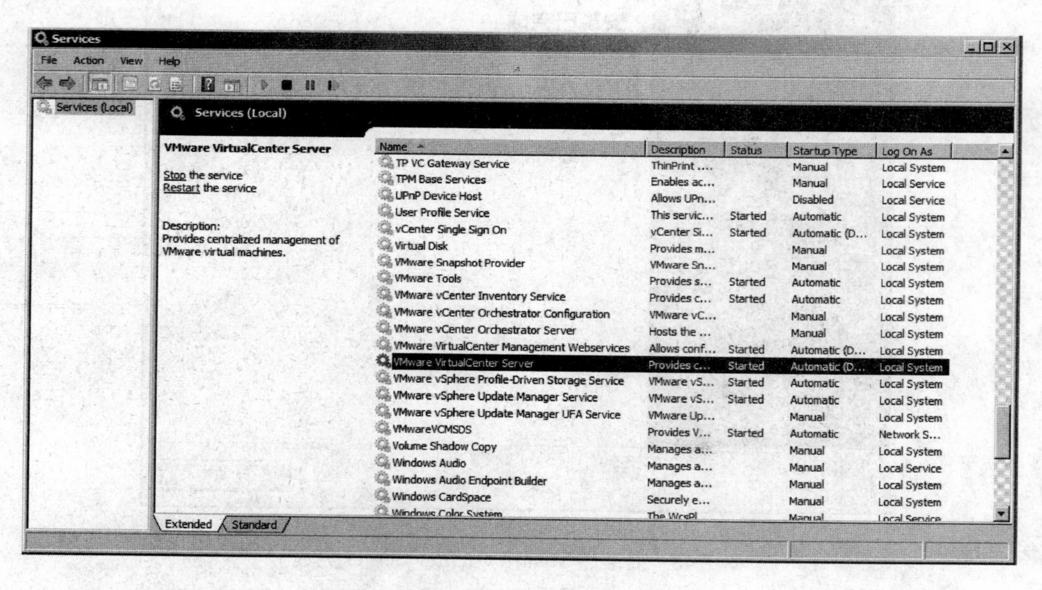

图 7-51　在服务里看到 VMware vCenter Serve

7.4.3　配置 vCenter

使用“vCenter Server 设置”对话框可以配置 vCenter Server，包括如许可、统计信息收集、日志记录等设置以及其他设置。

打开“vSphere Client”主页窗口，依次选择“Administration”→“vCenter Server Settings”，如图 7-52 所示。

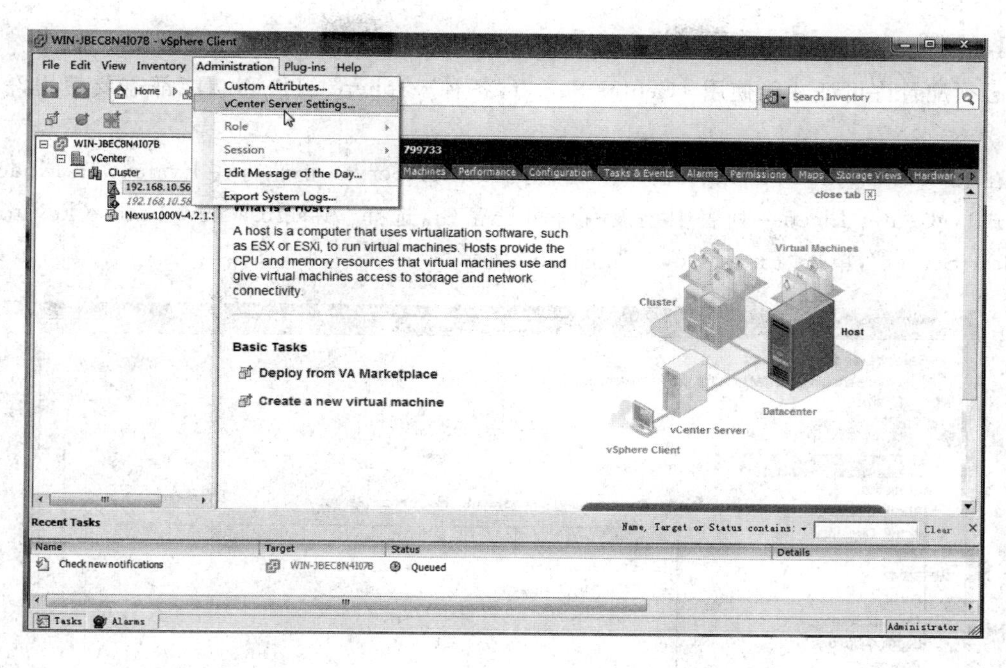

图 7-52　配置 VMware vCenter Serve

打开“vCenter Server Settings”窗口，在右侧选项栏中，提供了丰富的设置选项，如图 7-53 所示。

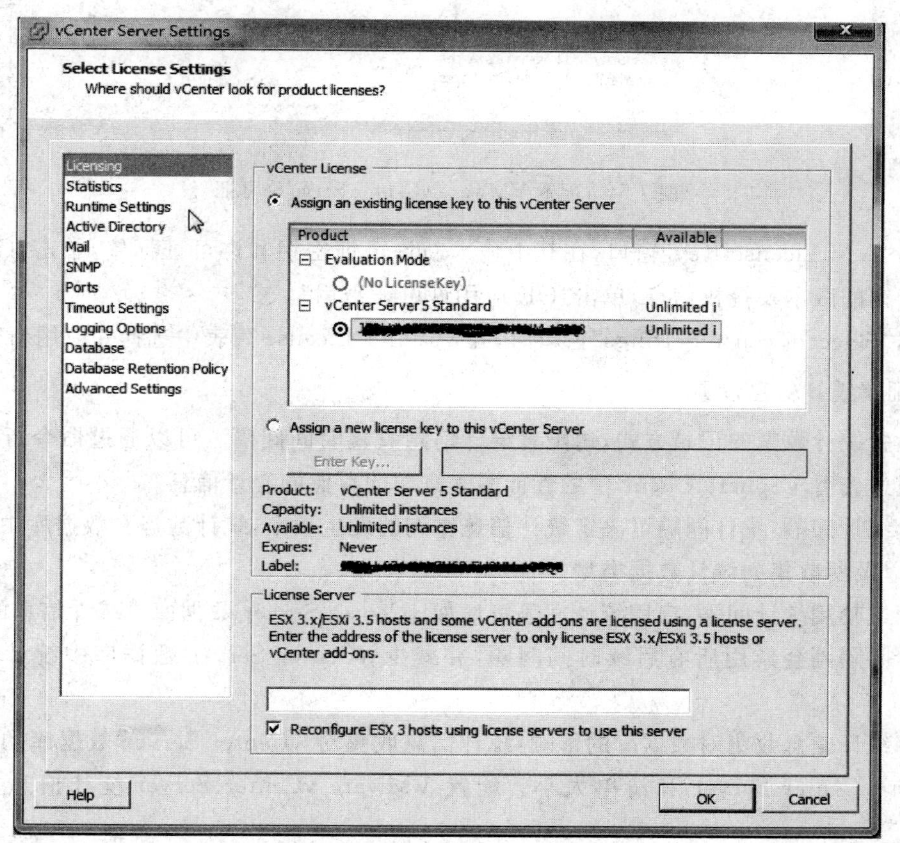

图 7-53　配置 VMware vCenter Server

1. 配置 vCenter Server 许可证

必须配置许可证才能使用 vCenter Server，各种 vSphere 组件和功能都需要使用许可证密钥。

在"Select License Settings"窗口中，当前 vCenter Server 正运行在 Evaluation Mode（评估模式），vCenter License 列表中还没有许可证密钥，选择"Assign a new license key to this vCenter Server"，单击"Enter Key"，如图 7-54 所示。

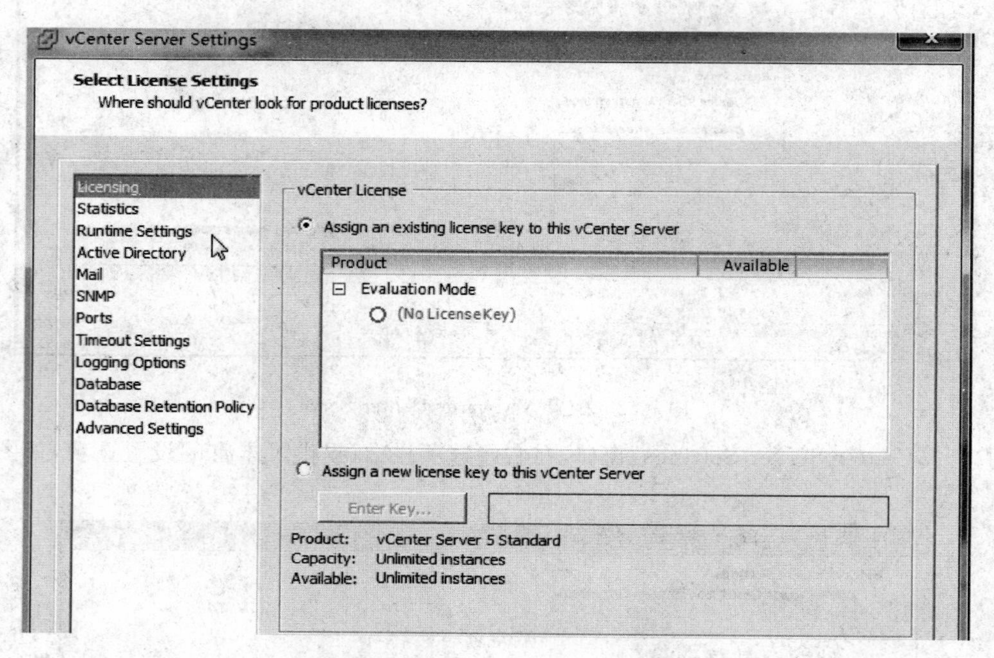

图 7-54　配置 VMware vCenter Serve 许可证

弹出"Add License Key"窗口，在其中输入新许可证密钥和许可证标签，单击"OK"，显示许可证密钥的版本及授权信息，单击"OK 应用许可证密钥"。

返回"Select License Settings"窗口时，在 vCenter License 列表中选择并应用新密钥。

2. 配置统计信息设置

要设置统计数据的记录方式，请配置统计信息收集时间间隔。可以通过命令行监控实用程序或通过查看 vSphere Client 中的性能图表来访问存储的统计信息。

配置统计间隔：统计间隔可决定统计信息查询的发生频率，统计数据在数据库中的存储时间长度，以及所收集的统计数据类型。

启用或禁用统计间隔：启用统计间隔可增加 vCenter Server 数据库中统计信息的存储量；禁用统计间隔则会禁用所有后续时间间隔，并减少 vCenter Server 数据库中统计信息的存储量。

估算统计信息收集对数据库的影响：统计信息收集对 vCenter Server 数据库的影响将取决于当前 vCenter Server 和清单大小。配置 VMware vCenter Server 统计信息如图 7-55 所示。

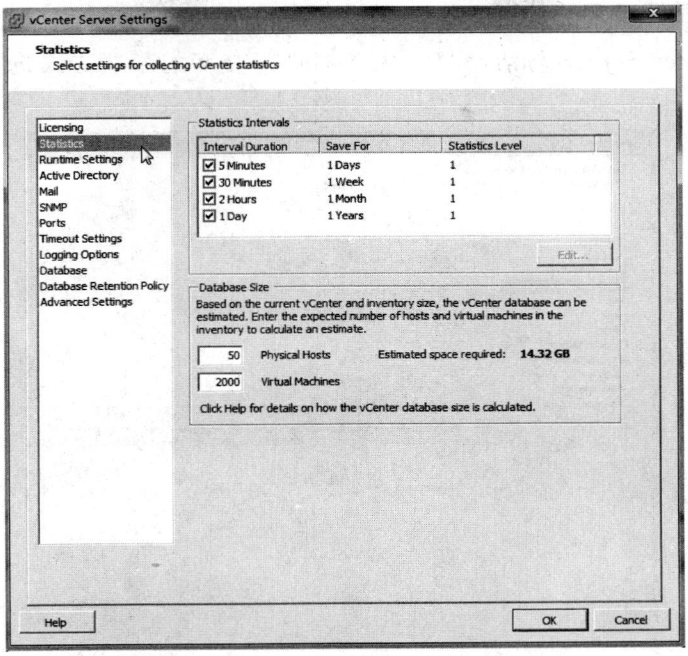

图 7-55　配置 VMware vCenter Server 统计信息

3. 配置运行时间设置

可以更改服务器用于通信的端口号，还可以更改 vCenter Server ID 和 vCenter Server 的受管 IP 地址。通常不需要更改这些设置，但如果在同一环境中运行多个 vCenter Server 系统，则可能需要进行更改。配置 VMware vCenter Server 运行时间如图 7-56 所示。

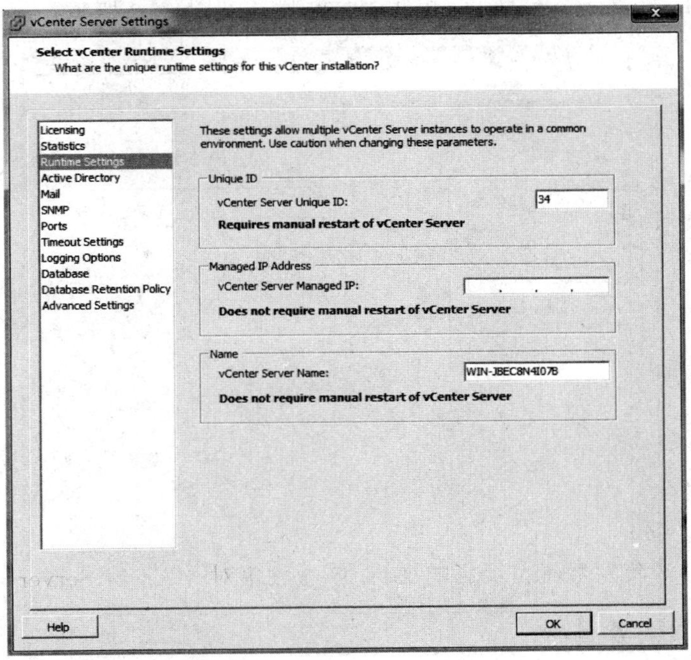

图 7-56　配置 VMware vCenter Serve 运行时间

4. 配置活动目录设置

可以对 vCenter Server 与活动目录服务器进行交互的某些方式进行配置，如图 7-57 所示。

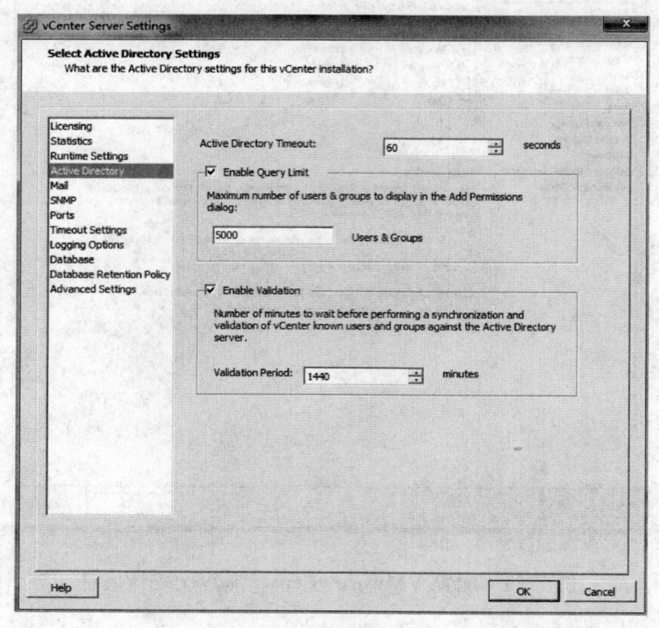

图 7-57 配置 VMware vCenter Server 活动目录

5. 配置邮件发件人设置

必须配置发件人账户的电子邮件地址，以便启用 vCenter Server 操作，如将电子邮件通知作为警报操作发送，配置 VMware vCenter Server 邮箱如图 7-58 所示。

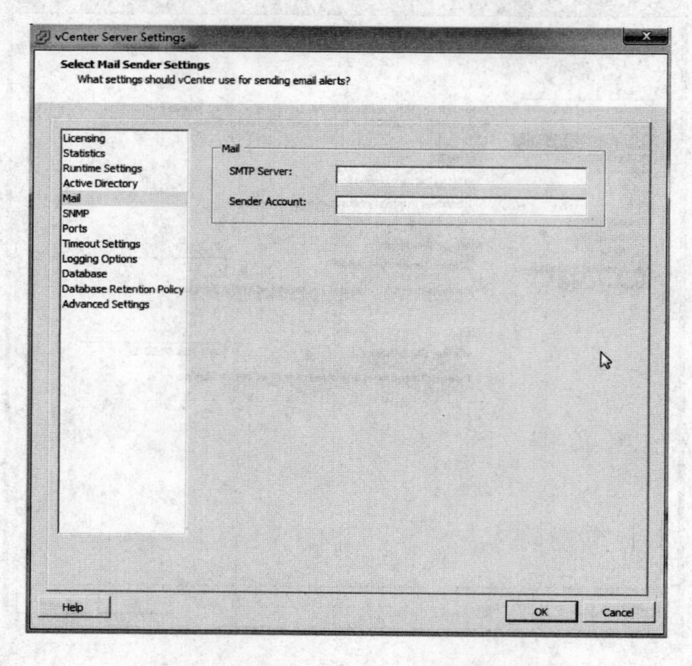

图 7-58 配置 VMware vCenter Server 邮箱

6. 配置 SNMP 设置

最多可以配置 4 个收件人从 vCenter Server 接收 SNMP 陷阱，对于每个收件人，请指定主机名称、端口和团体。配置 VMware vCenter Server 网管参数如图 7-59 所示。

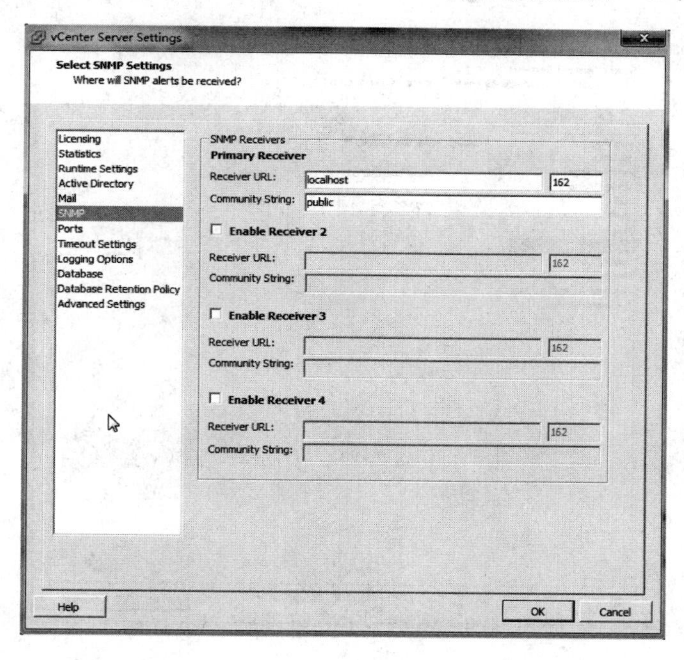

图 7-59　配置 VMware vCenter Server 网管参数

7. 查看端口设置

可以查看由 Web 服务使用的端口，如图 7-60 所示，与其他应用程序进行通信将无法再使用这些端口。

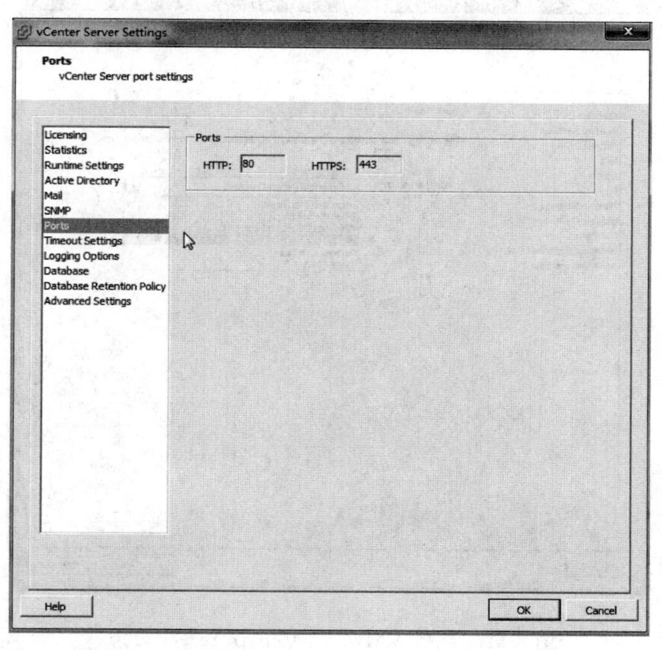

图 7-60　查看 VMware vCenter Server Web 端口

8. 配置超时设置

可以配置 vCenter Server 操作的超时时间间隔，如图 7-61 所示，这些时间间隔用于指定 vSphere Client 超时之后的时间量。

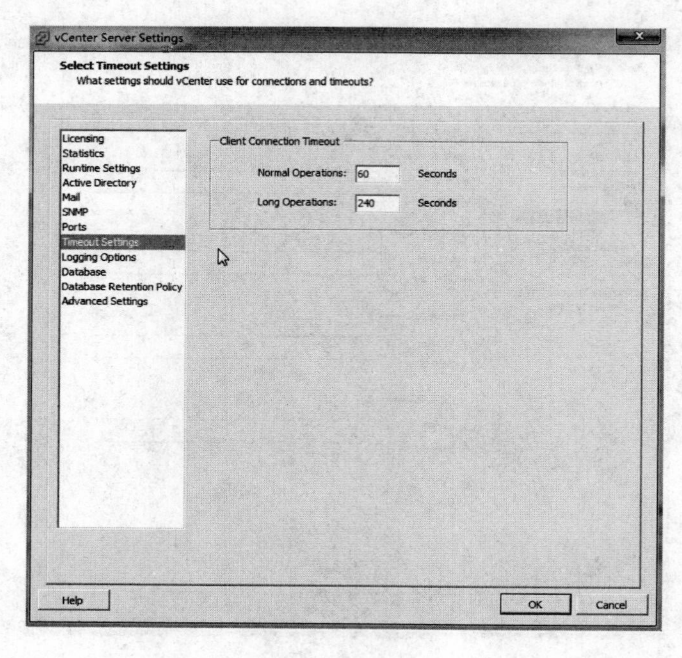

图 7-61　配置 VMware vCenter Server 超时间隔

9. 配置日志记录选项

可以对 vCenter Server 在日志文件中收集的详细信息的数量进行配置，如图 7-62 所示。

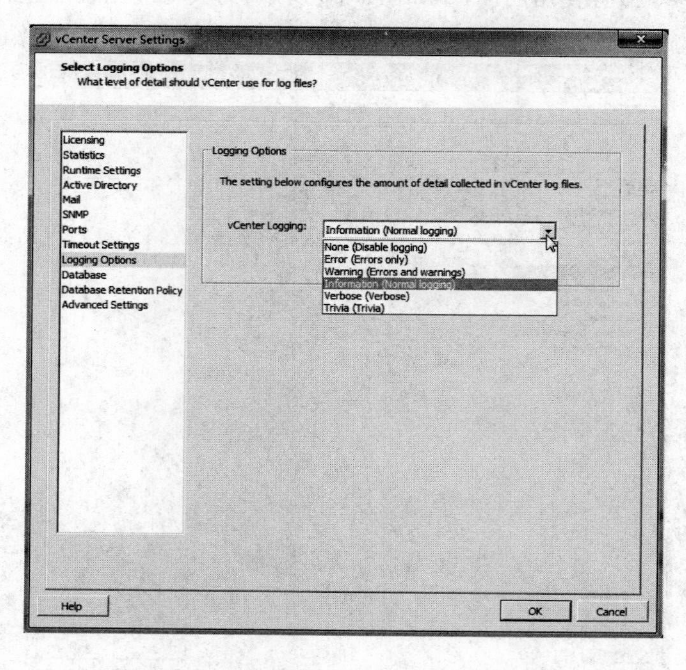

图 7-62　配置 VMware vCenter Server 日志

10. 配置最大数据库连接数

可以配置允许同时出现的最大数据库连接数,如图 7-63 所示。

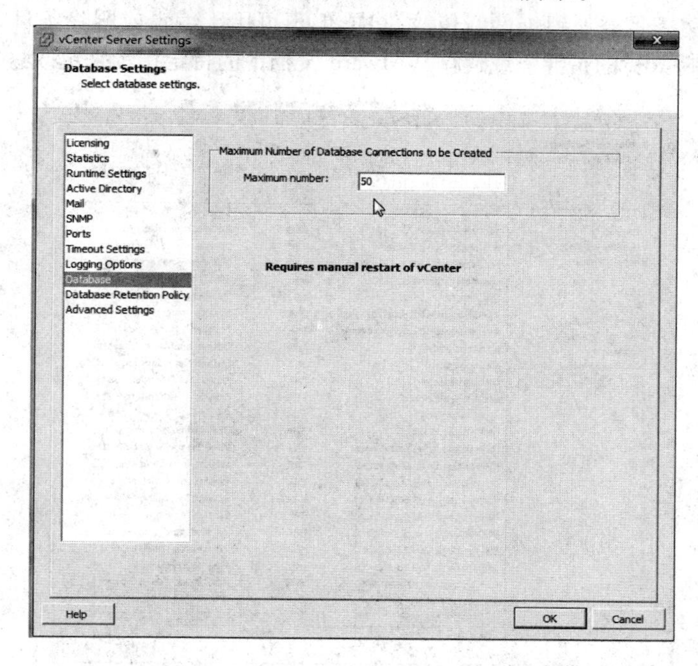

图 7-63　配置 VMware vCenter Server 数据库最大连接数

11. 配置数据库保留策略

为了限制 vCenter Server 数据库的增长并节省存储空间,可以将数据库配置为一段指定时间后放弃有关任务或事件的信息,如图 7-64 所示。

如果要保留 vCenter Server 的任务和事件的完整历史记录,请不要使用这些选项。

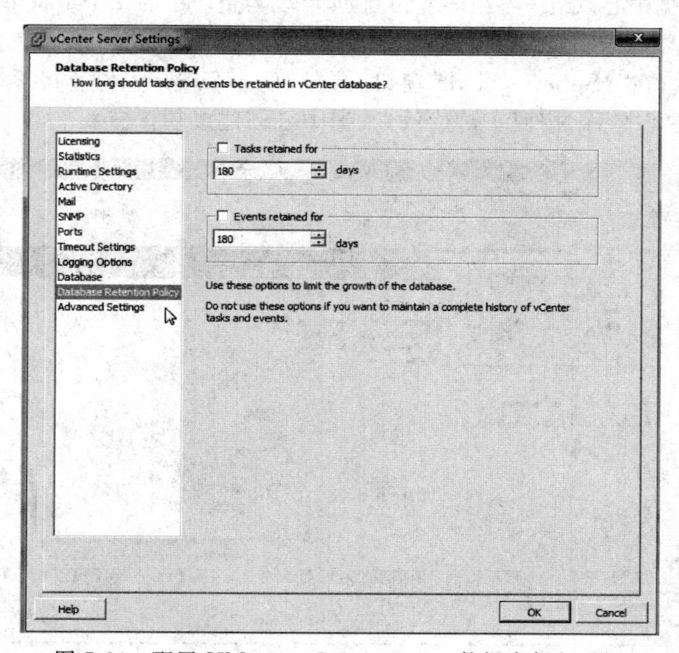

图 7-64　配置 VMware vCenter Server 数据库保留时间

12. 配置高级设置

可以在“高级设置”页面中修改 vCenter Server 配置文件 vpxd. cfg，如图 7-65 所示。

此页面可用于将条目添加到 vpxd. cfg 文件中，但不可用于编辑或删除条目。VMware 建议仅在 VMware 技术支持人员的指导下或遵循 VMware 文档中的特定指示来更改这些设置。

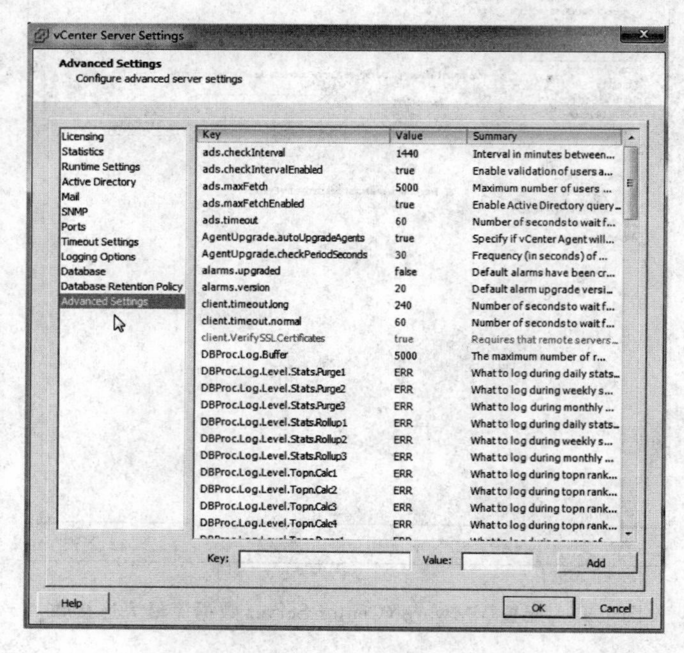

图 7-65　VMware vCenter Server 高级设置选项

13. 创建数据中心

虚拟数据中心是一种容器，其中包含配齐用于操作虚拟机的完整功能环境所需的全部清单对象，可以创建多个数据中心以组织各组环境。例如，可以为企业中的每个组织单元创建一个数据中心，也可以为高性能环境创建某些数据中心，而为要求相对不高的虚拟机创建其他数据中心。

如图 7-66 所示，在“vSphere Client”窗口，选择 vCenter Server 实例名称，单击“New Datacenter”，修改数据中心名称，按回车键确认，数据中心创建成功。

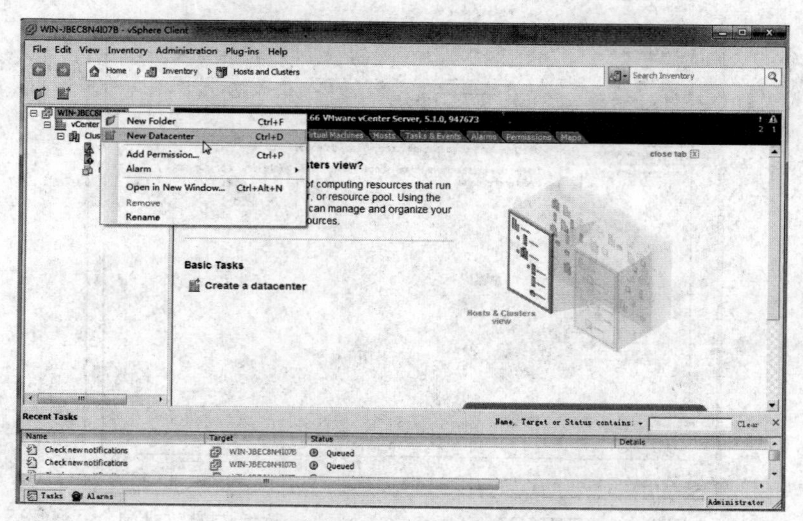

图 7-66　创建一个新的 Datacenter

14. 添加 ESXi 主机

可以在数据中心对象、文件夹对象或群集对象下添加主机，如果主机包含虚拟机，则这些虚拟机将与主机一起添加到清单。

在“vSphere Client”窗口，选择数据中心名称，单击“Add a host”，如图 7-67 所示。

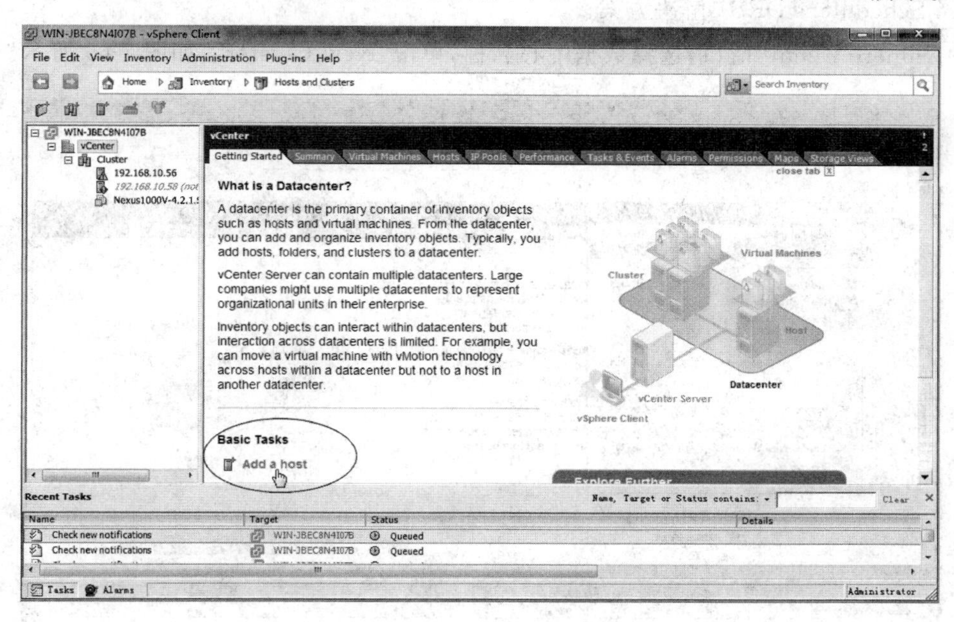

图 7-67　配置一个主机

打开“Add Host Wizard”窗口，在“Connection Settings”页，输入“Host”“Username”“Password”的内容，单击“Next”，如图 7-68 所示。

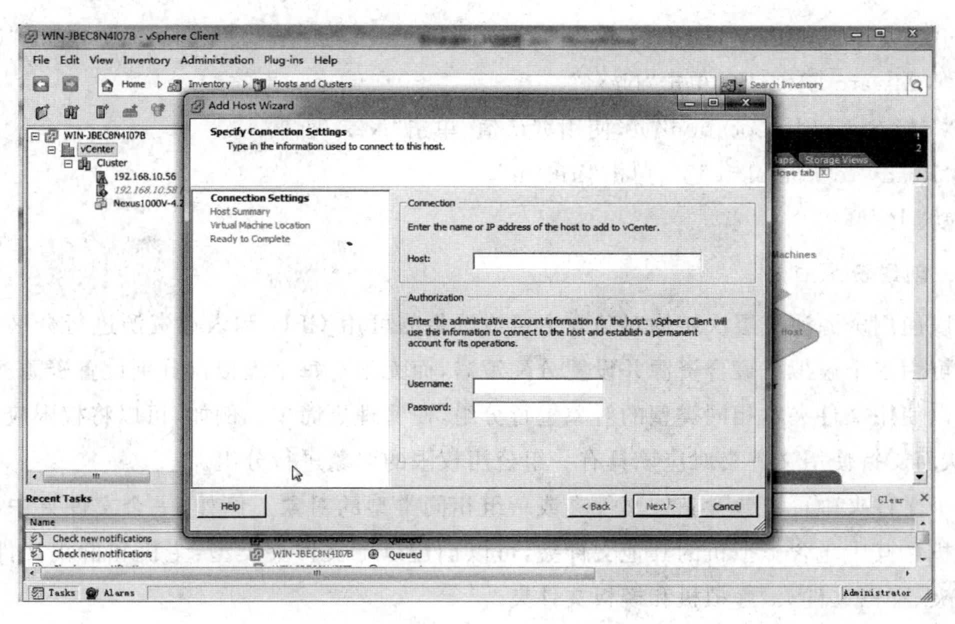

图 7-68　配置一个主机参数

根据向导完成主机的添加。

15. 创建群集

群集是一组主机，将主机添加到群集时，主机的资源将成为群集资源的一部分，群集管理其中所有主机的资源。群集启用 vSphere High Availability（HA）和 vSphere Distributed Resource Scheduler（DRS）解决方案。

在“vSphere Client”窗口，选择数据中心名称，单击“New Cluster”，如图 7-69 所示。

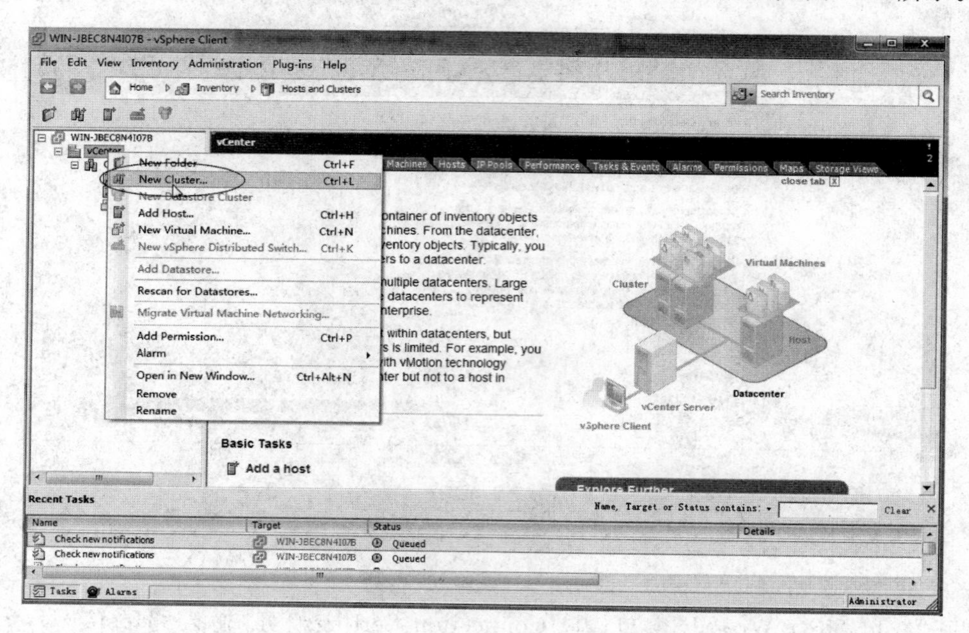

图 7-69　配置一个集群

打开“New Cluster Wizard”窗口，在“Cluster Features”页，输入群集的名称，单击“Next”。

在“VMware EVC”页，单击“Next”。

在“VM Swapfile Location”页，使用默认值，单击“Next”。

在“Ready to Complete”页，单击“Finish”。

成功创建群集。

16. 创建资源池

可以使用资源池按层次结构对独立主机或群集的可用 CPU 和内存资源进行分区。使用资源池可为多个虚拟机聚合资源并设置分配策略，而无须对每个虚拟机分别设置资源。

可以使用文件夹对相同类型的对象进行分组，使管理更简单。例如，可以将权限应用于文件夹，从而支持使用文件夹对应该具有一组公用权限的对象进行分组。

一个文件夹中可以包含其他文件夹或一组相同类型的对象。例如，一个文件夹中可以包含虚拟机和其中包含虚拟机的其他文件夹，可以创建以下文件夹类型：主机和群集文件夹、网络文件夹、存储文件夹、虚拟机和模板文件夹。

在“vSphere Client”窗口，选择 ESXi 主机名称，单击“New Resource Pool”，如图 7-70 所示。

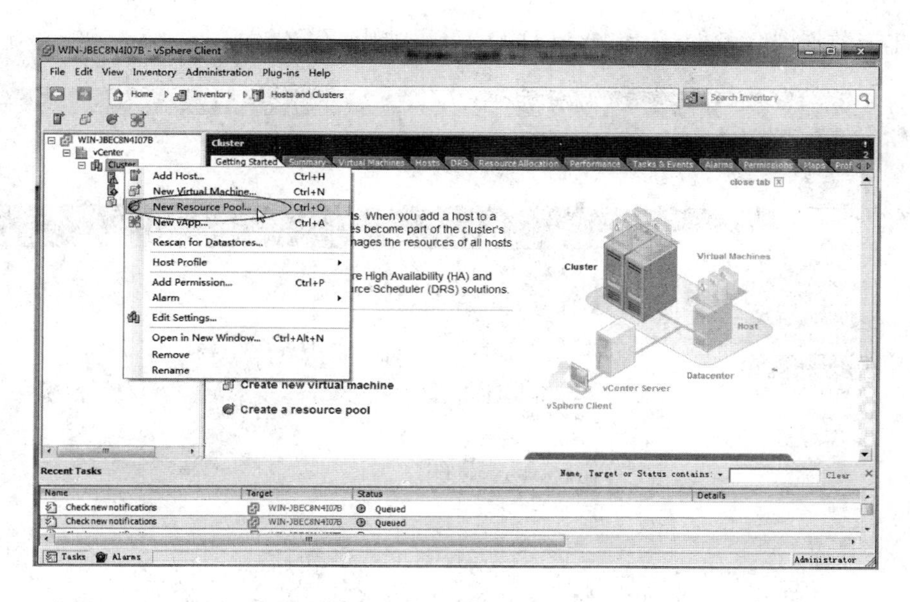

图 7-70　配置新的资源池

打开“Create Resource Pool”窗口，在“Name”处输入资源池名称，在“CPU Resource”和“Memory Resource”区域中设置资源池配置值，单击“OK”，直到资源池创建成功。

17. 创建数据存储

数据存储是用于保存虚拟机文件以及虚拟机操作所必需的其他文件的逻辑容器。数据存储存在不同类型的物理存储，包括本地存储、iSCSI、光纤通道 SAN 或 NFS。数据存储可以基于 VMFS，也可以基于 NFS。

依次选择“Home”→“Inventory”→“Datastores and Datastore Clusters”，选择数据中心下的一个存储，打开“Summary 选项”页，可以查看数据存储摘要信息。单击“Browse Datastore”，浏览数据存储中的文件，如图 7-71 所示。

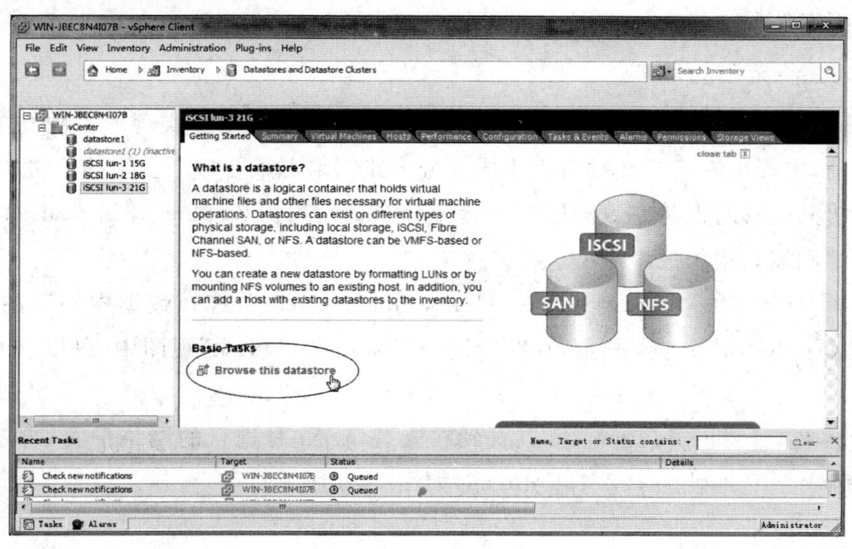

图 7-71　配置数据存储

右键单击数据中心名称，选择“Add Datastore”，如图 7-72 所示。

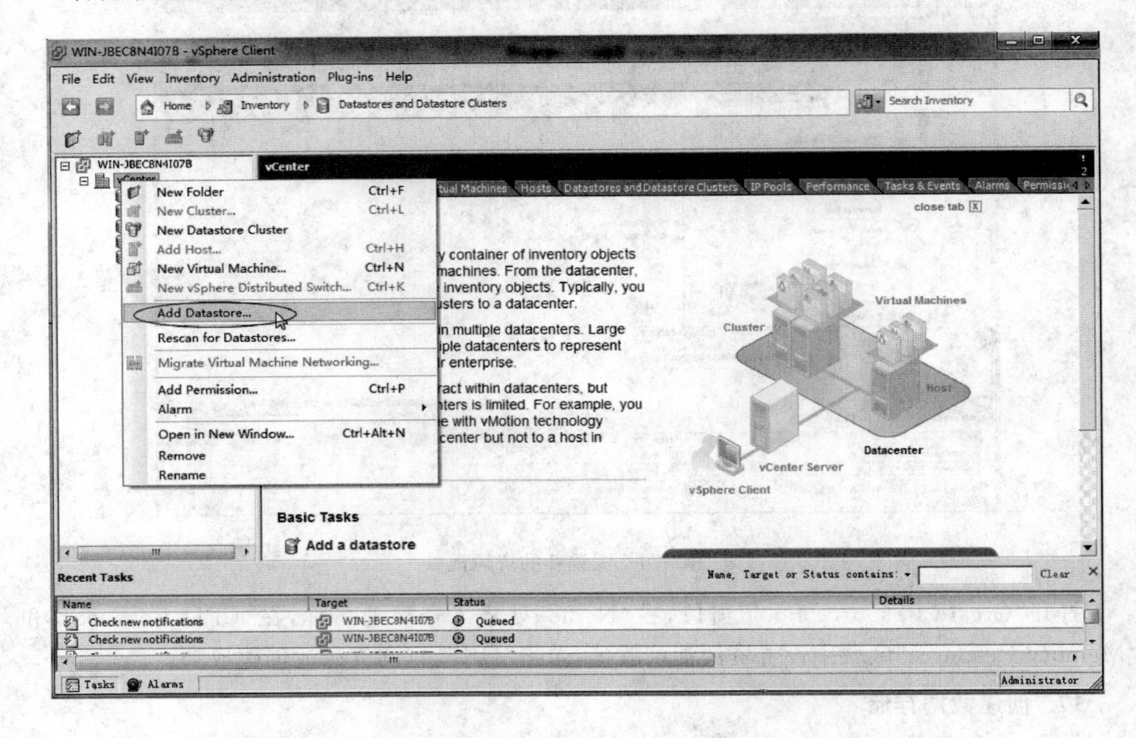

图 7-72　配置一个数据存储

打开“Add Storage”窗口，在“Select Host”页，选择一个 ESXi 主机，单击“Next”。在“Storage Type”中选择“Network File System”，单击“Next”。在 Properties 的“Server”处输入服务器名称，在“Folder”处输入共享文件夹的位置。在“Datastore Name”处输入数据存储名称，单击“Next”。查看配置摘要信息，单击“Finish”。

新的数据存储添加成功。

18. 创建主机范围的网络

在 vSphere 中，可以创建标准网络和分布式网络。标准网络可实现一个独立主机上虚拟机之间的通信，由标准交换机和端口组组成。分布式网络可聚合多个主机的网络功能，并允许虚拟机在主机间迁移时保持一致的网络配置。分布式网络由 vSphere Distributed Switch、上行链路端口组和端口组组成。

依次选择“Home”→“Inventory”→“Hosts and Clusters”，选择数据中心下的一个 ESXi 主机，打开 “Configuration 选项”页，在“vSphere Standard Switch”视图中，可以查看标准交换机配置信息。

第一个标准交换机 vSwitch0，可以看到在安装 ESXi 时进行的网络配置，绑定了两块网卡，默认管理网络 IP 地址为 192.168.10.56，如图 7-73 所示。

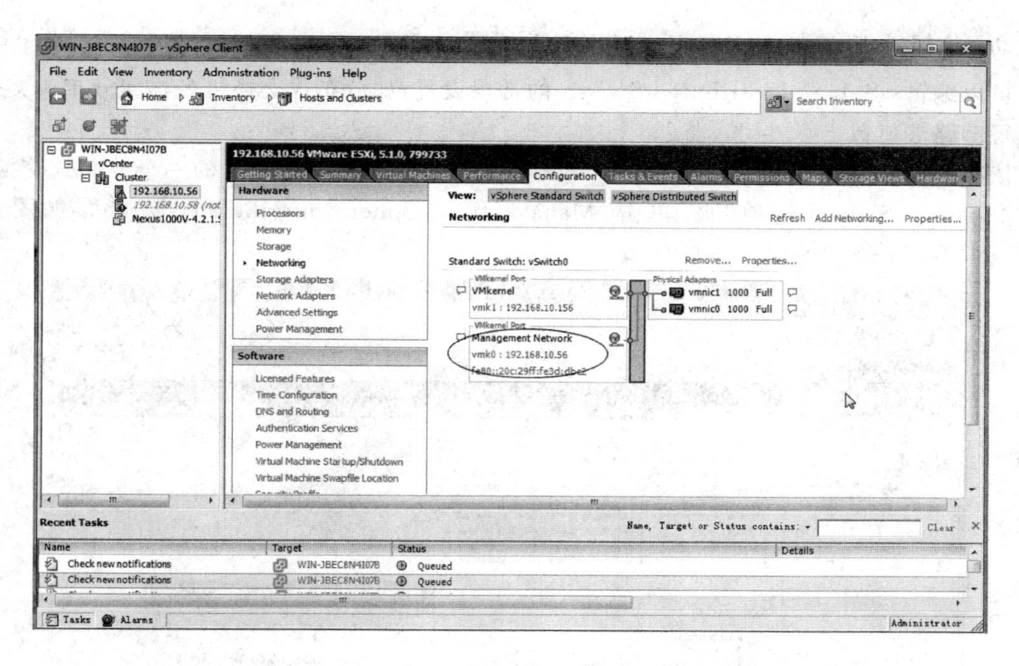

图 7-73　配置 vSwitch 参数

19. 创建数据中心范围的网络

完成主机范围的网络配置，下面是数据中心范围的网络配置。

依次选择“Home”→“Inventory”→“Hosts and Clusters”，选择数据中心下的一个 ESXi 主机，打开 “Configuration 选项”页，在“vSphere Distributed Switch”视图中，可以查看分布式交换机配置信息，如图 7-74 所示。

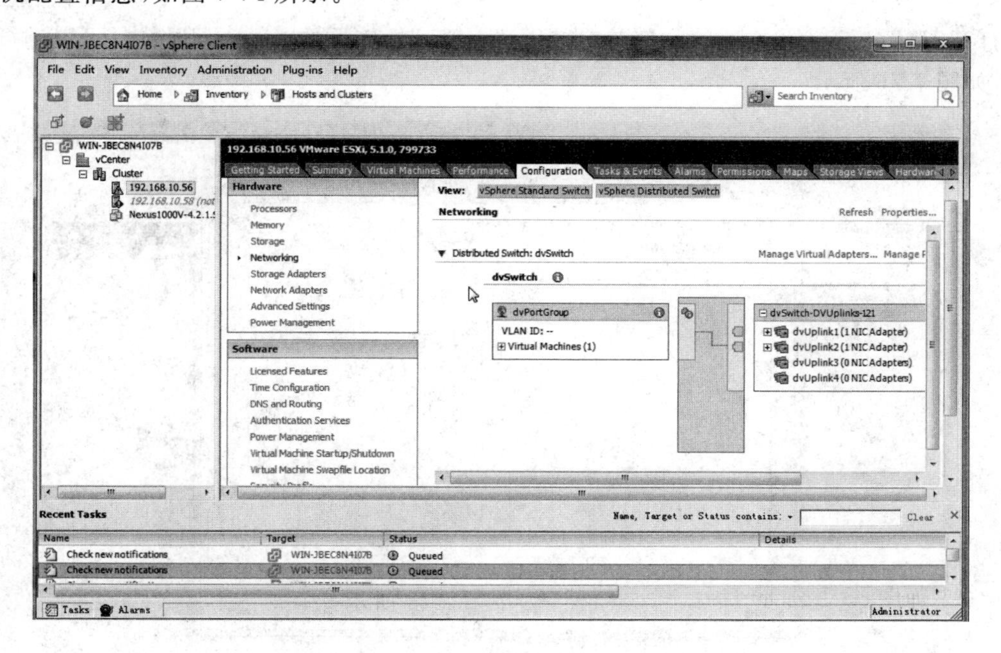

图 7-74　查看分布式交换机配置信息

20. 编辑常规 vSphere Distributed Switch 设置

可以编辑 vSphere Distributed Switch 的常规设置，如分布式交换机名称和分布式交换机的上行链路端口数。

依次选择“Home”→“Inventory”→“Networking”，选择数据中心下的一个分布式交换机，打开“Getting Started 选项”页，单击“Manage this vSphere distributed switch”，如图 7-75 所示。

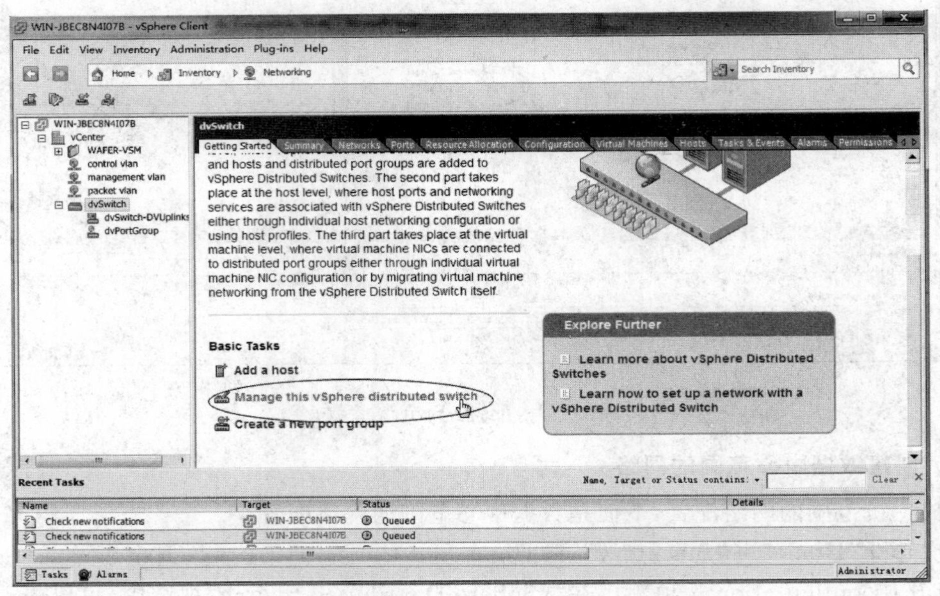

图 7-75 配置分布式交换机

打开 VM_dvSwitch Settings 窗口，在“Properties”标签页的“General”栏中，可以进行常规配置，如图 7-76 所示。

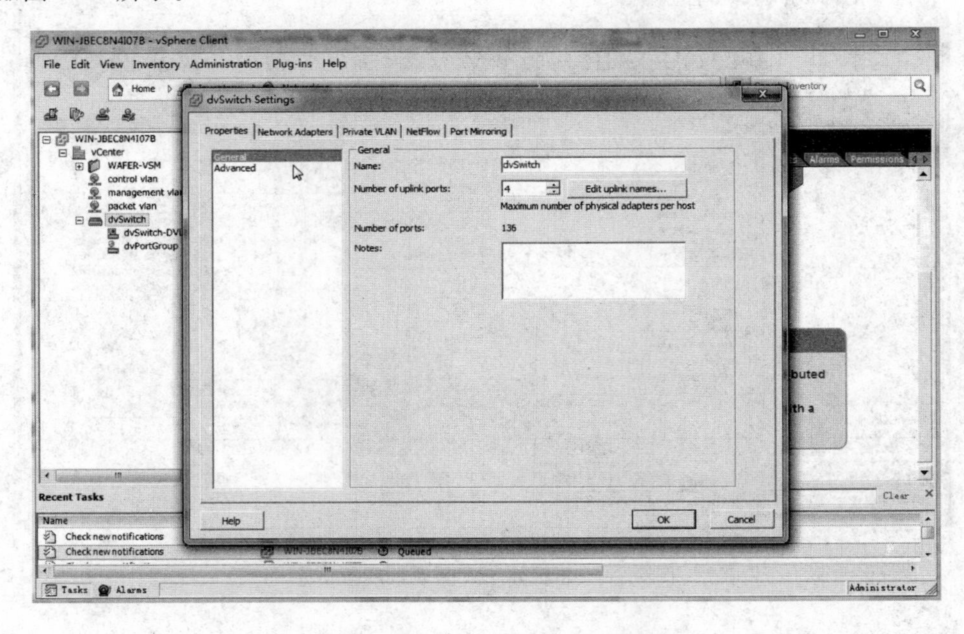

图 7-76 配置分布式交换机常规参数

21. 编辑 vSphere Distributed Switch 高级设置

可以更改高级 vSphere Distributed Switch 设置，如 vSphere Distributed Switch 的 Cisco 发现协议和最大 MTU，如图 7-77 所示。

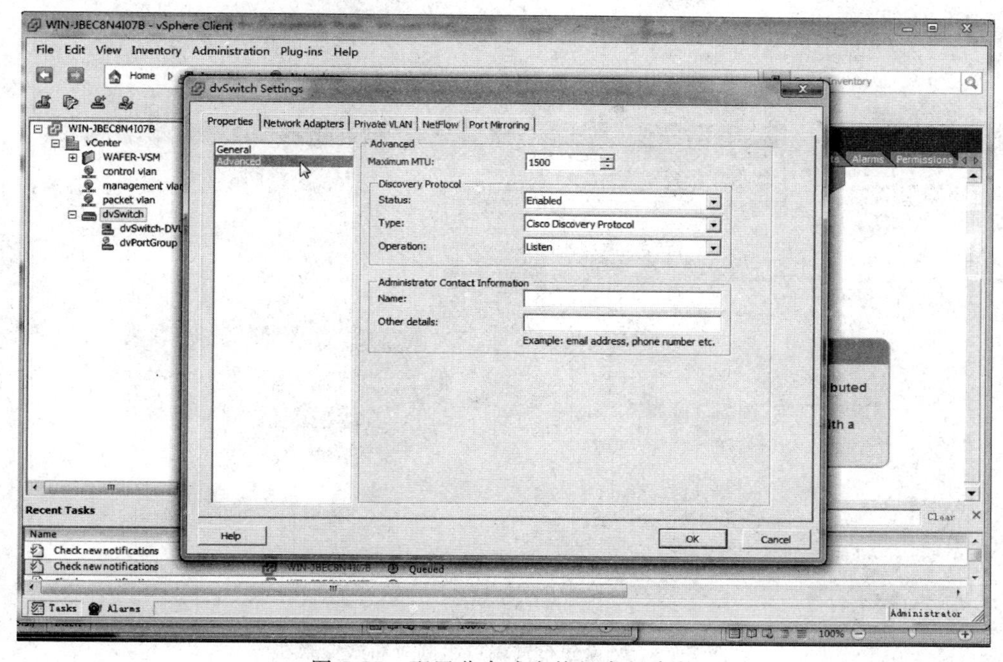

图 7-77　配置分布式交换机高级参数

22. 将主机添加到 vSphere Distributed Switch

创建 vSphere Distributed Switch 之后，可以在分布式交换机级别将主机和物理适配器添加到此 vSphere Distributed Switch，如图 7-78 所示。

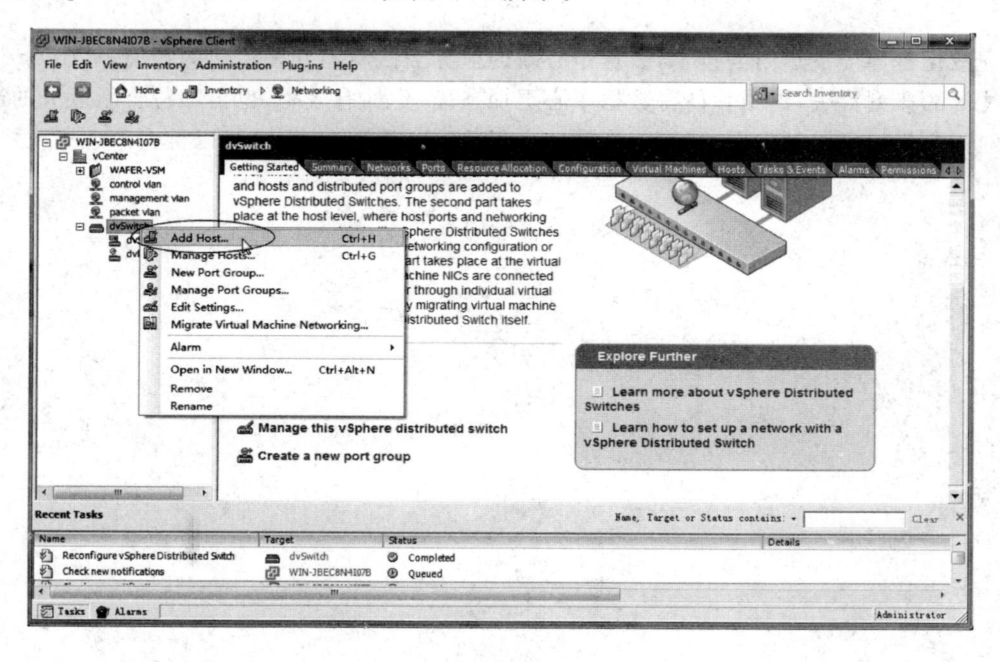

图 7-78　将主机添加到分布式交换机

23. 添加分布式端口组

将分布式端口组添加到 vSphere Distributed Switch 来为虚拟机创建分布式交换机网络，如图 7-79 所示。

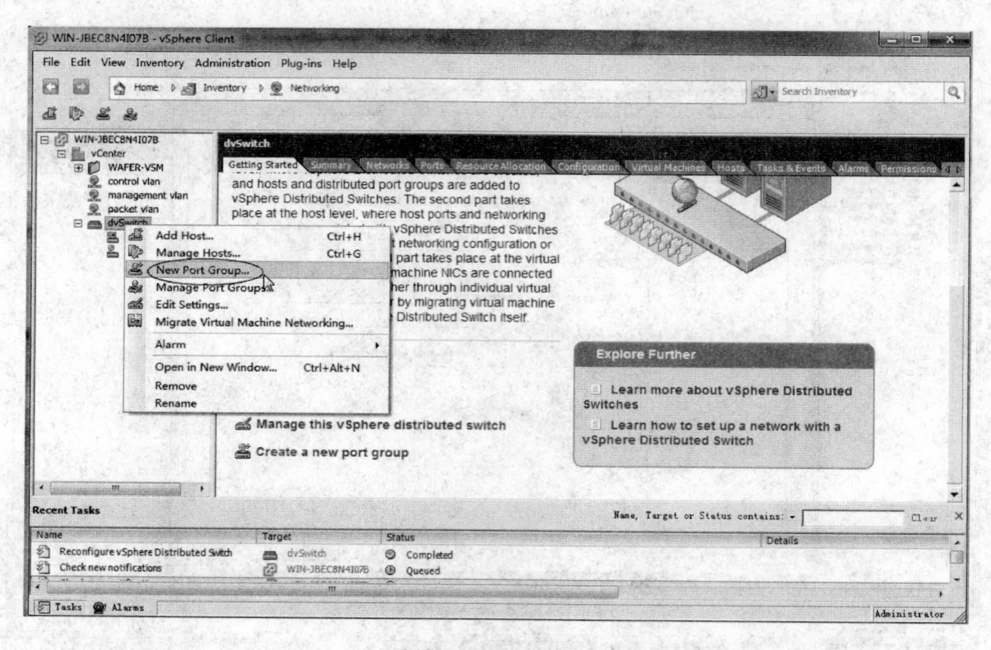

图 7-79 配置分布式端口组到分布式交换机

7.5 小 结

虚拟化是当今最热门的技术，数据中心新的业务当然离不开虚拟化的环境，VMware 是虚拟化技术的典型代表，虚拟化使得数据中心更加节能，更加有效率，也更加符合新业务的需求，互联网行业仍然是蓬勃发展的行业，新的技术仍然不断涌现，我们拭目以待。

第8章　思科虚拟数据中心产品系列安装

思科虚拟的设备，像虚拟交换机 Nexus 1000V、虚拟路由器 CSR 1000V、虚拟防火墙 ASA 1000V、虚拟应用加速和广域网优化设备 vWAAS，其功能就是物理世界在虚拟网络里的实现，所以本章的重点不在这些设备的配置，而是其在虚拟网络里的功能特性介绍及如何安装。

8.1　虚拟交换机 Nexus 1000V

8.1.1　Nexus 1000V 概念介绍

思科 Nexus 1000V 虚拟接入交换机是一款智能软件交换机，适用于 VMware ESXi 环境。思科 Nexus 1000V 在 VMware ESXi 管理程序中运行，支持思科 VN-LINK 服务器虚拟化技术，提供基于策略的虚拟机（VM）连接、移动 VM 安全保护和网络策略、对服务器虚拟化和联网团队运行无干扰的操作模式。

在数据中心部署服务器虚拟化时，虚拟服务器的管理方式一般与物理服务器不同。服务器虚拟化作为特殊部署对待，从而延长了部署所需时间，增加了服务器、网络、存储和安全管理员之间需要进行的协调工作。而采用思科 Nexus 1000V，就能拥有一个从 VM 到接入层、汇聚层和核心层交换机的统一网络特性集和调配流程。虚拟服务器能与物理服务器使用相同的网络配置、安全策略、工具和运行模式。虚拟化管理员能充分利用预先定义的、跟随 VM 移动的网络策略，重点进行虚拟机管理，这一全面的功能集能帮助我们更快地部署服务器虚拟化并从中受益。

思科 Nexus 1000V 是与 VMware 密切合作的结晶，它与 VMware 虚拟基础设施完全集成，其中包括 VMware Virtual Center，VMware ESX 和 ESXi。使用思科 Nexus 1000V 来管理 VM 连接，确定能保持服务器虚拟化基础设施的完整性。

思科 Nexus 1000V 交换机有两个主要组件，即在管理程序内部运行的虚拟以太网模块（VEM）和管理 VEM 的外部虚拟控制引擎模块（VSM），如图 8-1 所示。

8.1.2　虚拟以太网模块

思科 Nexus 1000V 虚拟以太网模块是 VMware ESX 或 ESXi 内核的一部分，能够取代 VMware 虚拟交换机的功能。VEM 充分利用了思科和 VMware 共同开发的 VMware 分布式虚拟交换机（DVS）API，为虚拟机提供高级网络功能。这一集成确保思科 Nexus 1000V 完全了解所有服务器虚拟化活动，如 VMware VMotion 和 Distributed Resource Scheduler（DRS）。VEM 从虚拟控制引擎模块获取配置信息，执行以下高级交换功能。

① 服务质量（QoS）。

② 安全：专用 VLAN、访问控制列表、思科 TrustSec 架构。

③ 监控:NetFlow、SPAN、ERSPAN。

思科 Nexus 1000V 交换机采用思科 NX-OS 软件,其设计旨在取代配备 VMware ESX 的基本交换机。思科 Nexus 1000V 提供了一个与 VMware ESX 内核紧密耦合的、丰富的 VM 管理特性集。思科 Nexus 1000V 适用于 1 Gbit/s 和 10 Gbit/s 架构,并且需与现有网络基础设施集成。

如果丢失了与虚拟控制引擎模块的连接,VEM 具有不间断转发功能,能根据最近了解的配置来继续交换流量。简而言之,VEM 为服务器虚拟化环境提供了高级交换功能和数据中心可靠性。

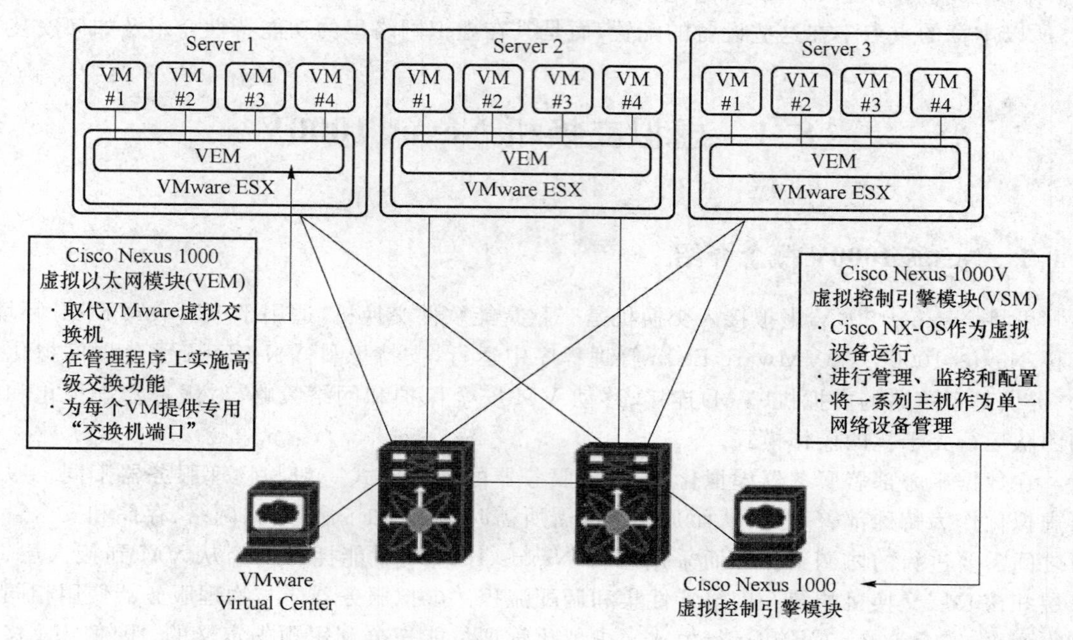

图 8-1 思科 Nexus 1000V 架构

8.1.3 虚拟控制引擎模块

思科 Nexus 1000V 虚拟控制引擎模块将多个 VEM 作为单一逻辑模块化交换机管理。用户不必再部署物理线卡模块,虚拟控制引擎模块支持服务器软件内部运行的 VEM。配置可通过虚拟控制引擎模块进行并自动传播到 VEM。管理员无须每次在一台主机上的管理程序内部配置软交换机,而是能够在所有由虚拟控制引擎模块管理的 VEM 上定义配置并立即使用。

思科 VN-LINK 的核心是一个称为端口简况的思科 NX-OS 特性。端口简况是一系列网络属性,如 VLAN 分配、服务质量(QoS)参数、访问控制列表(ACL)配置和其他过去应用于特定网络接口的网络特性。与需要分别配置各端口的传统网络调配不同,端口简况能够在一组端口上同时部署。

8.1.4 安装与配置

1. 安装环境

```
ESXi x 1 ip address: 192.168.0.10
```

```
vCenter Server x 1 ip address: 192.168.0.20
vSphere Client and RCLI x 1 ip address: 192.168.0.30
 (Cisco Nexus 1000V management ip: 192.168.0.50)
```

2. 安装步骤

① 首先，至少要先安装一个 VSM(实际是个 VM 虚拟机)。

② 安装完 VSM，确认 VSM-Vcenter 之间连接正常之后，需要在不同的 ESXi 主机上安装 VEM 模块，最后添加 ESXi 主机到 Nexus 1000V 里。

(1) 安装 VSM 过程

① 安装 VSM 的前期准备工作。

在 ESXi 上标记出 3 个 VLANS，一个用于 VSM 的 Control VLAN 3(用来检测及控制 VEMS)，一个用于 VSM 的 Management VLAN 101(远程管理 VSM)，一个用于 VSM 的 Packet VLAN 102(载有 CDP 和 IGMP 流量)，所属 VLAN ID 分别为 3，101，102(这些 VLAN 当然也要在物理交换机上创建)，如图 8-2 所示。

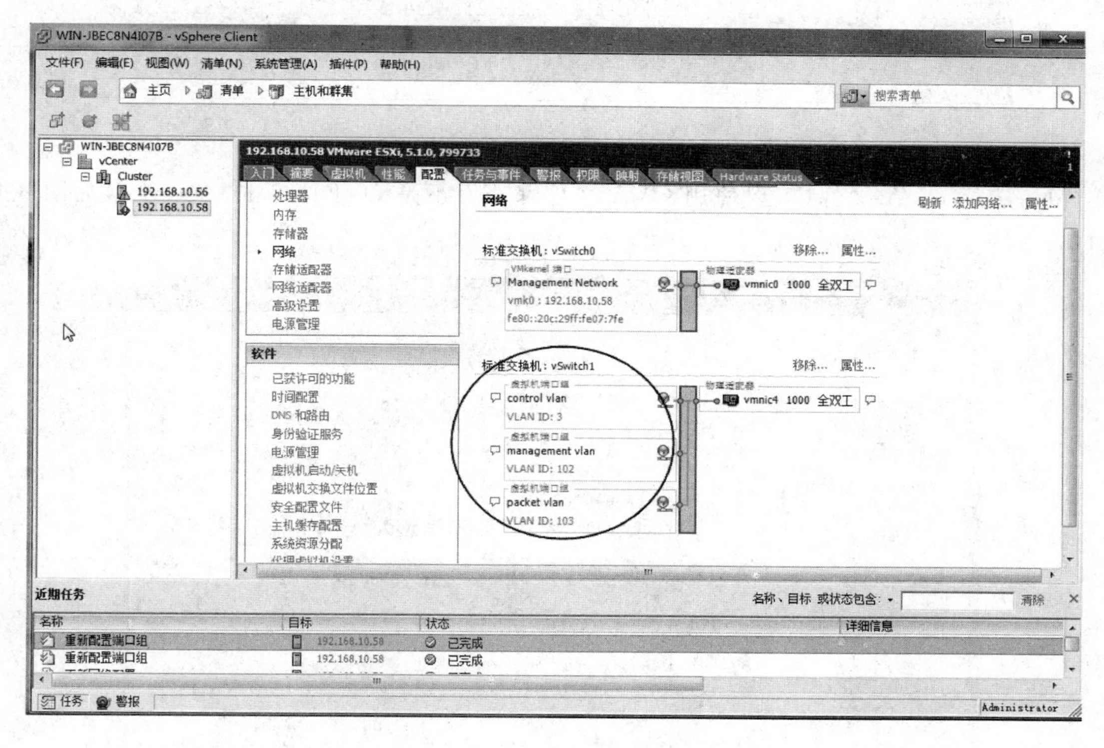

图 8-2 VSM 安装前的准备

注意: 这些 VLANS 区别于以后在 Nexus 1000V 分布式交换机上跑的生产流量，仅用于测试目的。

② 使用 OVA 模板安装 VSM 虚拟机。

a. 解压“Nexus 1000v. 5. 2. 1. SV1. 4a”安装文件到相应目录，然后执行如下操作，如图 8-3所示。

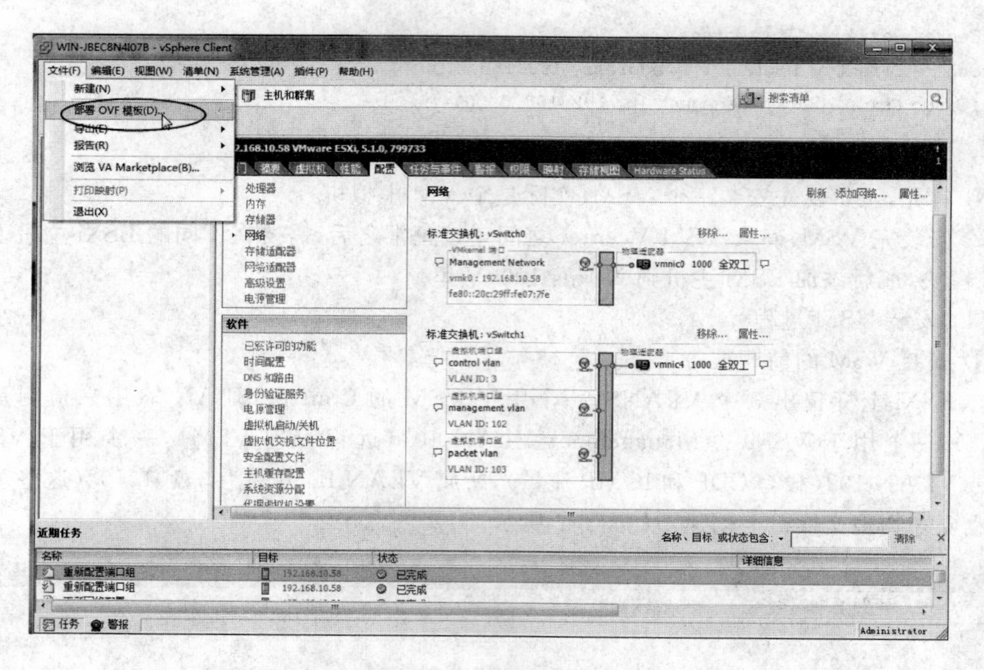

图 8-3 部署 OVA 模板

b. 选择 OVA 文件(建议使用 OVA 文件部署而不是 OVF)，单击“下一步”，如图 8-4 所示。

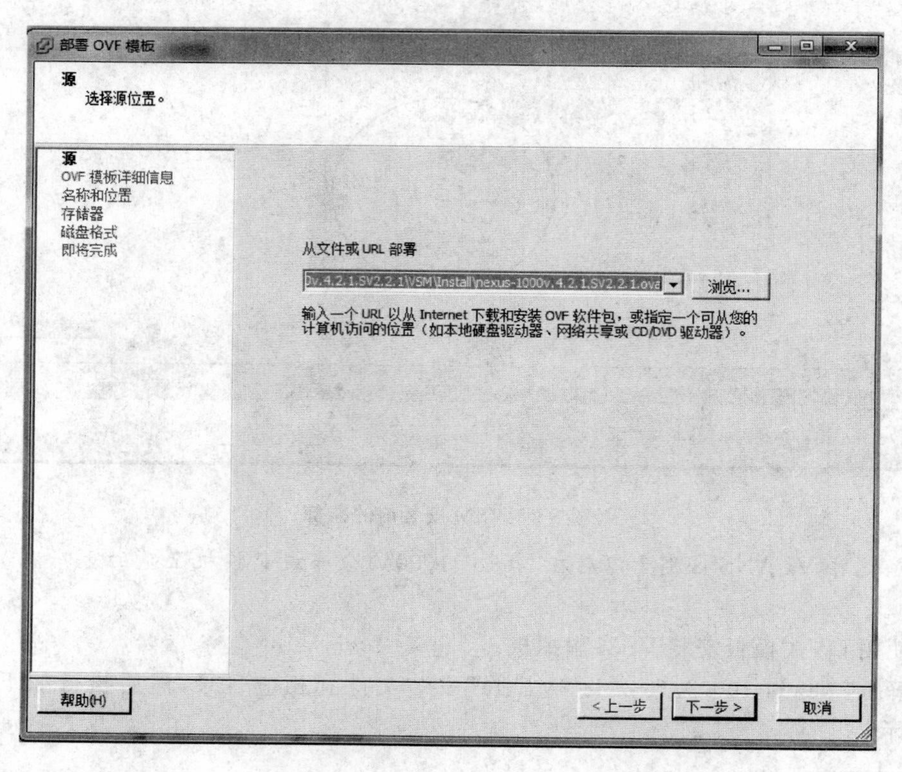

图 8-4 选择 OVA 文件路径

c. 配置选择 Nexus 1000V Installer,如图 8-5 所示。

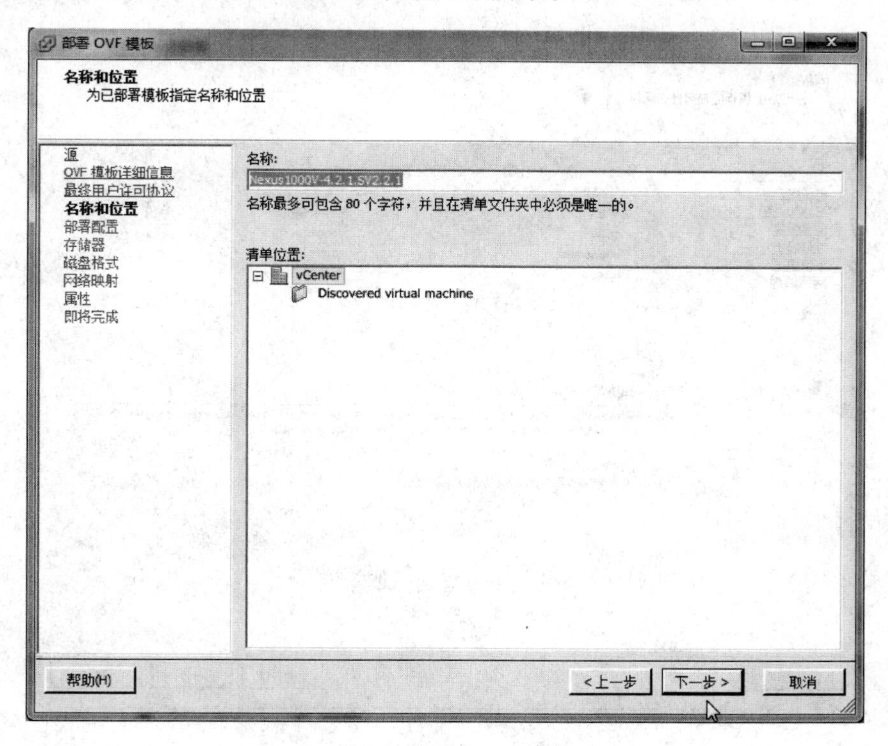

图 8-5　选择安装文件

d. 数据存储推荐选择“厚置备延迟置零”,如图 8-6 所示。

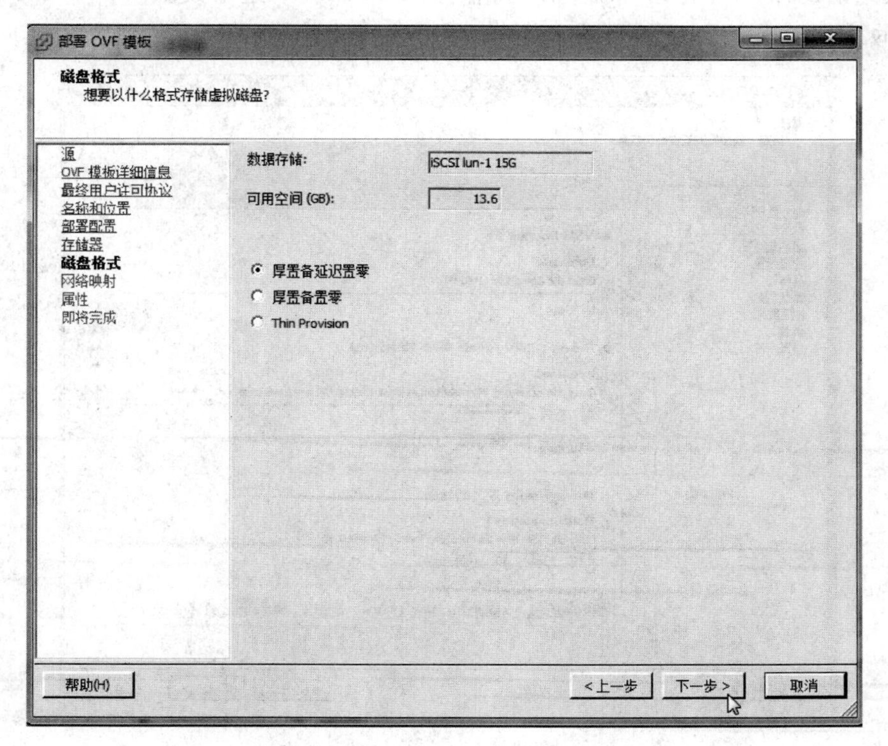

图 8-6　配置存储模式

e. 下一步关键设置好 3 个网卡对应的 3 个 VLANS，如图 8-7 所示。

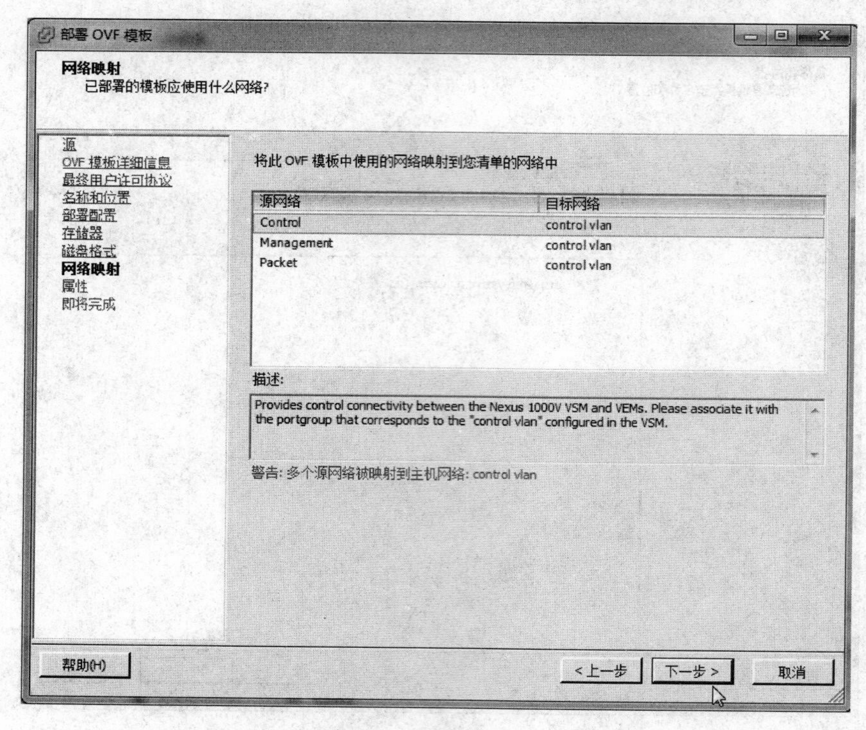

图 8-7　选择 VLAN

f. 设置 VSM 的初始网络参数(VSM 的 Domain ID 主要是用来标示 VSM 控制的不同网络环境，换句话说，同一个网络环境应使用相同的 Domain ID，这个数值在 1～4 095 之间)，如图 8-8 所示。

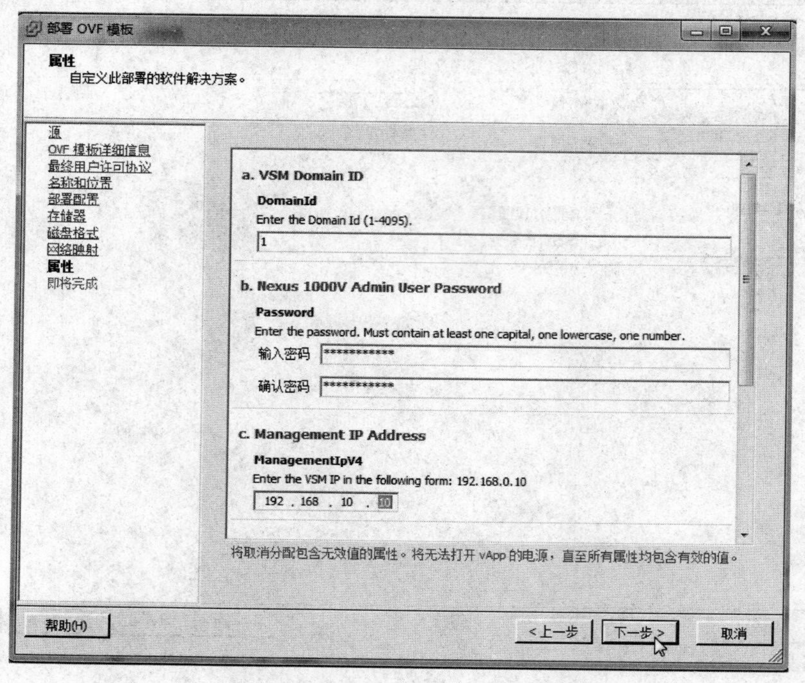

图 8-8　配置域 ID 及 1000V 账号

g. 部署设置完毕，如图 8-9 所示。

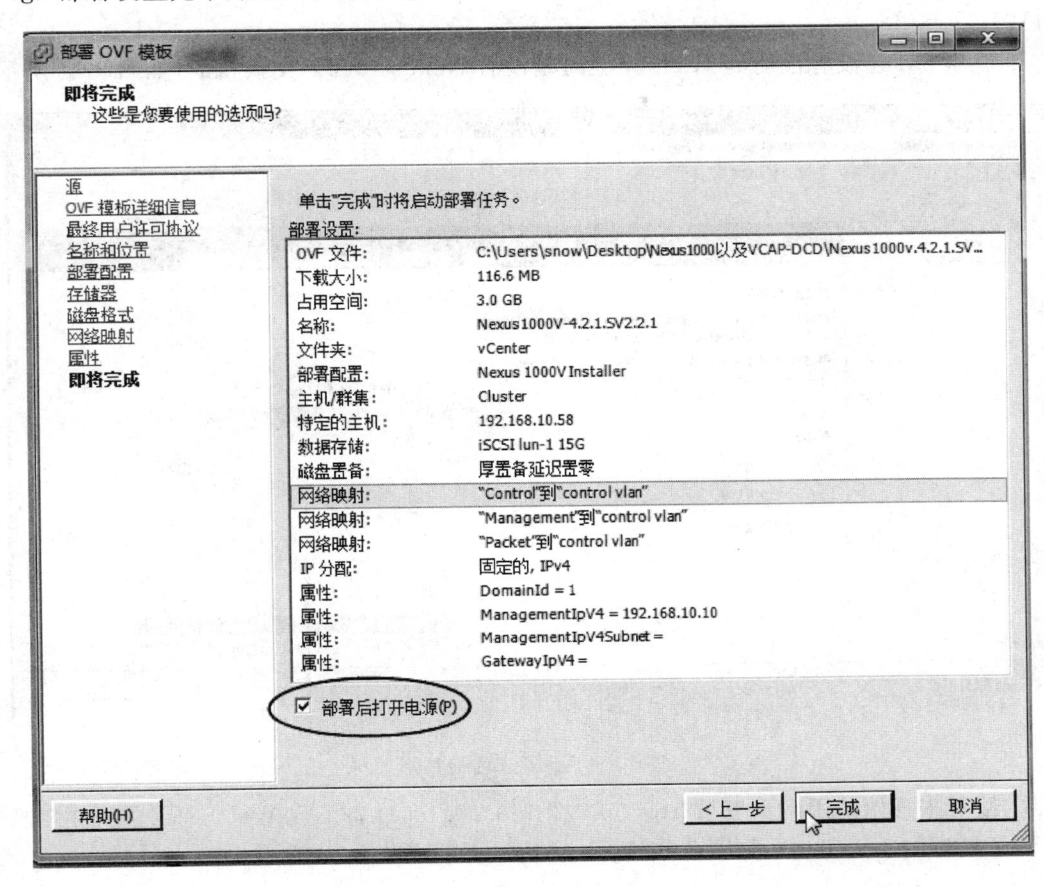

图 8-9 部署完成参数提示

h. 部署安装完之后使用 admin 账号登录测试网络连通性，Ping 下 vCeneter，如图 8-10 所示。

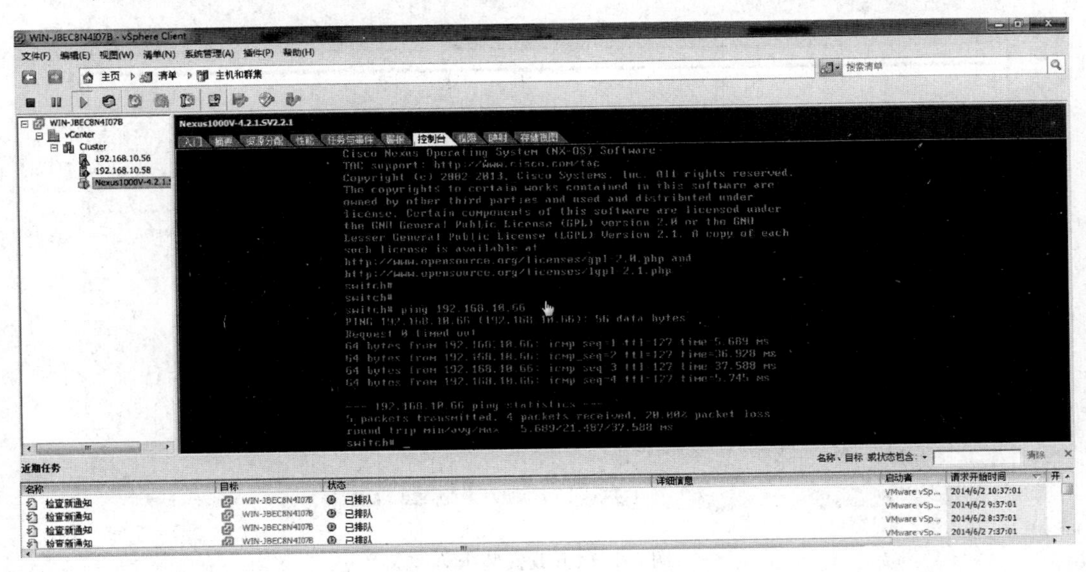

图 8-10 跟 vCenter 连通性测试

③ 在 vCenter 上使用 Web-based 工具配置 VSM，使 VSM 连接到 vCenter（也可以用命令行方式）。

a. 在 VSM 还没有注册到 vCenter 上时是没有 Nexus 1000V 交换机的，如图 8-11 所示。

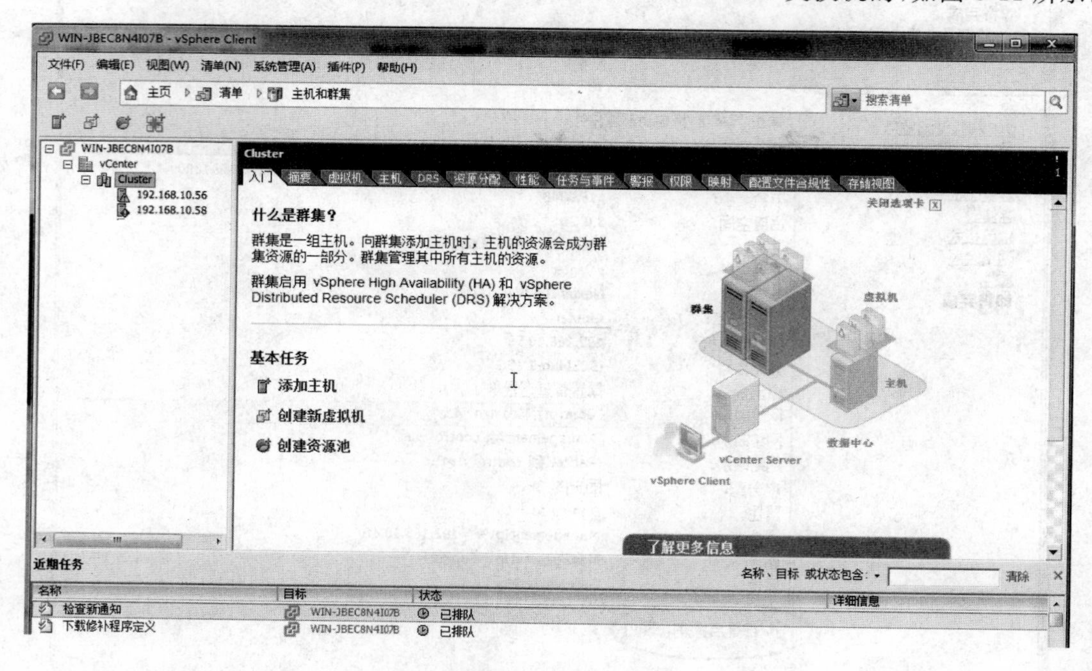

图 8-11　VSM 未注册情形

b. 接下来开始使用网页登录 http://192.168.10.10，准备下载 Web 工具（前提是需要下载安装了最近的 Java 6.0），如图 8-12 所示，会提示下载安装 Java 应用。

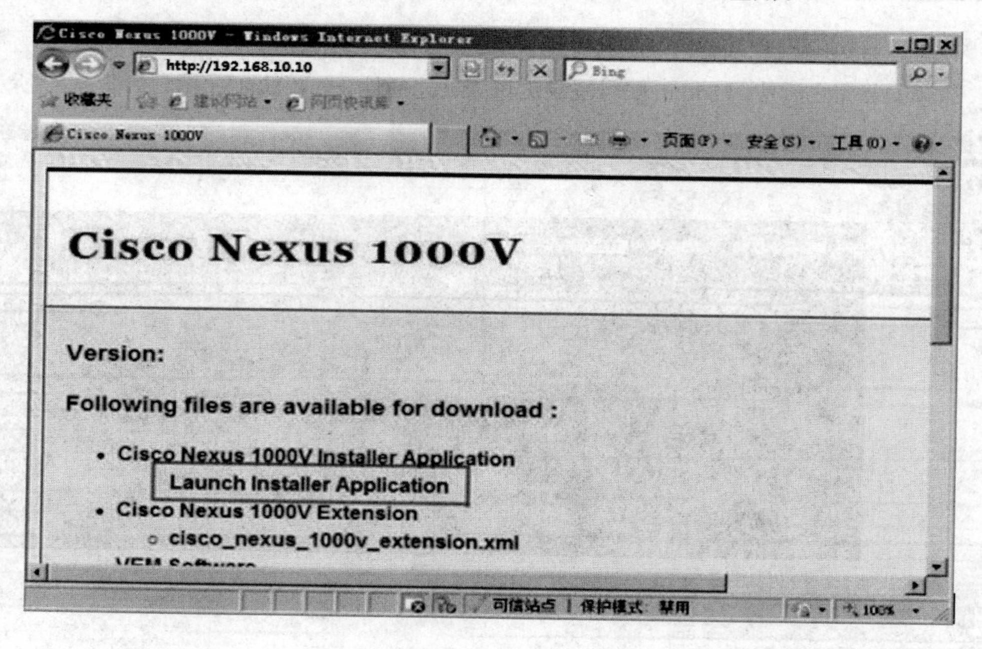

图 8-12　下载安装 Java

c. 连接上以后输入 vCenter 的 IP 地址及管理员账号和密码，下一步需要选择 VSM 虚拟机，这里配置选项使用 Advanced L2 方式，如图 8-13 所示。

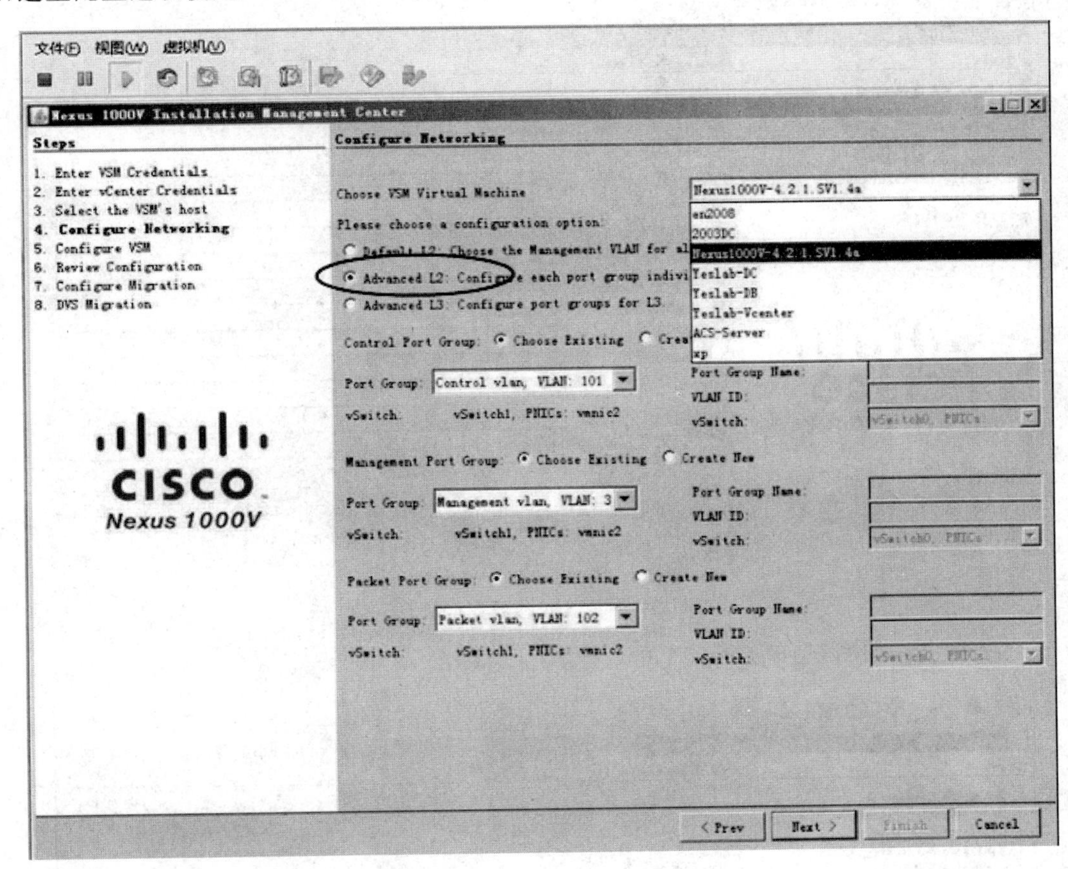

图 8-13　配置选项用 Advanced L2 方式

d. 单击“Next”，重新配置一下 VSM 的网络参数，如图 8-14 所示。

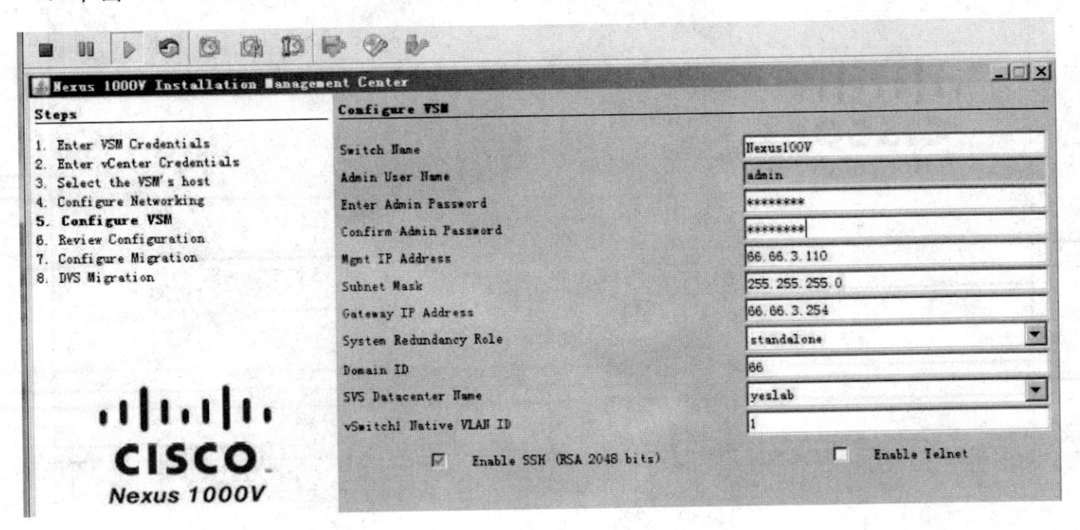

图 8-14　VSM 的网络参数设置

e. 下一步将显示全部配置信息，如图 8-15 所示。

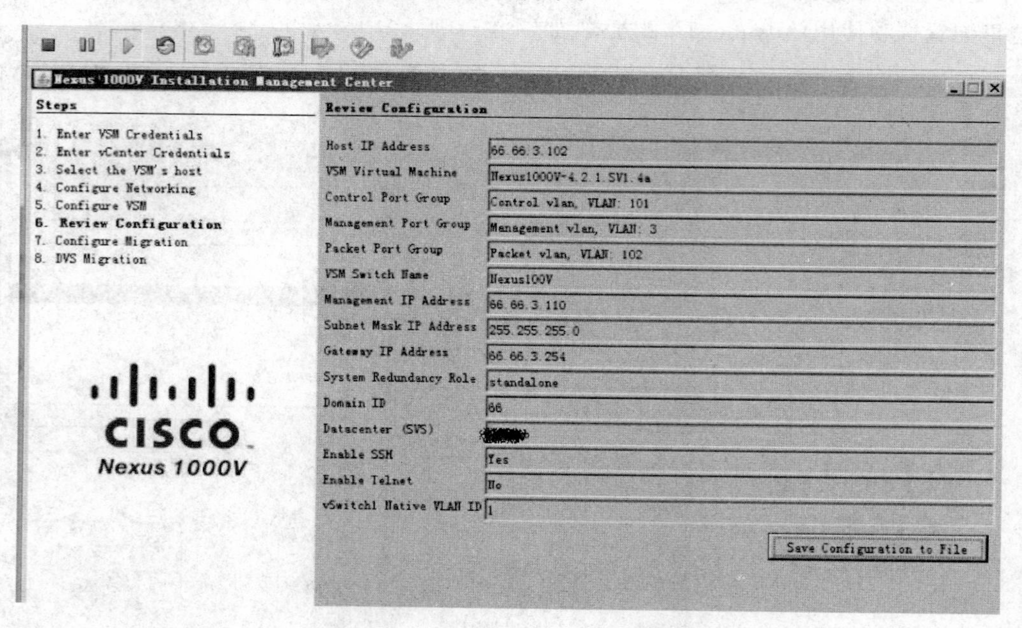

图 8-15　VSM 参数完成确认

f. 完成之后，如图 8-16 所示。

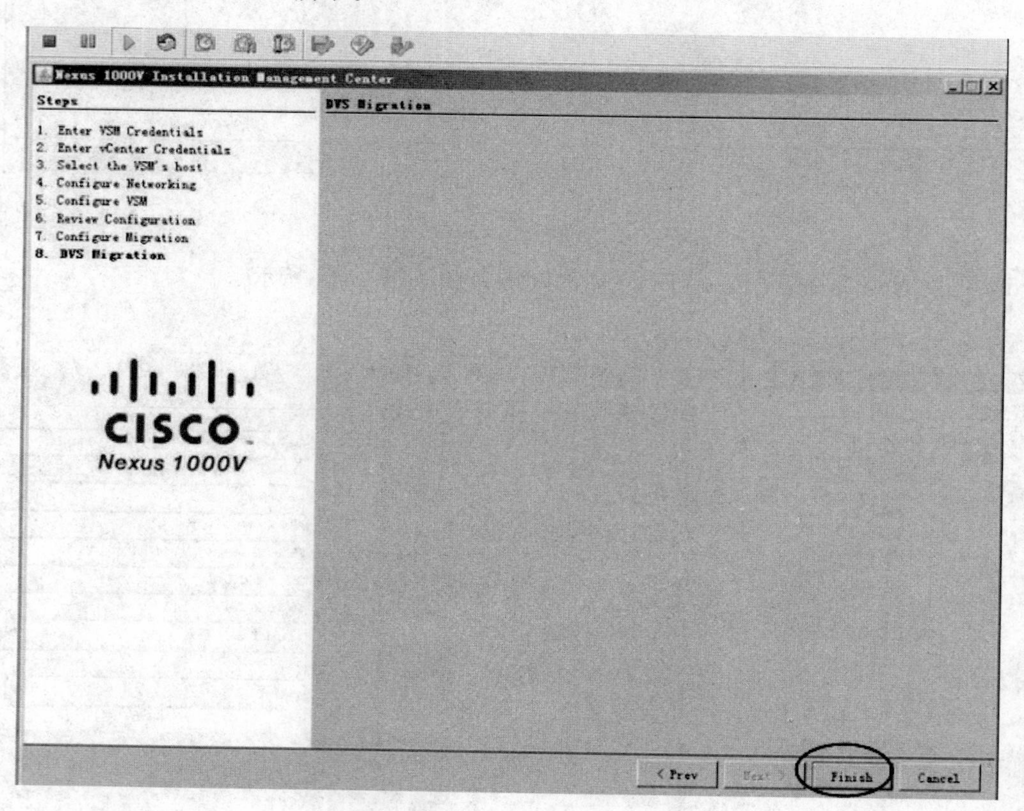

图 8-16　完成配置

g. 整个过程完成之后，VSM 需要重启一次，最终 VSM 会注册到 vCenter 上，Nexus 1000V 会添加到 vCenter 上，显示为两个默认的分布式交换机，如图 8-17 所示。

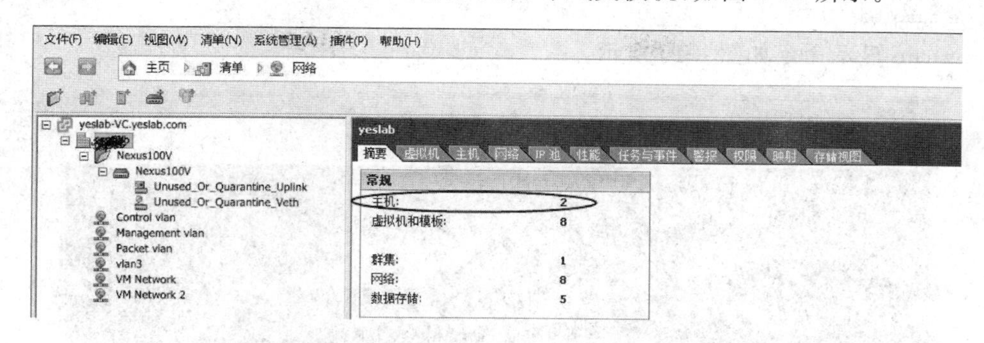

图 8-17 Nexus 1000V 配置完成的情形

h. 当然也可以在 VSM 上使用"show svs connections"查看连接状态，如图 8-18 所示。

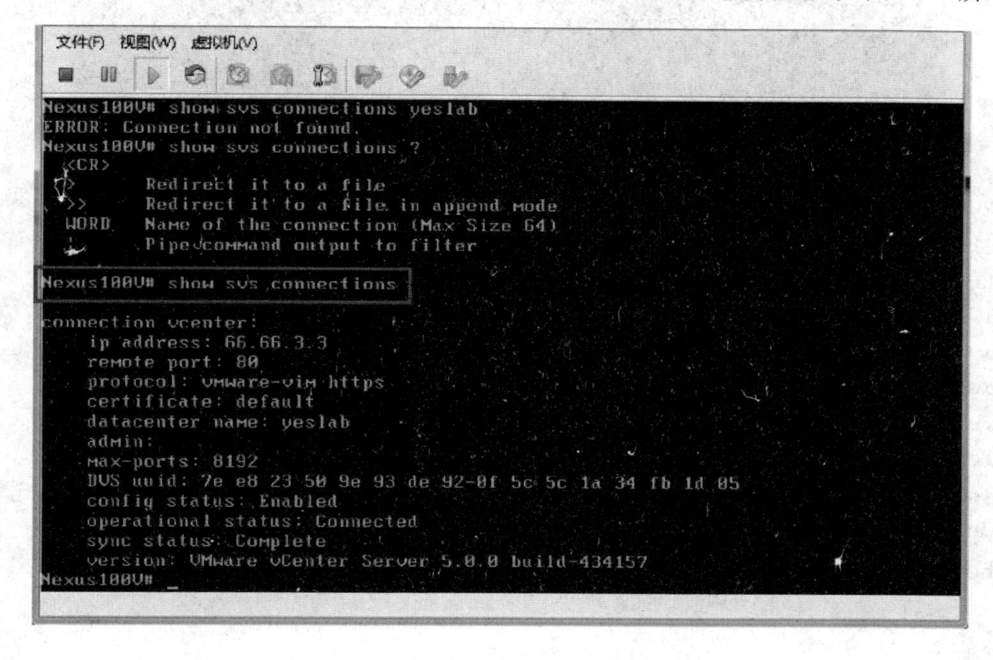

图 8-18 显示 Nexus 1000V 已经关联上

④ 在 VSM 上为 VEM 通信所需要配置 port-profile（此处需要理论性知识较强，注意的细节也很多，限于篇幅不作过多讲解，只给出配置命令，请注意命令的先后顺序）。

```
vlan 3
name Management-vlan
vlan 101
name Control-vlan
vlan 102
name Packet-vlan
port-profile type ethernet system-uplink
no shudown
switchport mode trunk
swithport trunk allow vlan all
```

sytem vlan 3,101,102

vmware port-group

state enabled

show run 显示命令如图 8-19 所示。

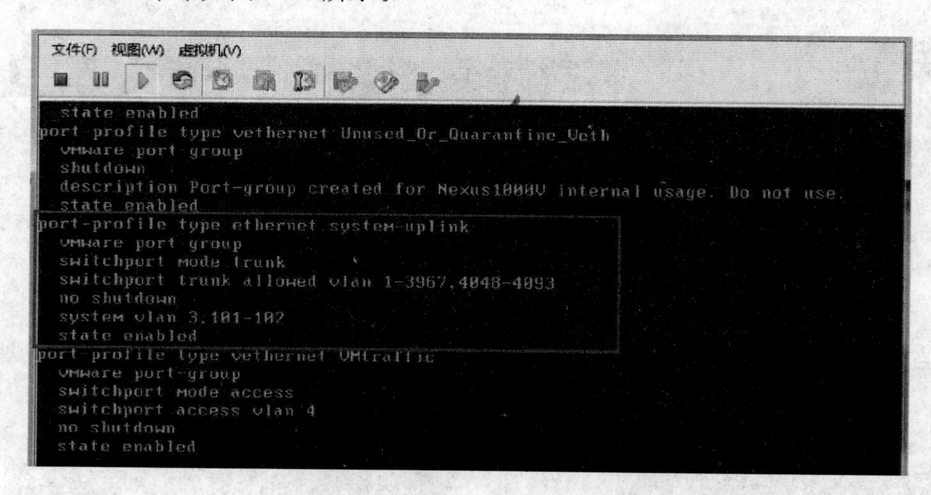

图 8-19　显示 VSM 配置

⑤ 在 VSM 上为 VM 通信所需要配置 port-profile(也是同样给出了命令,这步其实是为了最后的测试用的,在实际应用中,此内容会添加很多)。

vlan 4

name VMtraffic

port-profile type vethernet VMtraffic

no shutdown

switchport mode access

switchport access vlan 4

vmware port-group

state enabled

show run 显示命令如图 8-20 所示。

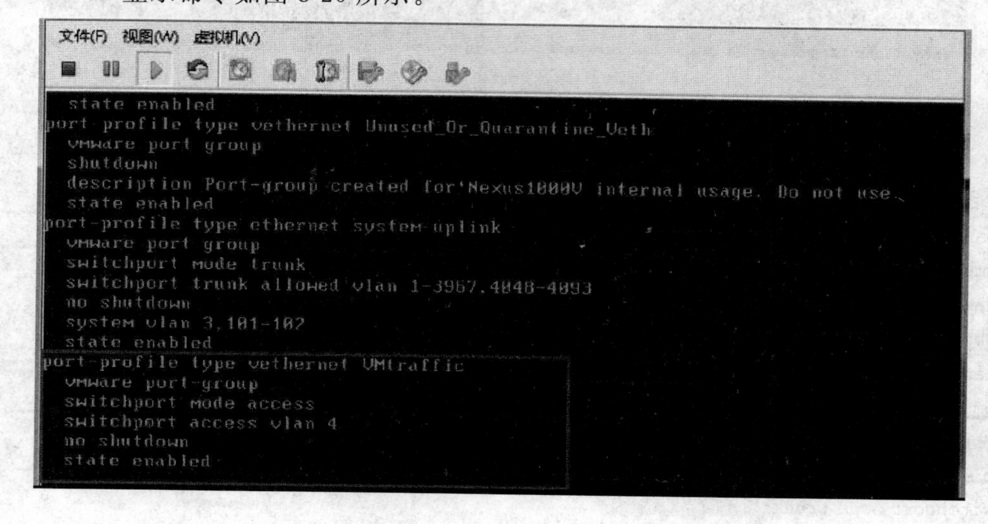

图 8-20　显示 port-profile 配置

新建的两个 port-profile 会在 vCenter 的 Nexus 分布式交换机里生成，如图 8-21 所示。

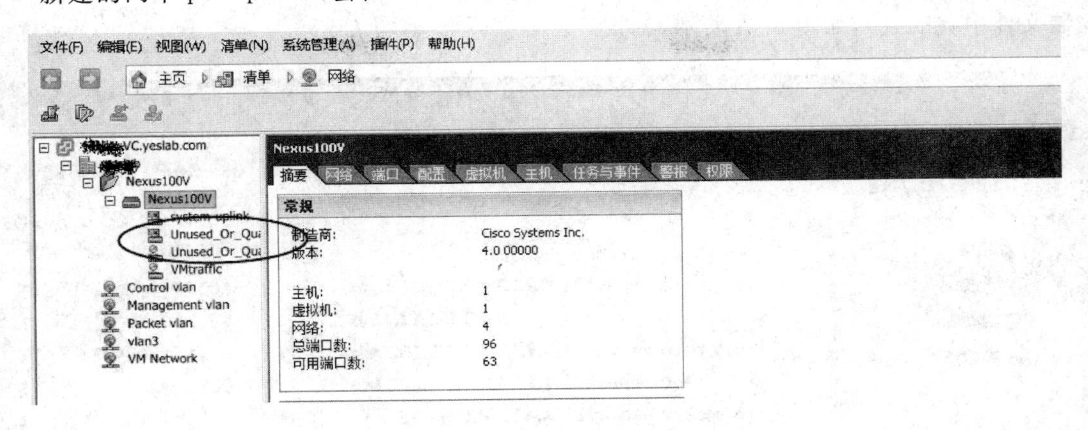

图 8-21　新建的两个 port-profile

至此，整个安装的第一步骤完成。

(2) 安装 VEM 到 ESXi 主机，添加 ESXi 主机到 Nexus 1000V 并最终测试。

① 使用 CLI 方式安装 VEM 模块到指定 ESXi 主机上（就是 ESXi2 这台）。

第一步，开启 ESXi 的远程 SSH，在“vSphere Client”中选中 ESXi 服务器，单击右面的“配置”→“安全配置文件”→“属性”，在“服务属性”画面中，选择“远程技术支持(SSH)”，然后单击“选项”，选择“自动启动”，并单击“确定”。

使用 SSHv2 远程登录 ESXi2(IP 地址为 192.168.10.56)的效果如图 8-22 所示。

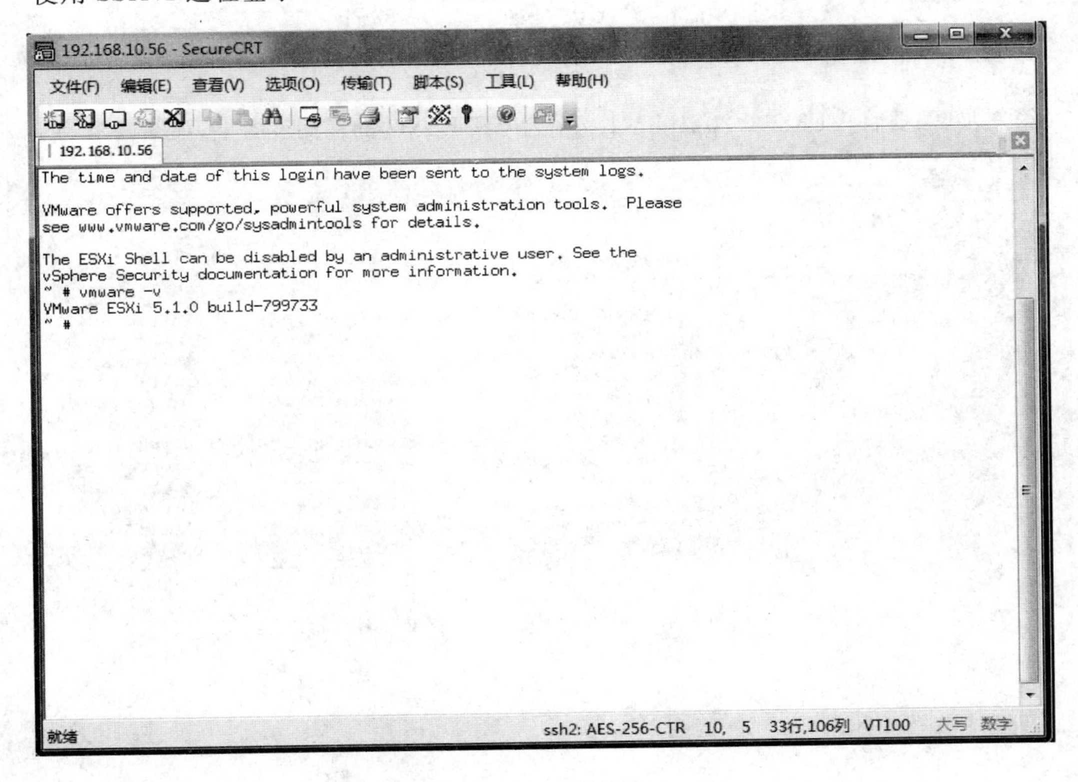

图 8-22　SSH 登录窗口

第二步，确定 VEM 的安装文件 VEM500-201108271.zip（在 Nexus1000v.5.2.1.SV1.4a 安装文件中找的到），如图 8-23 所示。

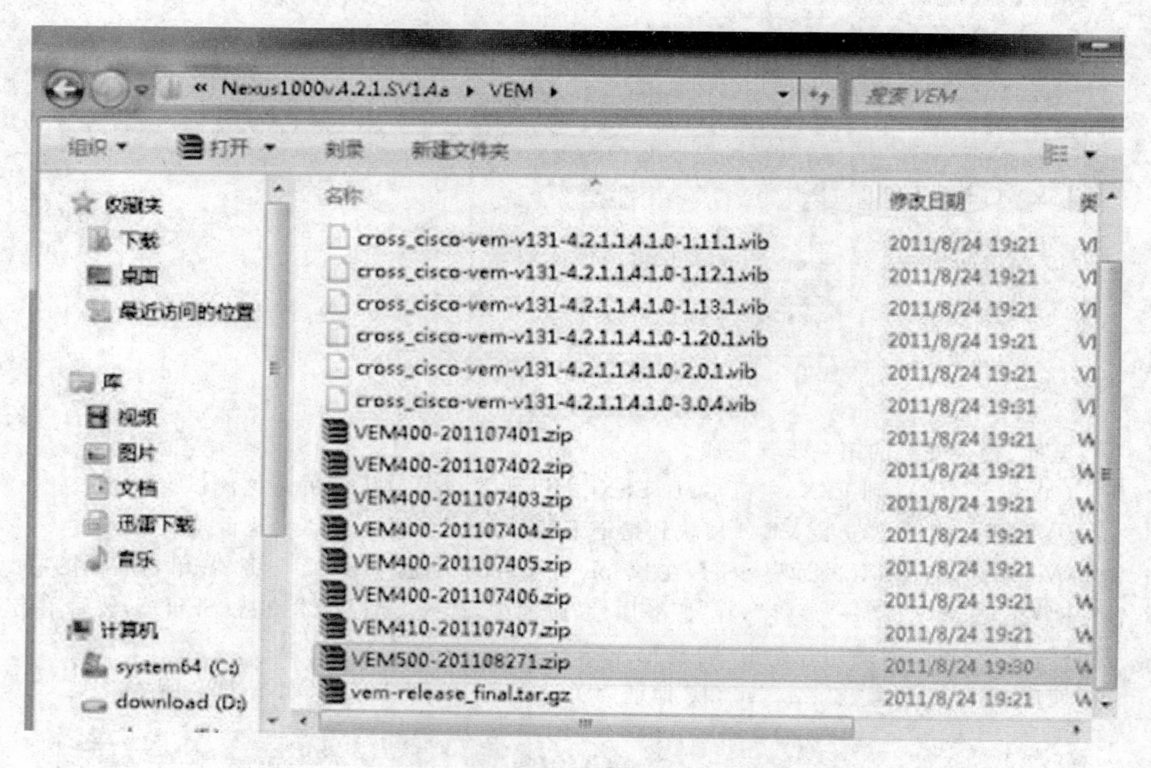

图 8-23　安装路径

第三步，需要将 VEM500-201108271.zip 上传到 ESXi2 的存储设备上，如图 8-24 所示。

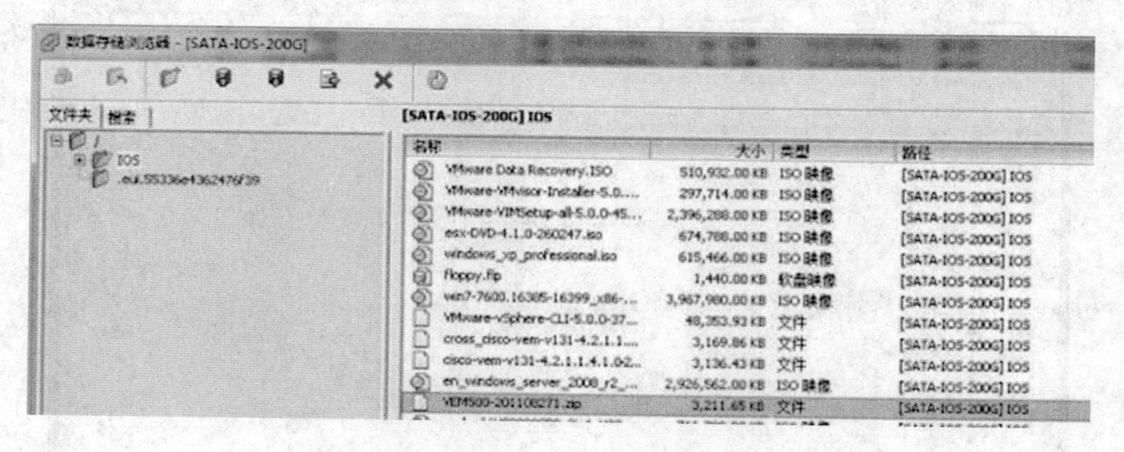

图 8-24　上传到 ESXi 共享存储文件

这样可以在 ESXi2 上找到此安装文件，查看命令如图 8-25 所示。

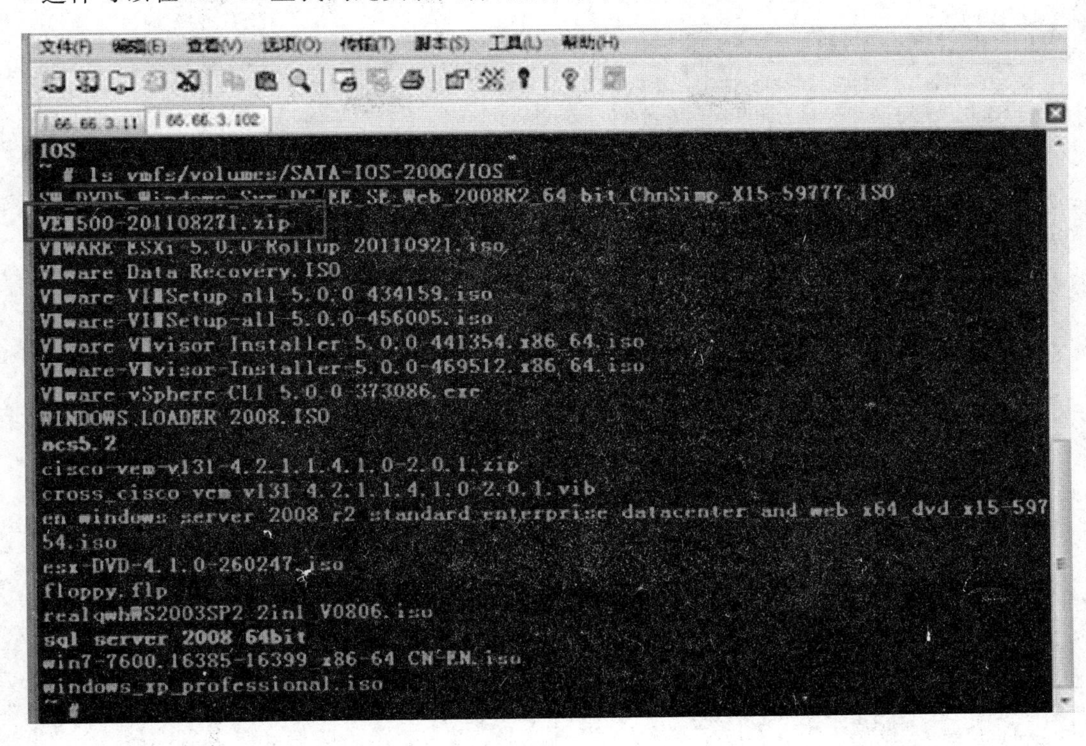

图 8-25　选择要安装的文件

执行命令“esxcli software vib install-d（VEM 文件的绝对路径）”，如图 8-26 所示。

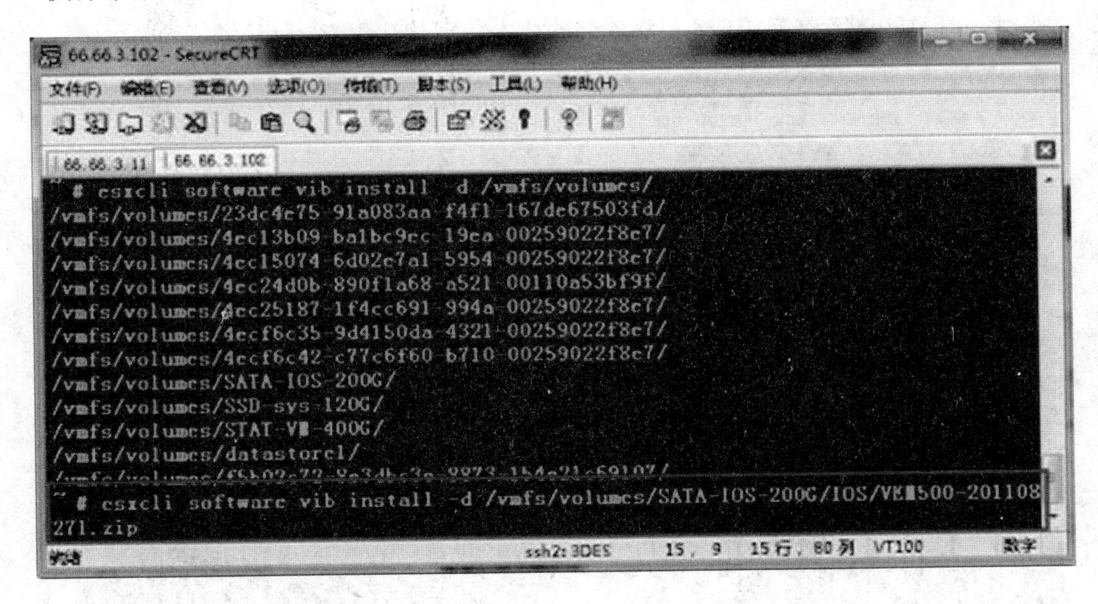

图 8-26　安装命令

安装完之后使用“esxcli software vib list | grep cisco”查看安装结果，如图 8-27 所示。

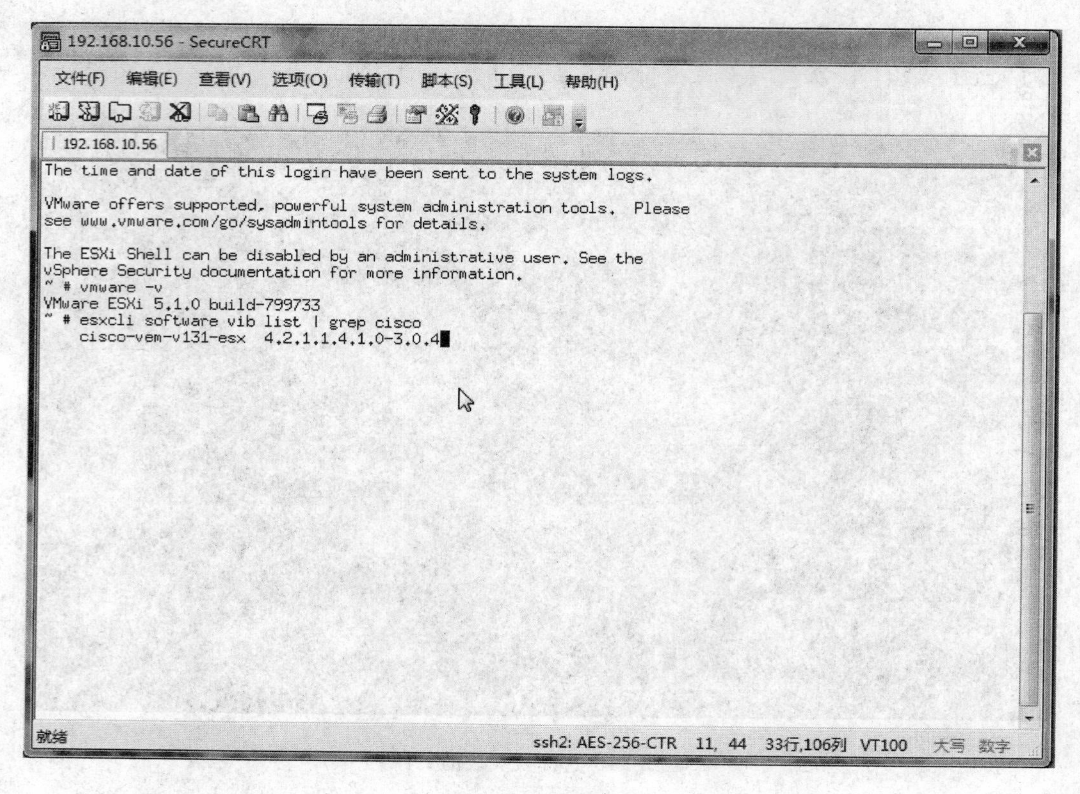

图 8-27 安装完成后确认

使用 vem status -v 查看 VEM 模块安装的版本信息，如图 8-28 所示。

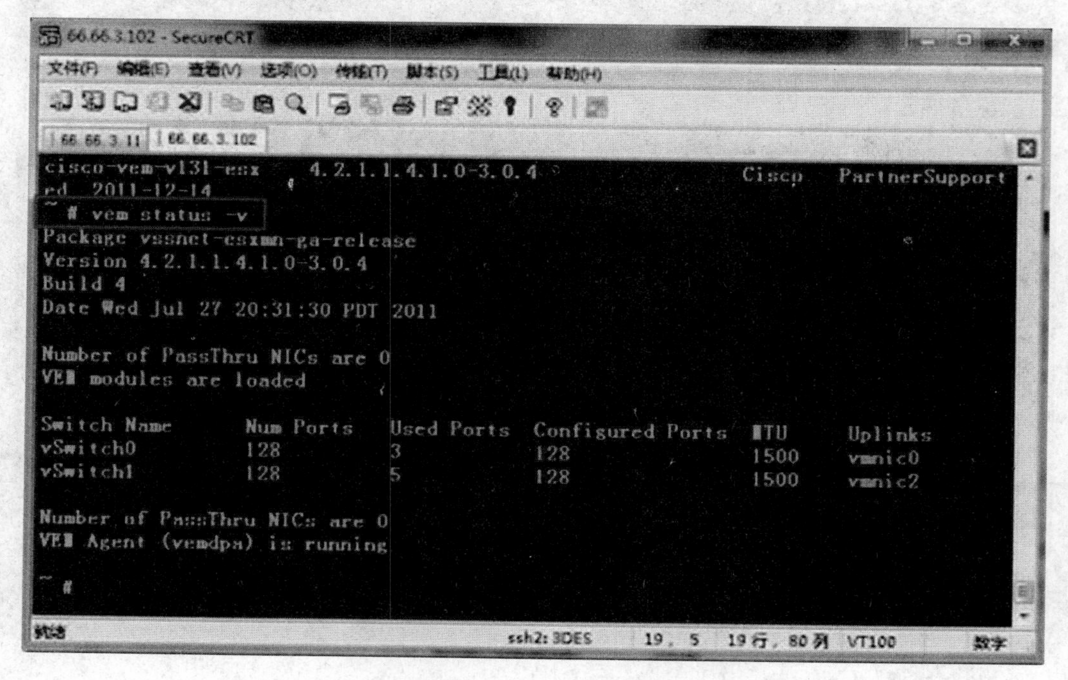

图 8-28 检查 VEM 版本信息

② 添加 ESXi2 主机到 Nexus 1000V。

这也是在整个演示中需要在理论知识上好好理解的地方，虚拟分布式交换机和真实物理交换机的整个通信过程理解起来有些复杂，如图 8-29 所示。

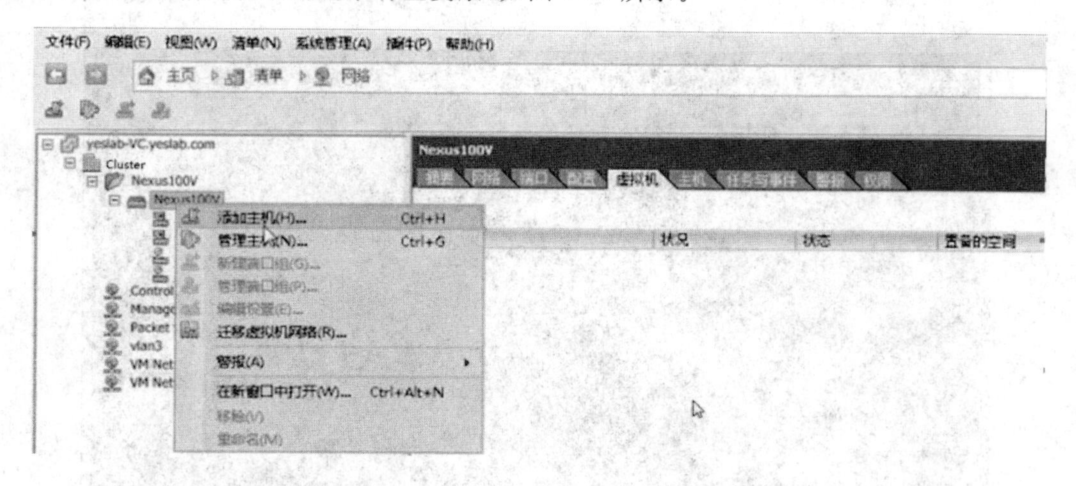

图 8-29　关联一个主机到 Nexus 1000V

a. 选择 ESXi2 网卡，选择新的未用的 vmnic1 网卡，如图 8-30 所示。

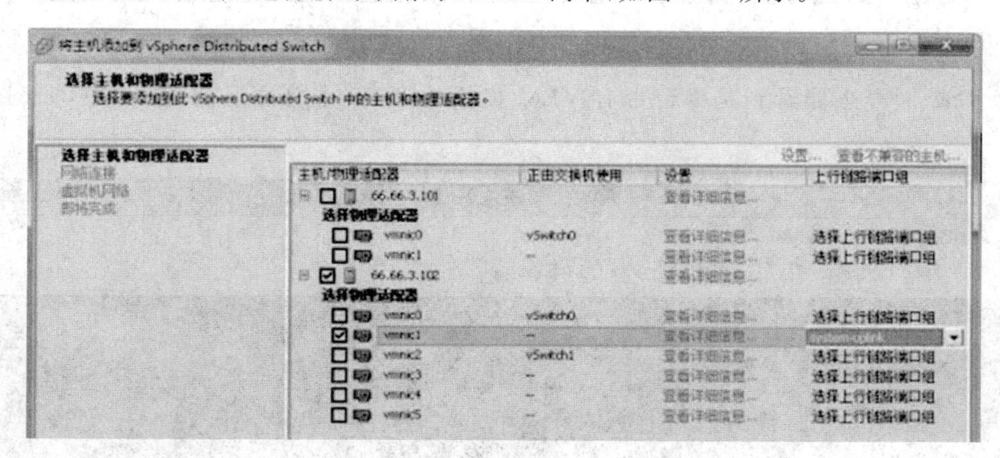

图 8-30　选择网卡

b. 添加完成之后在主机页面会看到 ESXi2，如图 8-31 所示。

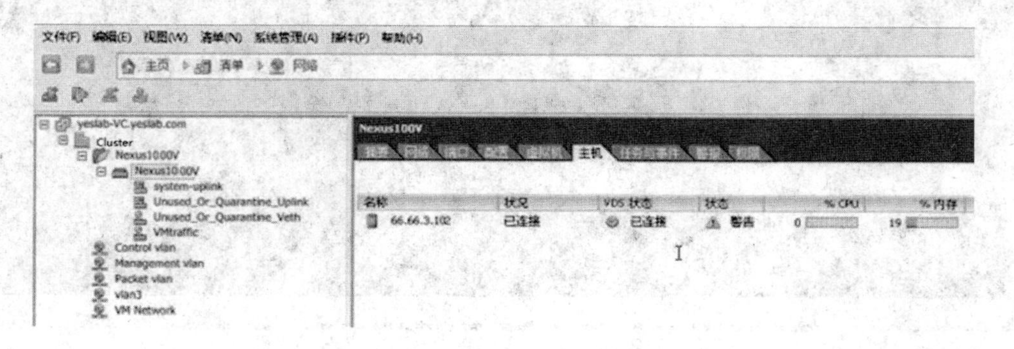

图 8-31　ESXi 已经连接到 Nexus 1000V

c. 在 VSM 上 show module 会看到 VEM 模块会添加到 VSM 中，如图 8-32 所示。

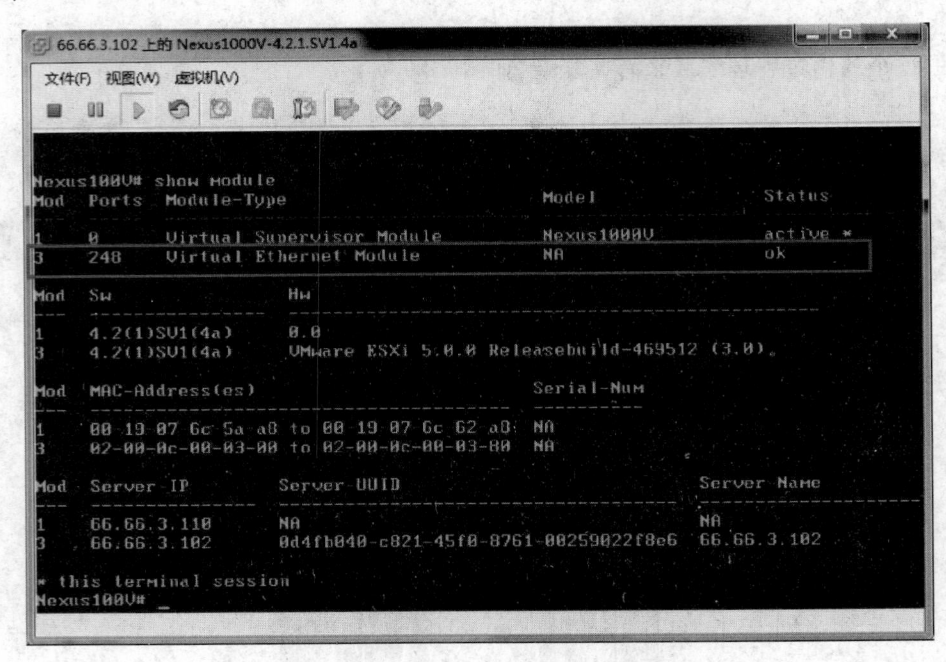

图 8-32　显示 VEM 模块

此命令同样也能看到是哪台带有 VEM 模块的 ESXi 主机注册到 VSM 中，如图 8-33 所示。

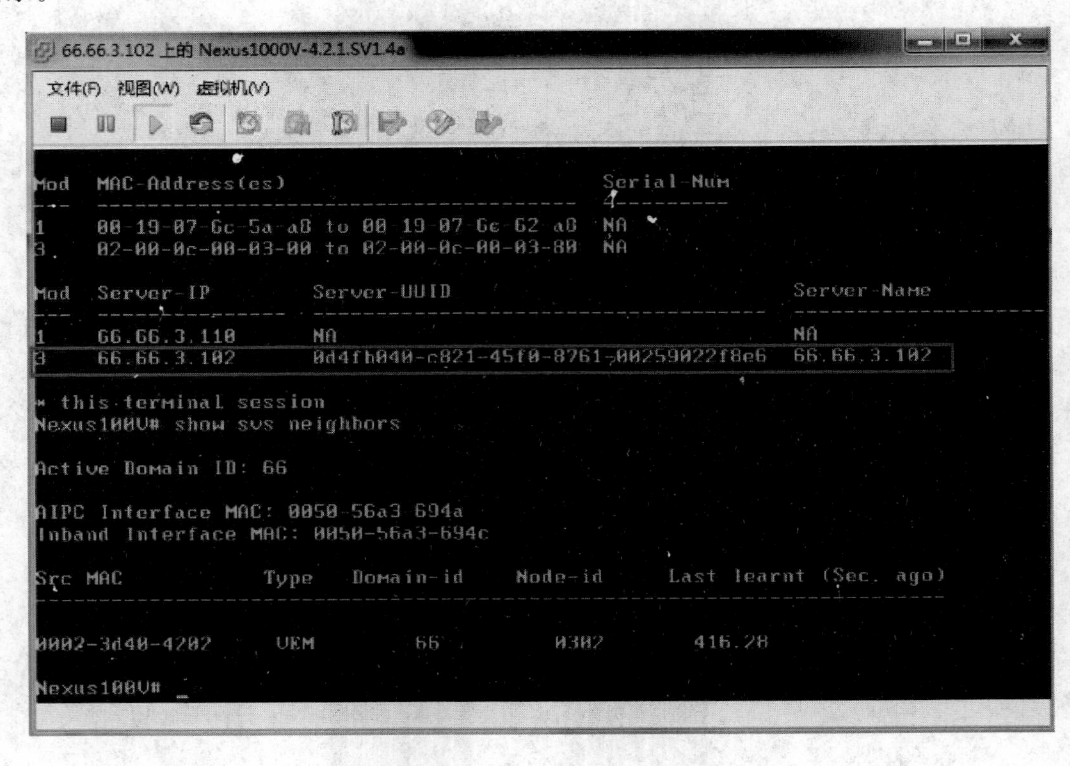

图 8-33　VEM 详细信息

至此，第二大步骤完成。

最终使用 Nexus 1000V 分布式交换机测试网络连通性。

有台 XP 虚拟机它的网卡使用的正是我们之前用的 VMtraffic 网络，在 ESXi Host 上 Ping Nexus 1000V，如图 8-34 所示。

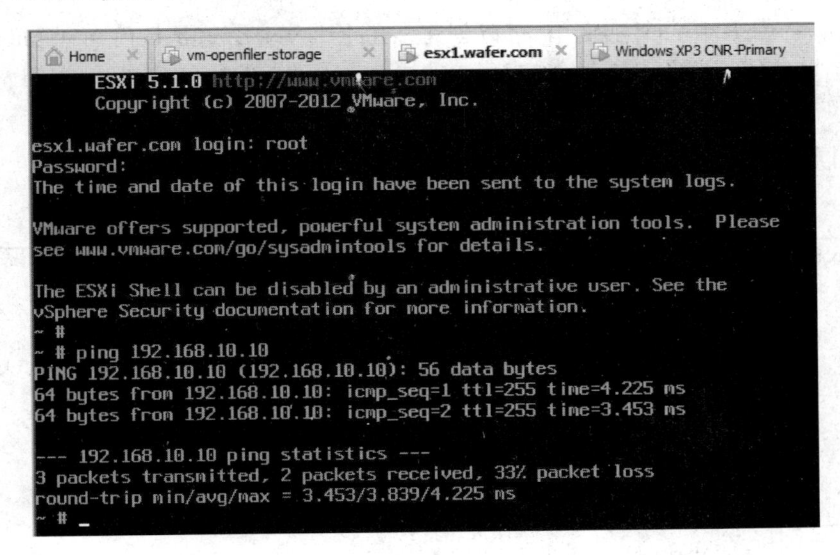

图 8-34 测试 Nexus 1000V 连通性

图 8-34 说明整个 Nexus 1000V 是正常工作的。

说明：Nexus 1000V 不是 Cisco 免费的虚拟化产品，投入生产网络使用需要购买相应授权。

8.2 虚拟路由器 CSR 1000V

8.2.1 CSR 1000V 介绍

思科云服务路由器(CSR)的 1000V 系列通过单一的软件，提供多租户、广域网网关等功能。它可以让企业的广域网扩展到云里和变成云服务提供商，以便为租户提供企业级的网络服务。

其主要特征包括：

① 为多租户提供灵活的虚拟外形设计，为供应商提供托管的主机云；

② 为每个租户提供完全的虚拟机管理程序隔离及多业务路由器；

③ 是经过验证的、熟悉的、企业级的 Cisco IOS 软件网络服务；

④ 功能和思科物理外形路由器操作一致；

⑤ 思科集成多业务路由器是端到端的广域网架构的组成部分。

主要应用场景包括：安全的 VPN 网关、MPLS 广域网终结、数据中心网络扩展、控制和流量重定向。

与其他思科虚拟设备相比，CSR 1000V 的安装非常简单，就是利用 OVF 或 ISO 文件。在撰写本文时，CSR 1000V 要求 VM guest 虚拟机操作系统是基于 Linux 2.6（就像 NX-OS，IOS XE 的内核都是基于此操作系统）的，4 个 vCPU，4 GB 内存，8 GB 虚拟硬盘，并映射到了 IOS XE ISO 文件的虚拟 CD/DVD 引导设备。

在一个单一 CSR 1000V 实例中，可以创建多达十个 vNIC（这是目前在 vSphere 中的虚拟机的限制）。当添加或删除一个虚拟网卡，需要重新加载，一个接口被删除后接口映射关系随之改变。

安装不同 CSR 1000V 版本对 VMware 虚拟机的要求不同，如表 8-1 所示。

表 8-1　不同 CSR 1000V 版本对 VMware 虚拟机的要求

Cisco CSR 1000V Release	VM Configuration Requirements
Cisco IOS XE Release 3.9S	• VMware ESXi 5.0 • Single hard disk • 8 GB virtual disk • 4 virtual CPUs • 4 GB of RAM • 3 or more virtual network interface cards
Cisco IOS XE Release 3.10S	• VMware ESXi 5.0 or 3.1 • Single hard disk • 8 GB virtual disk • The following virtual CPU configurations are supported: 1 virtual CPU, requiring 2.5 GB minimum of RAM 2 virtual CPUs, requiring 4 GB minimum of RAM 4 virtual CPUs, requiring 4 GB minimum of RAM • 3 or more virtual network interface cards
Cisco IOS XE Release 3.11S	• VMware ESXi 5.0 or 3.1 • Single hard disk • 8 GB virtual disk • The following virtual CPU configurations are supported: 1 virtual CPU, requiring 2.5 GB minimum of RAM 2 virtual CPUs, requiring 2.5 GB minimum of RAM 4 virtual CPUs, requiring 4 GB minimum of RAM • 3 or more virtual network interface cards
Cisco IOS XE Release 3.12S	• VMware ESXi 5.0, 3.1 or 3.5 • Single hard disk • 8 GB virtual disk • The following virtual CPU configurations are supported: 1 virtual CPU, requiring 2.5 GB minimum of RAM 2 virtual CPUs, requiring 2.5 GB minimum of RAM 4 virtual CPUs, requiring 4 GB minimum of RAM 8 virtual CPUs, requiring 4 GB minimum of RAM • 3 or more virtual network interface cards

注：VM 不支持多个硬盘的情形。

CSR 1000V 版本对 VMware 虚拟机的要求：思科 Application Visibility and Control

(AVC)需要 4 GB RAM,在主机 BIOS 里需要支持 64 位的 Virtualization Technology (VT)技术支持。当创建 OVA 时,会自动创建 3 个 vNIC,思科 CSR 1000V 启动后,可以手动增加 vNIC 到虚拟机。

8.2.2 CSR 1000V 的安装

CSR 1000V 的部署有两种方式,一个是 OVA 的方式,一个是思科的 CBD(Cisco Build Deploy)方式。通过 OVA 文件包格式部署就像轻松部署一个虚拟机,OVA 包括一个.ovf 的文件,带有缺省 IOS XE 配置,支持 hypervisor,所以这里以 OVA 的方式介绍 CSR 1000V 的安装。

注意: Cisco IOS XE 3.10S 和 3.11S 版本,OVA 只创建带有 4 个 vCPU 的虚拟机,要改变到 1~2 个 vCPU,必须首先部署 OVA 的模板,然后用 vSphere 去修改 vCPU 数量和需要的 RAM 值。如果虚拟 CPU 配置修改了,CSR 1000V 必须重启,而改变 RAM 不需要重启,从 Cisco IOS XE 3.12S 开始,OVA 包提供 vCPU 的配置选项。当部署 OVA 时,VM 需要两个虚拟 CD/DVD 驱动,一个是 OVF 的环境文件,一个是 ISO 文件。

下面是通过 VMware vSphere Client 执行的安装步骤:

① 运行 VMware vSphere Client 客户端软件;

② 从"VMware vSphere Client"菜单界面,选择"File"→"Deploy OVF Template";

③ 在 OVA 向导中,选要部署的 Cisco CSR 1000V OVA 源,单击"Next";

④ 在"Name"和"Inventory Location"中,说明 VM 的名字,单击"Next";

⑤ 在"Under Deployment Configuration"里,选择希望的硬件配置配置文件下拉菜单,单击"Next";(此步骤适用于 Cisco IOS XE 3.12S 之后的版本)

⑥ 在"Storage"里,选择用于虚拟机的 Datastore,单击"Next";

⑦ 在"Disk Format"里,选择磁盘格式:Thick Provision Lazy Zeroed 或 Thick Provision Eager Zeroed,选择后,单击"Next";

注意: Thin Provision 选项不被支持,Thick Provision Eager Zeroed 选项花费的安装时间较长,但是提供更好的性能。

⑧ 在 Network Mapping 下,在目的网络用下拉菜单选择一个或多个 virtual Network Interface Card (vNIC),选中 vNIC 连接时 Power On,单击"Next",当 CSR 1000V 用 OVA 模板部署完成时,会自动创建两个 vNIC,其他的 vNIC 需要手动创建,思科 CSR 1000V 最多支持 10 个 vNIC;

⑨ 为 VM 配置属性,不同的 IOS XE 版本可能有不同的属性;

⑩ 选择"Power on after deployment",自动在部署完成时为 VM 加电;

⑪ 单击"Finish"开始部署 OVA,开始安装过程。

CSR 1000V 显示通过 Console 连接情形,如图 8-35 所示。

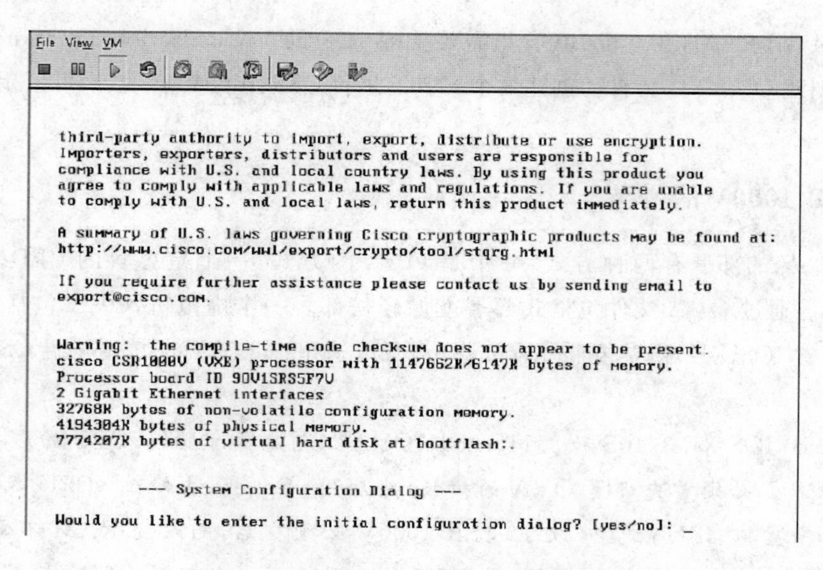

图 8-35　CSR 1000V 显示通过 Console 连接情形

8.3　虚拟防火墙 ASA 1000V

8.3.1　ASA 1000V 介绍

思科 ASA 1000V 云防火墙作为软件防火墙，运行在 VMware vSphere 平台上，部署于 VDC 和私有云的前端，图 8-36 为思科提供的架构图。

ASA 1000V 扩展了行之有效的自适应安全设备的安全平台，始终如一地保证在多租户私有云和公共云部署的租户边缘。补充了思科虚拟安全网关（VSG）的基于区域的安全功能，Cisco ASA 1000V 云防火墙提供了多租户边缘安全性，缺省网关功能，以及防止基于网络的攻击，是一个全面的云安全解决方案。Cisco ASA 1000V 云防火墙集成思科 Nexus 1000V 系列交换机，提供了多虚拟机管理程序功能的解决方案，使一个单一的 ASA 1000V 实例来保护多个 ESX 主机，有着卓越的部署灵活性和简化的管理。思科虚拟网络管理中心（Virtual Network Manager Center，VNMC）用于提供动态、政策驱动、多租户管理，也可以用命令行及 ASDM 方式进行管理和配置。

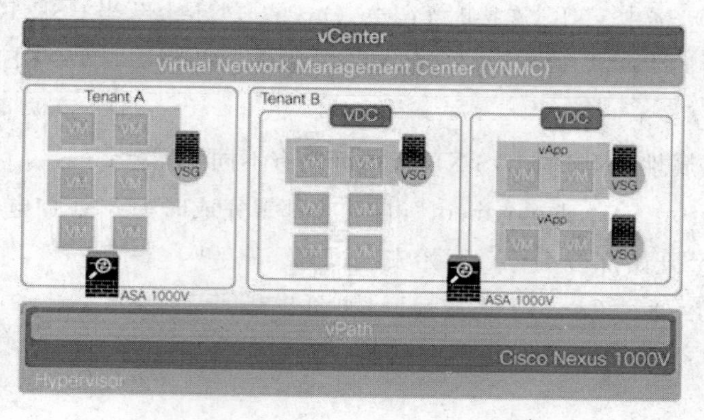

图 8-36　ASA 1000V 架构图

思科 ASA 1000V 提供单个 OVF 格式文件安装下载,当前版本为 ASA8.7,所以部署 ASA 1000V 就像部署一个虚拟机一样简单,还可以用克隆的方式部署多个 ASA 1000V。

Cisco ASA 1000V 云火墙部署需求,如表 8-2 所示。

表 8-2　Cisco ASA 1000V 云火墙部署需求

产　品	需　求
Cisco ASA 1000V 云火墙	Cisco ASA 1000V 云火墙作为一个虚拟应用 一个虚拟 CPU vRAM:1.5 GB vHard disk: 2.5 GB 网络数据接口:2 管理接口:1 高可用性:1
安装平台和管理程序	• VMware vSphere 5.1 or later releases with VMware ESX or ESXi • VMware vCenter 5.1 or later releases
分布式虚拟交换机	Cisco Nexus 1000V Series 软件版本 5.2(1)SV1(4) 及以上,包括 VEM(embedded in the VMware vSphere ESX or ESXi hypervisor); Essential edition or Advanced edition
网管	思科 CNMC 2.0 以上(作为一个虚拟应用部署)

注意:ASA 1000V 有 License 限制,进一步信息请查看思科在线文档。

8.3.2　ASA 1000V 详细安装步骤

部署 ASA 1000V,要用 VMware vSphere Client 和 OVF 模板文件。运行 ASA 1000V 向导文件之后,创建 ASA 1000V 需要的 VM。

注意:在 OVF 模板部署期间,ASA 1000V 主系统、配置、镜像文件占用 2 GB 的存储空间,这些文件显示在 disk0。

(1) 启动 VMware vSphere Client,选择“File”→“Deploy OVF Template”。

(2) 部署 OVF 文件路径,或 URL 区域,浏览已经下载的 ASA 1000V OVF 包,单击“Next”。

(3) 在 OVF Template Details 页,检查 ASA 1000V 包后,单击“Next”。

(4) 检查确认 End User License Agreement,单击“Next”。

(5) 在 Name 部分,输入 ASA 1000V VM 名称,选择 VM 菜单位置,单击“Next”。

(6) 为 ASA 1000V 选择下列部署模式(单台或 failover),单击“Next”。

- Deploy ASA as Standalone——单台 ASA 1000V 模式。
- Deploy ASA as Primary——ASA 1000V 被配置成 failover 主用模式。
- Deploy ASA as Secondary——ASA 1000V 被配置成 failover 从模式。

(7) 选择 ASA 1000V 运行的主机或族位置,单击“Next”。

(8) 选择 host 或 cluster 上存贮 ASA 1000V 文件的路径,单击“Next”,每个硬盘设备显示为一个 datastore。

(9) 选择存储格式,单击“Next”。

(10) 选择想要使用的OVF模板，作为ASA 1000V接口的端口配置文件。

注意：如果此时还没有创建端口配置文件，暂停ASA 1000V的部署，返回到VSM Console，为ASA 1000V接口创建需要的端口配置文件：inside，outside，management和HA(failover)，完成这些配置后，返回到ASA 1000V的部署，单击“Next”。

当部署完ASA 1000V后，网络适配器被创建为如下所示。

Network Adapter1——Management 0/0

Network Adapter2——GigabitEthernet 0/0 (used as the inside interface)

Network Adapter3——GigabitEthernet 0/1 (used as the outside interface)

Network Adapter4——GigabitEthernet 0/2 (used as the failover interface)

端口配置文件是通过VMware vCenter Server连接到Nexus 1000V VSM获得。

(11) 重启前面设置好的ASA 1000V属性，单击“Next”。

设置的参数有：管理接口DHCP模式、管理地址、管理地址的掩码、管理地址的备用地址、选择设备管理员、思科VNMC地址。

(12) 查看ASA 1000V总体配置，单击“Finish”，之后ASA 1000V实例会显示在指定的数据中心。

(13) 运行ASA 1000V。

从VMware vSphere Client，右击主机或族中部署的ASA 1000V实例，选择“Power”→“Power on”，在菜单右边有ASA 1000V Console向导做进一步的运行配置。

8.3.3 ASA 1000V配置

ASA 1000V可以像配置ASA 5500系列防火墙一样配置，既可以通过命令行配置，也可以通过ASDM软件配置，或兼而有之。

1. ASA 1000V命令行配置

(1) 启动VMware vSphere Client，选择“Home”→“Inventory”→“Hosts and Clusters”，单击安装的ASA 1000V实例，加电引导。

(2) 在菜单右边，单击“Console”项，如图8-37所示。

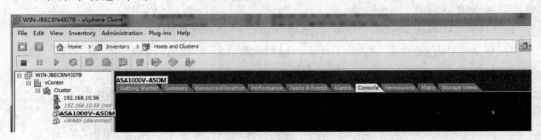

图8-37 通过Console安装ASA 1000V

(3) 可以看到：hostname>。这个指示在用户执行模式下，这种模式只能执行基本的命令。

(4) 进入特权模式，键入命令：hostname> enable。则出现提示：Password。

(5) 提示后面键入enable需要的密码，缺省密码是空，可以直接按回车键进入，进入后可以修改用户名和密码。提示改变成：Hostname#。

这是在特权模式，可以进入全局配置模式，也可以键入disable或exit退出特权模式。

(6) 进入全局配置模式，进入下列命令：hostname# configure terminal。提示变成：Host-

name(config)#。

至此,可以在全局模式下开始配置 ASA 1000V,要退出全局模式,用 exit,quit 或 end 命令。

2. 应用 ASDM 配置 ASA 1000V

可以通过两种方式启动 ASDM,一种是 ASDM-IDM Launcher(只在 Windows),另一种是 Java Web Start。从 ASDM-IDM Launcher 启动,首先需要下载 ASDM-IDM Launcher 软件,ASDM Demo Mode 可以在离线方式下配置 ASA 防火墙。

图 8-38 为连到 ASA 1000V 的画面。

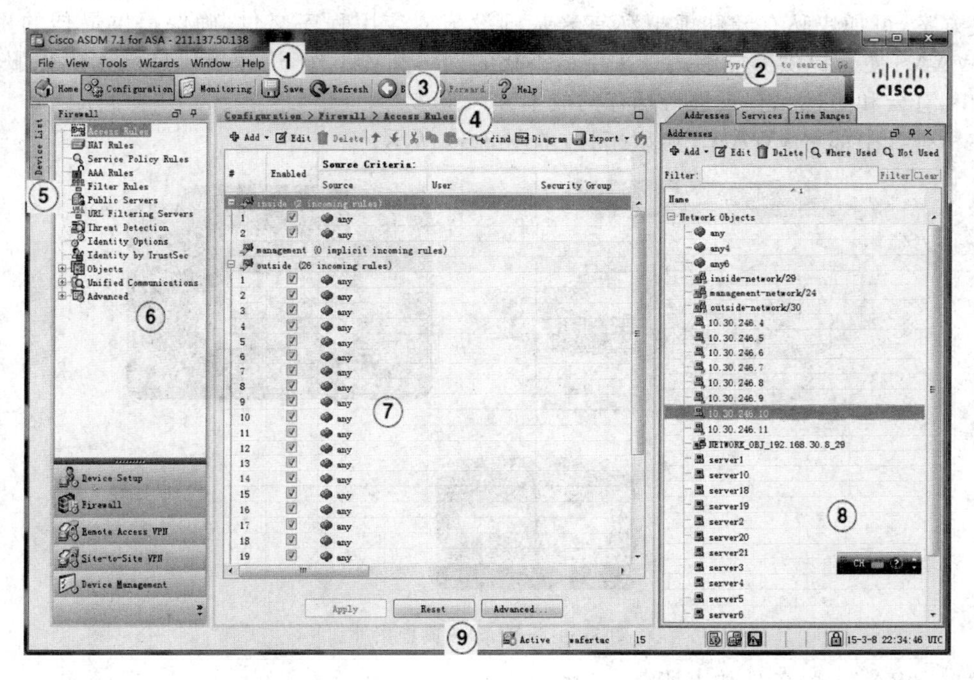

图 8-38　连到 ASA 1000V 的画面

熟悉 ASA 5500 防火墙的读者一眼可以看出,ASDM 连到 ASA 1000V 与连接到 ASA 5500 是一样的,如表 8-3 所示。

表 8-3　菜单功能项的解释

数字序列	描　述
1	菜单栏
2	搜寻区域
3	工具项
4	引导路径
5	设备列表
6	左向导面板
7	内容区域
8	右向导面板
9	状态栏

8.4　虚拟应用加速 vWAAS

8.4.1　vWAAS 介绍

思科虚拟广域应用服务(vWAAS)可通过运行在 VMware ESX/ESXi 管理程序、思科统一计算系统 TM 或其他 x86 服务器上的虚拟设备提供服务。它是第一款支持云的广域网优化解决方案,可加快私有云和虚拟私有云基础设施的应用程序交付,确保提供最佳的用户体验。vWAAS 使云提供商能够通过采用最小化的网络配置工作,快速提供广域网优化服务,并降低向私有云和虚拟私有云进行迁移的成本和风险。

在私有云中部署 vWAAS 典型拓扑,如图 8-39 所示。

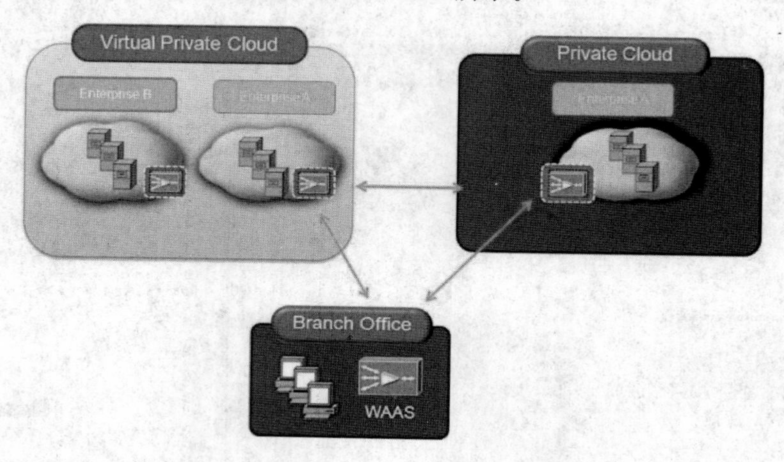

图 8-39　vWAAS 典型拓扑

部署 vWAAS 的条件如下所示。

(1) Cisco UCS 或 x86 服务器:VMware 兼容的 64 bit CPU;BIOS 支持的 Intel VT(虚拟化技术)。

(2) ESX/ESXi 版本: VMware ESX/ESXi 5.0 以上;Cisco ISR G2 VMware ESX/ESXi 5.0.1 以上。

不同型号 vWAAS 参数指标如表 8-4 所示。

表 8-4　各 vWAAS 型号参数

vWAAS 型号	内　存	硬　盘	虚拟 CPU
vWAAS-200	2 GB	160 GB	1
vWAAS-750	4 GB	250 GB	2
vWAAS-1300	6 GB	300 GB	2
vWAAS-2500	8 GB	400 GB	4
vWAAS-6000	8 GB	500 GB	4
vWAAS-12000	12 GB	750 GB	4
vWAAS-50000	48 GB	1 500 GB	8
vCM-100N	2 GB	250 GB	2
vCM-2000N	8 GB	600 GB	4

8.4.2 vWAAS 安装步骤

(1) 在“vSphere Client”窗口，选择“File”→“Deploy OVF Template”，如图 8-40 所示。

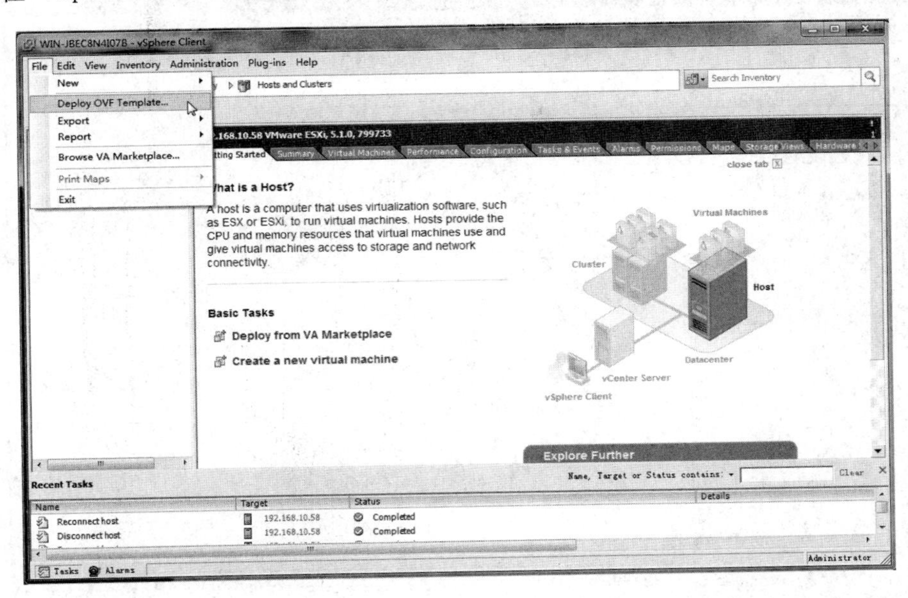

图 8-40 部署 vWAAS

(2) 单击“Browse”，打开一个 Window 窗口。

(3) 查到 vWAAS OVA 文件位置，单击“Open”。

(4) 单击“Next”，接受选择的 OVA 文件，名字和位置窗口将显示。

(5) 键入 vWAAS VM 名称，选择相应的数据中心，单击“Next”，如图 8-41 所示。

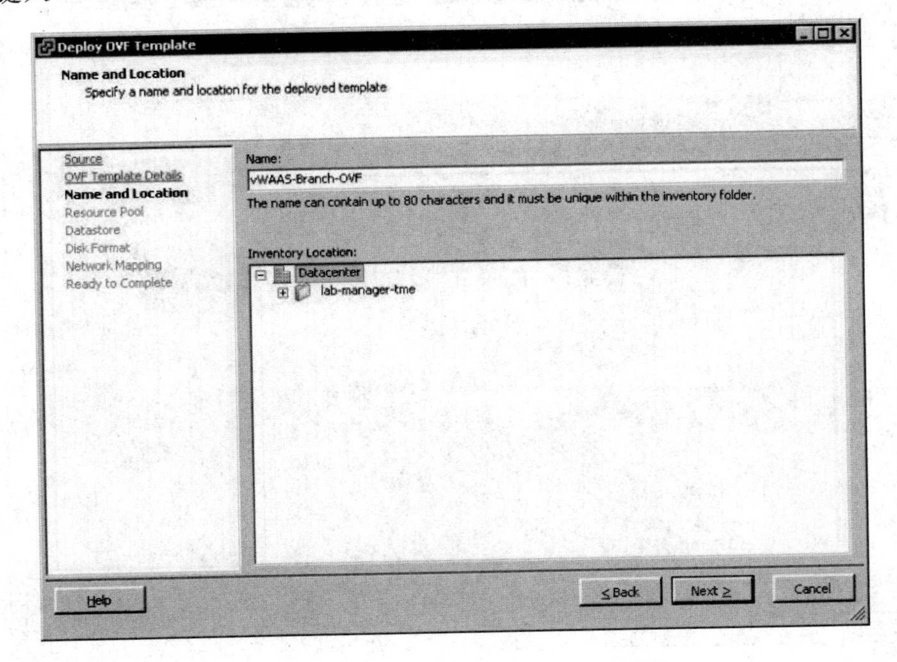

图 8-41 命名 vWAAS 名称

(6) 根据已经建立的数据中心或资源池配置情况,选择"Next"。

(7) 为虚拟机选择 Datastore,单击"Next",如图 8-42 所示。

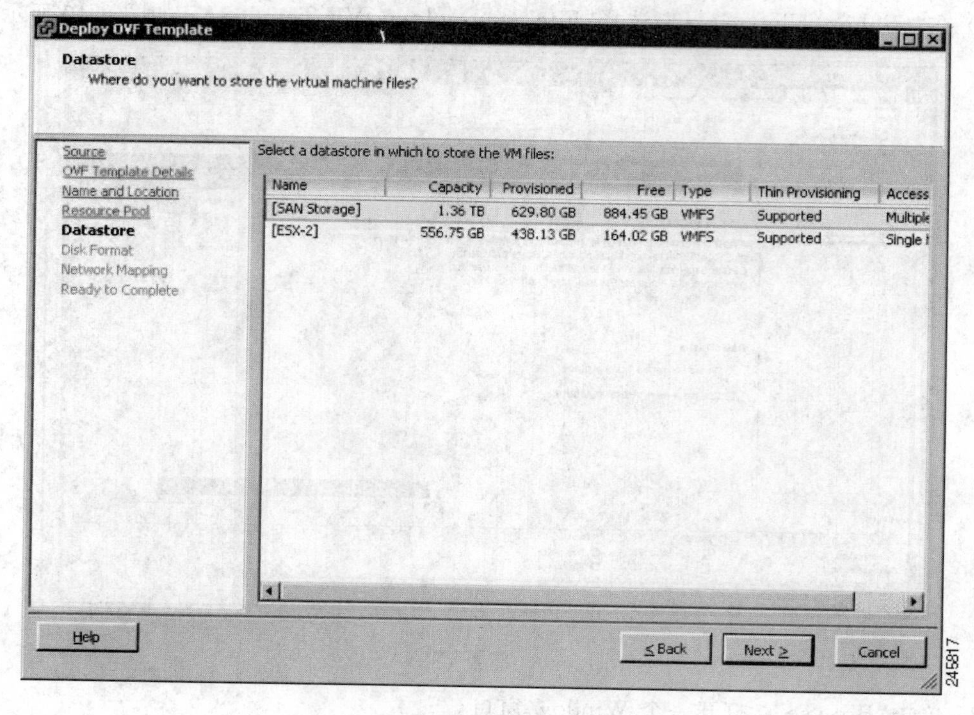

图 8-42　选择虚拟机 Datastore

(8) 选择"Thick Provision format"磁盘格式,单击"Next",如图 8-43 所示。

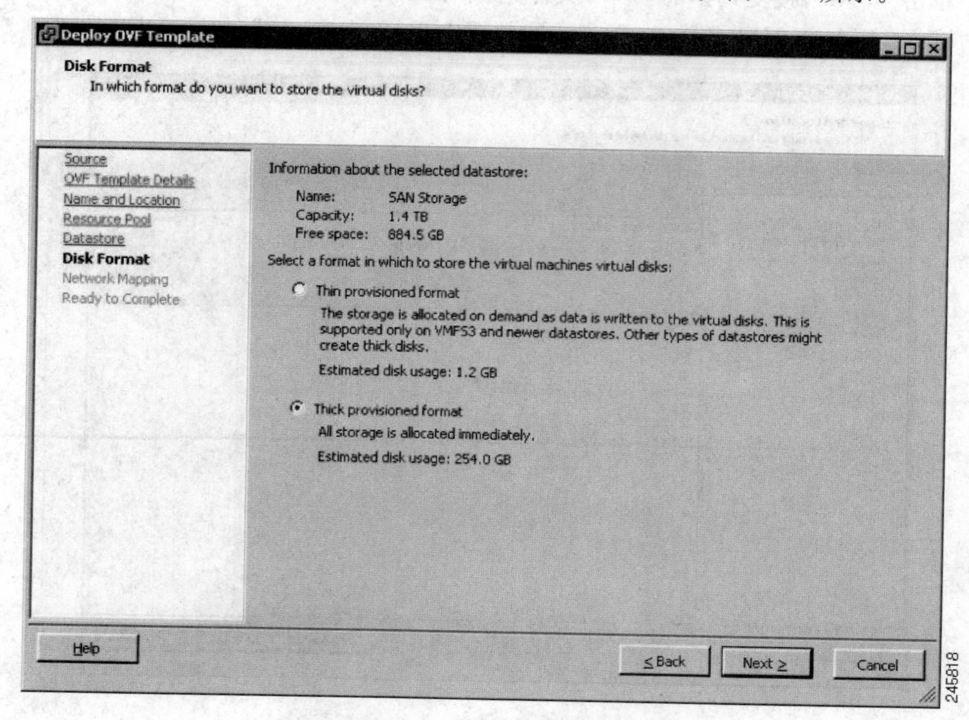

图 8-43　选择磁盘格式

(9) 选择 ESXi 网络映射，单击“Next”，也可以选择后面再修改，如图 8-44 所示。

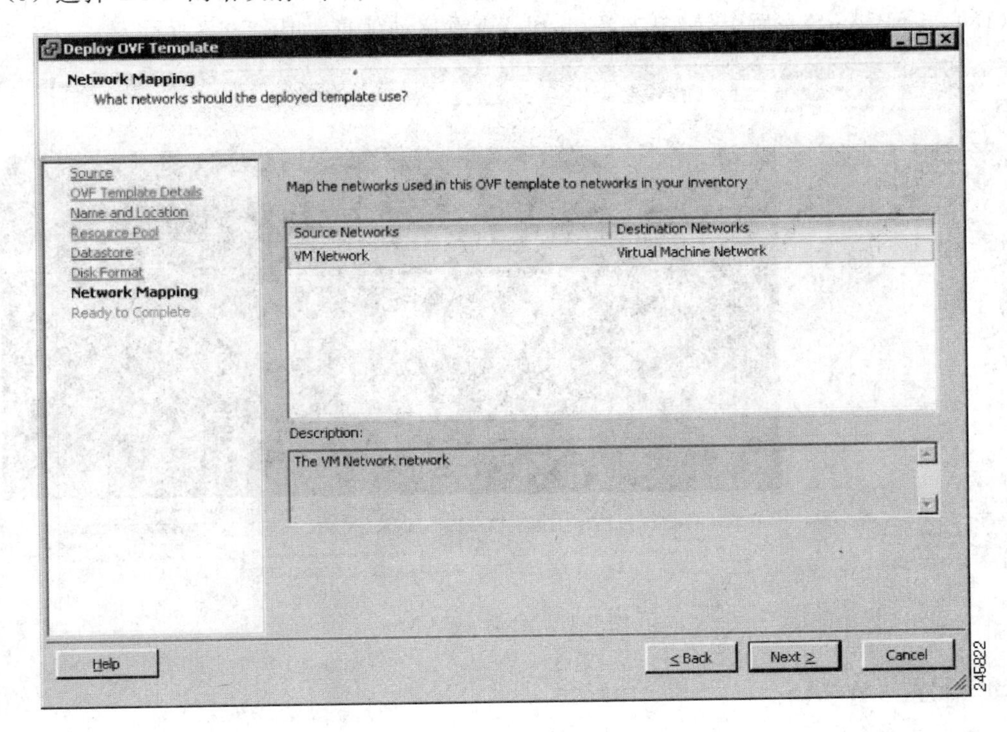

图 8-44　EXSi 网络映射

(10) 单击“Finish”完成安装，图 8-45 显示 OVA 正在部署。

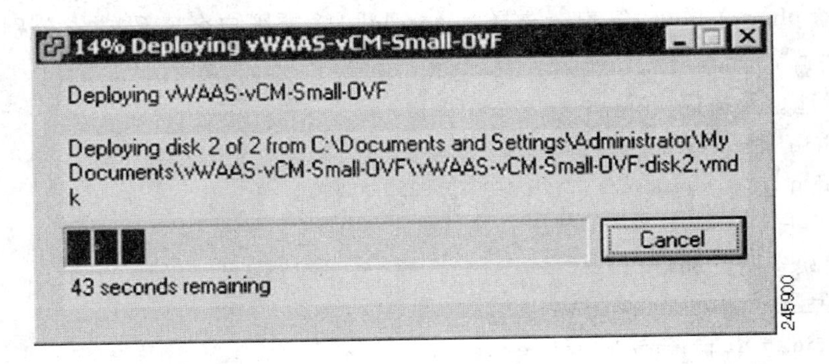

图 8-45　OVA 安装过程

(11) 部署完成时显示成功窗口，单击“Close”，如图 8-46 所示。

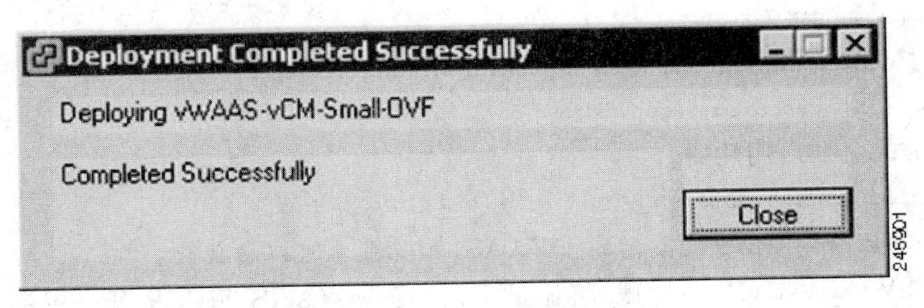

图 8-46　vWAAS 安装成功

(12) 现在已经能够启动 vWAAS 的虚拟机,单击此虚拟机 Power on Virtual Machine。

(13) 当 vWAAS 完后引导,单击“Console”查看启动信息,如图 8-47 所示。

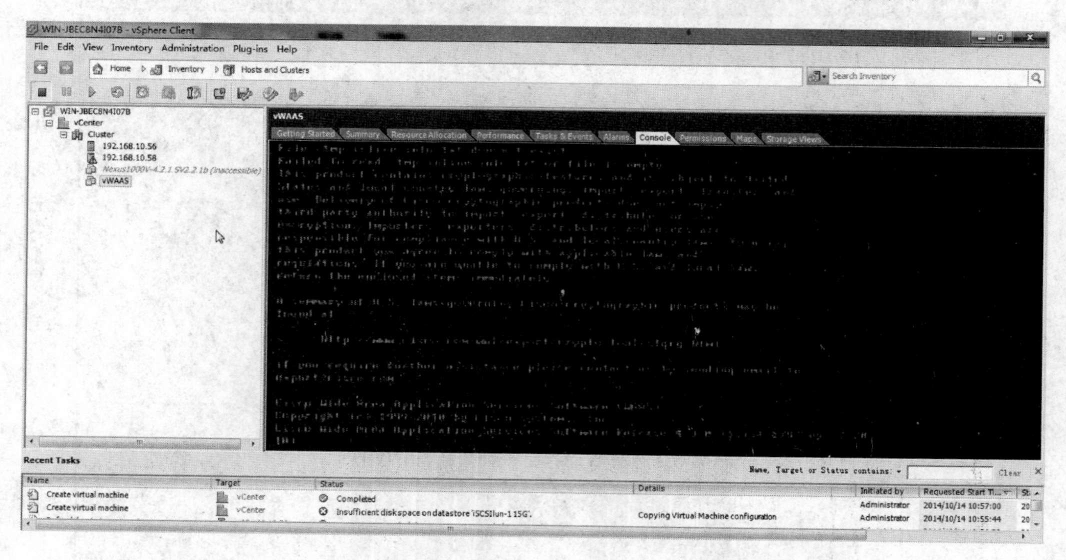

图 8-47 启动 vWAAS

8.4.3 配置 vWAAS

一旦 vWAAS VM 安装完毕,至少需要配置的内容:地址及掩码;缺省网关和基本接口;企业级 License;中心管理地址;CM(Center Manager);侦听(WCCP 或其他)。

(1) 在“vSphere Client”窗口,选择“Console”,登录用户名密码缺省分别为 admin、default。

(2) 配置地址及掩码。

```
VWAAS(config)# interface virtual 1/0
VWAAS(config-if)# ip address 2.1.6.111 255.255.255.0
VWAAS(config-if)# exit
```

(3) 配置缺省网关和接口。

```
VWAAS(config)# ip default-gateway 2.1.6.1
VWAAS(config)# ip primary-interface virtual 1/0
```

进行下一步前,Ping 缺省网关地址来验证连通性。

(4) 用命令行加入企业 License。

```
VWAAS# license add Enterprise
```

(5) 通过 center-manager 命令增加 Center Manager 地址。

```
VWAAS(config)# central-manager address 2.73.16.100
```

(6) 用 cms 命令启用注册。

```
VWAAS(config)# cms enable
```

注意:只有注册到 CMS 服务器的 vWAAS,其流量才可以被优化。

(7) 配置 WCCP 或 vPath 拦截,将流量重定向到 vWAAS,WCCP 可以用带有 WCCP 功能的路由器或三层交换机,vPath 重定向要利用 Nexus 1000V 虚拟交换机。

8.5 小 结

思科在数据中心有一系列的虚拟化产品，可以理解为把物理世界的网络搬到了数据中心、路由、交换、防火墙、应用加速和广域网优化等，本书只是介绍这些虚拟设备的概念和安装，深入的配置跟对应的物理设备一样，需要参考思科相关产品专门的配置文档。